T0317482

BIOMIMETIC PRINCIPLES AND DESIGN OF ADVANCED ENGINEERING MATERIALS

BIOMIMETIC PRINCIPLES AND DESIGN OF ADVANCED ENGINEERING MATERIALS

Zhenhai Xia
Department of Materials Science and Engineering, Department of Chemistry, University of North Texas, Denton, USA

This edition first published 2016
© 2016 John Wiley & Sons, Ltd

Registered Office
John Wiley & Sons, Ltd, The Atrium, Southern Gate, Chichester, West Sussex, PO19 8SQ, United Kingdom

For details of our global editorial offices, for customer services and for information about how to apply for permission to reuse the copyright material in this book please see our website at www.wiley.com.

The right of the author to be identified as the author of this work has been asserted in accordance with the Copyright, Designs and Patents Act 1988.

Wiley also publishes its books in a variety of electronic formats. Some content that appears in print may not be available in electronic books.

Designations used by companies to distinguish their products are often claimed as trademarks. All brand names and product names used in this book are trade names, service marks, trademarks or registered trademarks of their respective owners. The publisher is not associated with any product or vendor mentioned in this book.

Limit of Liability/Disclaimer of Warranty: While the publisher and author have used their best efforts in preparing this book, they make no representations or warranties with respect to the accuracy or completeness of the contents of this book and specifically disclaim any implied warranties of merchantability or fitness for a particular purpose. It is sold on the understanding that the publisher is not engaged in rendering professional services and neither the publisher nor the author shall be liable for damages arising herefrom. If professional advice or other expert assistance is required, the services of a competent professional should be sought.

Library of Congress Cataloging-in-Publication Data

Names: Xia, Zhenhai, 1963–
Title: Biomimetic principles and design of advanced engineering materials /
    Zhenhai Xia, Department of Materials Science and Engineering, Department of Chemistry,
    University of North Texas, Denton, TX 76203, USA.
Description: Chichester, West Sussex, United Kingdon : John Wiley & Sons, Inc., 2016. |
    Includes bibliographical references and index.
Identifiers: LCCN 2016009192 (print) | LCCN 2016022001 (ebook) | ISBN 9781118533079 (cloth) |
    ISBN 9781118926246 (pdf) | ISBN 9781118926239 (epub)
Subjects: LCSH: Biomimicry–Materials. | Bionics–Materials. | Biomimetic materials.
Classification: LCC TA164 .X527 2016 (print) | LCC TA164 (ebook) | DDC 620.1/1–dc23
LC record available at https://lccn.loc.gov/2016009192

A catalogue record for this book is available from the British Library.

Set in 10/12pt Times by SPi Global, Pondicherry, India

1   2016

Contents

Preface

Material designs by nature are quite different from traditional engineering concepts, and offer a vast reservoir of elegant solutions to engineering problems. Nature uses ingenious methods to create a large variety of materials with outstanding physical and mechanical properties. These materials are built at ambient temperature and pressure from a fairly limited selection of components. Thus, their extraordinary properties stem not from what or how they are made, but from their unique microstructures. Mimicking these biological materials designs and processes could create advanced engineering materials useful for various applications ranging from portable electronics to airplanes. However, natural designs of biological materials have proven difficult to mimic synthetically mainly because of a lack of knowledge of materials structure–property relationship and process methods. As a result, there is a growing requirement for the academic, research, and industrial communities to understand biomimetic materials design principles and look for innovative ways of addressing these issues.

This book explores novel biomimetic materials design concepts and materials structure–property relationships, as well as their implementation from a materials science and engineering perspective. It starts by understanding the microstructures of natural materials (e.g., squid beak, gecko footpad, butterfly wings, etc.) and then extracts biomimetic strategies on how to create advanced structural and functional materials though examining the microstructure relationship of these biological materials. These bioinspired design concepts and strategies, together with examples of how they are implemented, are then applied to synthetic materials. It is believed that considerable benefits can be gained by providing an integrated approach using bioinspiration with materials science and engineering.

This book was initiated when I attended the ASME conference as the session chair of the Bioinspired Materials and Structures Program in November 2012. I acknowledge that the topics covered in this book only scratch the surface of a considerable amount of both completed and ongoing research in a wide variety of disciplines. Yet, efforts have been made to provide fundamental understanding of the biomimetic principles and draw general viewpoints across the different topics. I hope this book will empower the reader to think beyond the current paradigms of biomimetic materials science and engineering when translating bioinspired design concepts to engineering reality.

I am indebted to a large number of colleagues for discussions and inputs that led to the writing of this book. I gratefully acknowledge my family, especially my wife, Shuqin Zhu, and my daughter, Serena Xia, for all their patience, help, and understanding while I completed this book.

Zhenhai Xia
Professor
Department of Materials Science and Engineering
Department of Chemistry
Center for Advanced Scientific Computing and Modeling
University of North Texas, Denton, TX 76203

1

General Introduction

1.1 Historical Perspectives

Living organisms in nature have evolved over billions of years to produce a variety of unique materials that possess extraordinary abilities or characteristics, such as self-cleaning, self-healing, efficient energy conversion, brilliant structural colors, intelligence, and so on. These biological materials are made by nature using earth-abundant elements at ambient temperature, pressure, and neutral pH. Mimicking these biological materials structures and processing could lead to the development of a new class of advanced engineering materials useful for various applications ranging from transportation (e.g., aircraft and automobiles) to energy production (e.g., turbine blades, artificial photosynthesis), to biomedical products (e.g., implants, drug delivery). Some of these solutions provided by nature have inspired humans to achieve outstanding outcomes. For example, artificial dry adhesives mimicking gecko foot hairs have shown strong adhesion, 10 times higher than what a gecko can achieve,[1] and the strength and stiffness of the hexagonal honeycomb have led to its adoption for use in light-weight structures in airplane and other applications.[2]

The idea of mimicking nature's materials design has been around for thousands of years. Since the Chinese attempted to make artificial silk over 3000 years ago[2] there have been many examples of humans learning from nature to design new materials and related products. One of history's great inventors, Leonardo da Vinci, is well known for his studies of living forms and for his inventions, which were often based on ideas derived from nature.[3] Although the lessons learned by da Vinci and others were not always successful, as seen in the countless efforts throughout the ages by humans to fly like a bird, these explorations provided some clue for the Wright brothers, who designed a successful airplane after realizing that birds do not flap their wings continuously, rather they glide on air currents.[4] Perhaps the most common and successful product developed based on bioinspiration is Velcro, a fastener. In the 1940s a Swiss engineer, George de Mestral, noticed how

Biomimetic Principles and Design of Advanced Engineering Materials, First Edition. Zhenhai Xia.
© 2016 John Wiley & Sons, Ltd. Published 2016 by John Wiley & Sons, Ltd.

the seeds of an Alpine plant called burdock stuck to his dog's fur. Under a microscope, he saw that the seeds had hundreds of tiny hooks that caught on the hairs. This unique biological material structure inspired him to invent the nylon-based fastener that is now commonly used.

Although the idea of learning from nature has been around for a long time, the science of biomimetics has gained popularity relatively recently. This approach, which uses nature's blueprints to design and fabricate materials, dates back to the 1950s, when the term "biomimetics" was first introduced by Schmitt in 1957.[5] Biomimetics is derived from *bios*, meaning life (Greek), and *mimesis*, meaning to imitate.[6] The term "bionics" was introduced by Steele[7] as "the science of systems, which has some function copied from nature, or which represents characteristics of natural systems or their analogues". The term "biomimicry", or imitation of nature, coined by Janine Benyus in 1997, refers to "copying or adaptation or derivation from biology".[8] From a materials science and engineering perspective, the science of biomimetic materials is thus the application of biological methods and principles found in nature to the study and design of engineering materials. This "new" science is based on the fundamentals of materials science and engineering, but takes ideas and concepts from nature and implements them in a field of technology. While the term "biomimetic" is frequently used in this book to describe mimicking the microstructure of biological materials, "bioinspired" is also employed to describe more general inspiration from nature.

The variety of life is huge; many things fascinate us. Leaves use sunlight, water, and carbon dioxide to produce fuel and oxygen. Geckos keep their sticky feet clean while running on dusty walls and ceilings. Some kinds of bacteria thrive in harmful environments by producing enzymes that break down toxic substances. Materials scientists are increasingly interested in how these phenomena work, and applying this knowledge to create new materials for clean energy conversion and storage, reusable self-cleaning adhesives, cleaning up pollution, and much more. Once the biomimicking succeeds, the impact is enormous.

1.2 Biomimetic Materials Science and Engineering

1.2.1 Biomimetic Materials from Biology to Engineering

Applying materials design principles taken from nature's design to engineering materials can create a new paradigm in materials science and engineering. The term "biomimetic materials science and engineering" is defined here as the study and imitation of nature's methods, mechanisms, and processes for the design and engineering of materials. Materials science, also commonly known as materials engineering, is a vibrant field creating various materials with specific properties and functions, and applying the materials to various areas of science and engineering. The knowledge, including physics and chemistry, is applied to the process, structure, properties, and performance of complex materials for technological applications. Many of the most pressing scientific problems that are currently faced today are due to the limitations of the materials that are currently available. As a result, breakthroughs in this field are likely to have a significant impact on the future of human technology. While humans make great efforts to look for better materials for technological applications, nature has already provided a vast reservoir of solutions to engineering problems, ready for us to exploit. Thus, it is necessary to extend materials science into biomimetic fields where scientists and engineers create materials with properties and performance beyond those of existing materials by mimicking nature-designed structures, and discover new routes for manufacturing materials

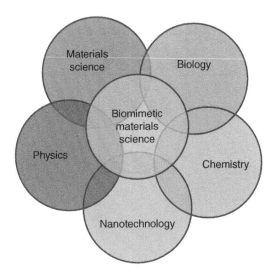

Figure 1.1   Scope of biomimetic materials science and engineering, and its relationship with other disciplines.

by imitating biological processes. The integration of biology, material sciences, chemistry, and physics together with nanotechnology and information technology has brought the subject of biomimetic materials to the science and engineering frontier (Figure 1.1); it represents a major international competitive sector of research for this new century.

1.2.2   Two Aspects of Biomimetic Materials Science and Engineering

Biomimetic materials science and engineering advocates looking at nature in new ways to fully appreciate and understand how it can be used to help solve problems related to materials design and processing. This is achieved by considering nature as model, measure, and mentor in two ways (Figure 1.2). The most obvious and common type of biomimetic materials is the emulation of natural material structures or functions. In this aspect, artificial materials that mimic both the structural form and function of natural materials are designed and fabricated using modern technology. With better understanding of the microstructure, chemistry, and function of biological systems, artificial materials with more precisely controlled microstructure and better function can be designed and produced by following biomimetic principles. With advances in nanotechnologies, biological materials can now be characterized at the level of atoms and molecules, and the biomimetic design of materials can be carried out on the same atomic and molecular scale. Computer modeling and simulations can further optimize the biomimetic design and even create new materials based on biological prototypes.

Emulating nature in the process is another aspect of the biomimetic design of engineering materials, which involves learning from the way nature produces things or evolves. Traditionally biomimetics has involved making artificial materials that replicate biological systems by conventional methods, but now it is possible to utilize biomolecules (nucleic acids, proteins, glycoproteins, etc.) and microbes (archaea, bacteria, fungi, protista, viruses, and symbionts) to actually fabricate artificial materials. This development has the potential to

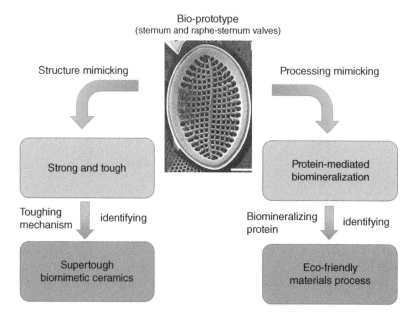

Figure 1.2 Two aspects of biomimetic materials science and engineering: structure and process mimicking. Image shows the structure of sternum and raphe-sternum valves of *Cocconeis scutellum* var. *scutellum* (scale 5 μm). Source: De Stefano *et al.* (2009).[9] Reproduced with permission of Elsevier.

revolutionize materials processing because biosystems synthesize inorganic materials like apatites, calcium carbonate, and silica with nanoscale dimensions.[3] Unlike the traditional materials processes that involve high temperature and high pressure with emission of toxic substances, biological systems produce materials under ecofriendly environments. Beyond the synthesis of nanomaterials, biological systems possess the ability to assemble nanoparticles into larger structures (e.g., bones and shells), effectively performing large-scale integration of nanoparticles. As opposed to the traditional engineering approach, biological materials are grown without final design specifications, using the recipes and recursive algorithms contained in their genetic code.[3] This provides new approaches for materials scientists and engineers to scale up nanoparticles into bulk materials or large structures with desired properties or functions, although this is more challenging than making nanoparticles. Mimicking these bio-assembly processes promises to be an enormously fruitful area of biomimetic manufacturing for advanced engineering materials.

1.2.3 Why Use Biomimetic Design of Advanced Engineering Materials?

Although tremendous progress has been made in the field of advanced materials beyond traditional materials, there still remain technological challenges, including the development of more sophisticated and specialized materials, as well as the impact of materials production on the environment. Many scientists and engineers, whether mechanical, civil, chemical, or electrical, will be looking for new and better designs involving materials. Over some 150 million years, nature has created and tested materials structures from nano and micro to macro and mega, using the principles of physics, chemistry, mechanics, materials science, and many other fields that we recognize as science and engineering. These materials, or "products", have been ruthlessly prototyped, market-tested, upgraded, refined, and otherwise improved as the world around them changed. Each of these fragile specimens is a package of innovation waiting to be understood and adapted as a biological prototype for advanced engineering materials. This evolution produced sophisticated materials and structures which rarely overlap with the methods and products made by humans.

In addition to new materials that could be fabricated based on biological design principles, biomimetic materials could be created that are better than the biological prototypes themselves in some aspects since the bioprototypes are optimized based on the elements available in their environments. Compared to nature, a multitude of synthetic materials with diverse properties are available for selection. Nature has achieved various functions or performances via microstructures restricted to limited kinds of biological materials, mainly collagen and minerals. In contrast, there are abundant artificial materials, including metals, ceramics, and polymers, with various properties facilitating the design and fabrication of microstructures. For example, artificial materials can provide refractive indexes of up to 2.0 and higher for building optical structures while most biomaterials are restricted to an index below 1.5.[10] Besides refractive index contrast, metamaterials with properties that do not exist in nature can be employed to create unique optical effects. To transfer the sophisticated design in nature, materials scientists need to design a broad range of fabrication approaches and adopt various artificial materials to fabricate microstructures with desirable features based on biological design principles.

Many biological materials have remarkable properties that cannot be achieved by conventional engineering methods, for example a spider can produce huge amounts (comparing with the linear size of his body) of silk fiber, which is stronger than steel, without access to high temperatures and pressures. Through biomimetics, it is possible to produce synthetic fibers with properties similar to those of natural fibers. These properties are achieved by mimicking the composite structure and hierarchical multiscale organization of the natural fibers.[11]

Biological materials are different from traditional engineering materials in a number of interrelated ways (Table 1.1). These differences may provide excellent opportunities for biomimetic materials science and engineering to create advanced engineering materials for various engineering applications. In terms of structures, the differences include the following:[12]

1. *Hierarchy.* Biological materials with different organized scale levels (nano to macro) exhibit distinct and translatable properties from one level to the next. A systematic and quantitative understanding of this hierarchy could provide a new route to building more complex synthetic materials with desirable properties and functions.
2. *Multi-functionality.* While many synthetic materials are designed for one function, most biological materials serve more than one purpose. For example, feathers provide flight capability, camouflage, and insulation, whereas the coating on moth eyes provides anti-reflection, self-cleaning, and protection functions.
3. *Self-healing capability.* Unlike synthetic materials in which damage and failure occur in an irreversible manner, biological materials often have the capability to heal damage or injury because of the vascular systems embedded in the structure.

Table 1.1 Comparison between biological materials, traditional engineering materials, and biomimetic materials (adapted from Fratzl (2007)[12]).

Materials	Biological materials	Engineering materials
Chemical compositions	Mostly earth-abundant elements: C, H, O, N, Ca, P, S, Si, etc.	Large variety of elements: Fe, Cr, Ni, Al, Si, C, N, O, etc.
Formation/ fabrication	Growth by genetically guided self-assembly (approximate design)	Fabrication from melts, powders, solutions, etc. (exact design)
Processing	Ambient temperature, pressure, neutral pH	Involve high temperature, high pressure, strong acid/base
Microstructure	Hierarchical structures at all length scales	Mostly microstructures at single length scale
Functions	Adaption of form and structure to the function, multifunctionality	Selection of materials according to function
Design criteria	Modeling and remodeling capability of adaption to changing environmental conditions	Secure design (consider large safety factor)
Failure prevention	Healing: capability of self-healing	Component replacement
Environmental impact	Biodegradable	Biodegradable/ non-biodegradable

4. *Evolution.* Biological structures are not necessarily optimized for all properties but are the result of an evolutionary process leading to satisfactory and robust solutions. "Living" materials (e.g., bone) have evolved in response to their environments during their lifetime.
5. *Environmental constraints.* Biological materials are limited in the elements they are composed of (e.g., C, H, O, N, Fe, etc.) and the availability of these elements dictates the morphology, properties, and functions of the materials.

The differences in processing between biological materials and traditional engineering materials could include the following:

1. *Self-assembly.* In contrast to many synthetic processes, most biosystems assemble structures from the bottom up, rather than from the top down.
2. *Mild synthesis conditions.* The majority of biological materials are synthesized at ambient temperature and pressure as well as in an aqueous environment, a notable difference from synthetic materials fabrication.
3. *Macromolecule-mediated processes.* Most biological processes involve macromolecules as templates, transporters, and catalysts for templating, guiding, and catalyzing the nucleation and growth of biomaterials, especially biominerals.

Biomimetic materials science and engineering also contribute to economy. Some examples found in nature that are of commercial interest include self-cleaning materials, drag reduction in fluid flow, energy conversion and conservation, high and reversible adhesion, materials and fibers with high mechanical strength, biological self-assembly, and antireflection. The applications of these biomimetic materials could generate an enormous market for new products. It is estimated that activity in the field of innovation based on nature increased seven-fold from 2000 to 2013, and papers published around the field increased eight-fold. Between 2012 and 2013, biomimetic patent issuance increased by 27% and scholarly articles jumped by 28%. By 2030, bioinspiration will generate $425 billion of US GDP and $1.6 trillion of global GDP.[13] Several universities have launched biomimicry disciplines and design courses, and biomimicry design challenges are also gaining popularity.[13]

1.2.4   Classification of Biomimetic Materials

Materials science encompasses various classes of materials, each of which may constitute a separate field. Materials can be classified in several ways, for instance by the type of bonding between the atoms or functions. Traditionally, materials are grouped into ceramics, metals, polymers, and composites based on atomic structure and chemical composition. Although biomimetic materials can also be classified in a traditional way, it is more convenient to divide them into classes according to materials properties, as follows: (1) structural materials, (2) functional materials, and (3) process (or procedure). This book follows this classification for biomimetic materials. The typical biomimetic materials described in this book are summarized in Table 1.2.

Structural materials are materials whose primary purpose is to transmit or support a force. The key properties of materials related to bearing load are elastic modulus, yield strength, ultimate tensile strength, hardness, ductility, fracture toughness, fatigue, and creep resistance. Unlike traditional structural materials, biomimetic materials could simultaneously have high strength and high toughness. On the basis of nature's design, for example gecko footpads with strong adhesion and controllable friction, biomimetic materials could be fabricated to have the ability to generate controllable friction and reversible adhesion. In addition, these materials could possess adaptive capabilities that could change their mechanical properties and/or self-shaping under external stimuli, or have self-healing capabilities that can recover their mechanical properties upon damaged.

Functional materials display particular native physical properties and functions of their own. There is a huge range of functional materials, including optical (e.g., structural color and antireflection), stimuli-responsive (e.g., electromechanical, photomechanical, mechanical induced and photomechanical materials), self-cleaning (wet, dry, and under water), catalytic (oxygen reduction, oxygen evolution, hydrogen evolution and artificial photosynthesis), tissue engineering materials, etc.

1.3   Strategies, Methods, and Approaches for the Biomimetic Design of Engineering Materials

Materials scientists and engineers usually take biological materials with remarkable properties or functions as a source of inspiration for the design of advanced engineering materials. While in most cases it is not possible to directly borrow solutions from living nature and to apply them in engineering, it is often possible to take biological systems as a starting point and a source of inspiration for engineering design. Biomimetic engineering materials do not result from the observations of natural structures alone but require a thorough investigation of structure–property relationships in biological materials, and the application of these relationships to engineering materials.

Table 1.2 Typical biomimetic materials, their prototypes, properties, and applications.

Materials	Properties/ functions	Biological prototypes	Biomimetic materials	Potential applications
Structural biomimetic materials	Strength, toughness	Spider silk, abalone shell, bone	Strong and tough materials	Super-tough CNT yarns, tough ceramic composites
	Hardness, wear resistance	Enamel, DEJ	Tough ceramic composites	Cutting tools, wear-resistant coatings
	Impact	Horns, hoof	Damage-tolerant composites	Impact resistance
	Adaptive (stiffness, shape)	Sea cucumber dermis, squid beak	Adaptive nanocomposites with reversible stiffness change capability	Self-shaping, morphing structures
	Self-healing	Soft tissue, bone, plants	Self-healing composites, concretes	Self-healing composites, roads
	Friction and adhesion	Gecko, tree frogs, pitcher plant, shark skin	Dry adhesive, low friction surface	Surface friction control, drag reduction
Functional biomimetic materials	Stimuli-responsive	Muscle, nastic action, sun-tracking plants	Artificial muscle, smart materials	Actuator, control, sensing
	Self-cleaning	Lotus leaf, gecko feet, pitcher plant, shark skin	Self-cleaning materials, anti-fouling coating	Self-cleaning and anti-fouling coating
	Photonics	Beetle, butterfly, moth eye, feather	Structural color materials, anti-reflective materials	Monitoring, sensing, anti-reflectivity
	Catalysis	Leaf, hydrogenase enzyme, blood cells	Catalyst for oxygen, hydrogen evolution, oxygen evolution	Fuel cells, metal–air batteries, water splitting
Biomimetic materials process	Biomineralization	Protein-mediated mineralization	Processing at ambient temperature and neutral pH	Formation of ceramic, metal nanoparticles

1.3.1 General Approaches for Biomimetic Engineering Materials

A successful biomimetic transfer of technology from nature to actual engineering materials can be broken down into four steps, as schematically shown in Figure 1.3. Each step must be brought to a reasonable level of completion to ensure a successful technology transfer.[14]

1. *Identify a high-performance natural model.* At this initial stage, the structure and function of biological systems are studied as prototypes for the design and engineering of materials and structures, but it is not always obvious which should serve as models for manmade designs. In fact, the most successful biomimetic designs have a similar function in nature and in the engineering application. To identify an appropriate model from a great variety of natural materials, one must keep in mind structure–performance–function relationships, in particular multifunctionality.

2. *Abstract key mechanisms, structures, and design principles.* At this stage, the principles and abstract ideas of natural phenomena and model systems are extracted. Mimicking is not copying. The intrinsic relationship between the features of natural materials and their attractive properties should be identified, understood, and abstracted from the natural model so they can be successfully implemented into engineering designs. However, this process is not always straightforward: natural materials are usually extremely complex and hierarchical over several length scales.

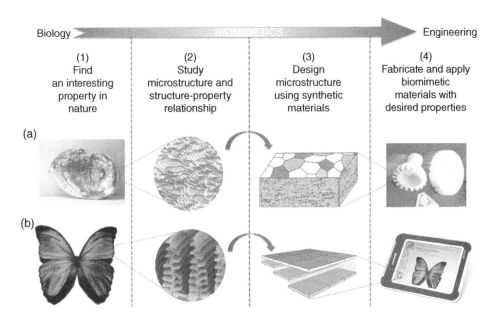

Figure 1.3 From interesting observation to advanced engineering materials and devices. Examples of transfer routes. (a) From abalone shell to super-tough engineering ceramics.[15,16] Source: Barthelat *et al.* (2007)[15] and Launey *et al.* (2010).[16] Reproduced with permission of Elsevier. (b) From butterfly to structural color-based flat panels. Source: Vukusic & Sambles (2003).[17] Reproduced with permission of Nature Publishing.

3. *Transfer and design biomimetic materials structures.* The principles and abstract ideas of natural phenomena and model systems are applied to technical applications and design. This technology implementation includes the choice of proper synthetic materials for the biomimetic structures and often computer modeling is useful to determine and optimize the synthetic systems based on the biomimetic design principles.

4. *Fabricate and implement biomimetic materials.* Once the previous three steps are achieved to a reasonable degree, an actual biomimetic material can be fabricated. However, there is a reason that nature is always one step ahead: we often find that cost or resources limit our ability to exactly replicate nature's efficiency. Nevertheless we are able to come close enough to produce some very interesting things.

As an example, scientists developed super-tough ceramics, drawing their inspiration from abalone shells, which are composed of 95% mineral calcium carbonate ($CaCO_3$), but are a thousand times tougher than their components. They followed the process delineated in Figure 1.3a. After recognizing the shell's outstanding toughness, in step 2 above, the scientists took a closer look at the chemical compositions and structures at micro/nanoscale, learning how they integrate hierarchically. They uncovered how these different features in the shell interacted with one another to produce its unique toughness. This included looking at how the thin mineral layers pile up and the role of organic molecules in the mineral tablets, and how these hard and soft materials work together to resist to brittle crack growth and increase energy assumption. This understanding of how the structure properties of shells work allowed the scientists to form a basis from which to design a synthetic version. It was the interplay of these complex structures that inspired the scientists to formulate super-tough ceramics. In step 3, synthetic ceramic (Al_2O_3) thin tablets and polymers were used to fabricate the brick-like structures. In the end, the scientists were able to develop biomimetic ceramics that leverage the super-tough and super-strong capabilities borrowed from the abalone shells.

Similar approaches have been taken for functional materials. Structural color butterfly wings are one example of a biological structural design that has been successfully transferred into products (Figure 1.3b). Brilliant iridescent coloring in male butterflies enables long-range conspecific communication and it has long been accepted that microstructures, rather than pigments, are responsible for this coloration. Although the final products are not a "material", rather a structure (display), the principles used in creating high-performance electronic color displays are the same: actively varying the interspatial distances of light-interacting layers (e.g., for cell phones), which can change colors rapidly. These flat panels show vivid colors even under low-light conditions, and require less energy than other electronic display methods.

1.3.2   Special Approaches for Biomimetic Engineering Materials

In practice, the biomaterials (steps 1 and 2 in the previous section) and synthetic materials fabrication and characterization (steps 3 and 4) are usually done by biologists and materials scientists, respectively. A systematic approach of biomimetics is to store the biomimetic solutions, once they have been uncovered in the analysis of biological materials (steps 1 and 2), in large databases, from where they can be retrieved by engineers in search of technical solutions. Similar to the selection of engineering materials and processes,[18] initial attempts have been made to establish a system into which all known biomimetic solutions can be placed, classified in terms of function.[8] Such tools would be extremely valuable for the development of bioinspired materials and processes. With the documentation, the biological mechanisms can be verified by following biomimetic manufacture and characterization. This will lead to an iterative process between biology and engineering in which the understanding gained from engineering may be fed back into biology. This mostly unexplored pathway offers the possibility that engineers can also contribute to biological sciences.[19]

Finally, computational methods could play an important role in developing biomimetic materials. Biological materials usually have distinct hierarchical levels in their structure, which leads to an increased diversification in microstructures and multifunctionalities, and enhancement in material properties.[20] A highly controllable assembly strategy can be applied using relatively simple building blocks to create complex structures. It is therefore possible tailor functional materials to match relevant requirements by designing the hierarchical structures using the basic blocks. Computational techniques have been developed to allow us to simulate the material structures and properties at the length scale from nano to micro and macro scales. At each level of hierarchy, computational methods can be applied to simulate the structure–property relationship, while multiscale modeling approaches can be used to link the properties at the macroscale to the phenomena at the nanoscale, outlining an overall picture over whole levels of hierarchy. In addition, computational concept generation is an effective route to generate several conceptual design variants to optimize biomimetic structures. With computer-based design tools such as interactive evolutionary algorithms based on the *de novo* design concept in drag design,[21] it is possible to carry out, based on biomimetic prototypes, a full search of the chemical space and fine optimization of biomimetic molecular structures with targeted properties/functions.

Bioinspired materials can also be designed for specific applications by fine tuning multiple design variables in materials synthesis and processing that pertain to the functional outcomes required.[20] With the aid of computer design and simulation, such an approach would produce a more rationally directed design based on model predictions. More cost- and time-efficient design processes could be achieved by controlling structural features on multiple hierarchical levels via integrated computational modeling and processing in the early stages of the material design.[22]

References

1. Qu, L.T., Dai, L.M., Stone, M., Xia, Z.H. & Wang, Z.L. Carbon nanotube arrays with strong shear binding-on and easy normal lifting-off. *Science* **322**, 238–242 (2008).
2. Eadie, L. & Ghosh, T.K. Biomimicry in textiles: past, present and potential. An overview. *Journal of the Royal Society, Interface/The Royal Society* **8**, 761–775 (2011).
3. Mann, A.B., Naik, R.R., DeLong, H.C. & Sandhage, K.H. Introduction. *Journal of Materials Research* **23**, 3137–3139 (2011).
4. Eggermont, M.J. Biomimetics as problem-solving, creativity and innovation tool in a first year engineering design and communication course. *WIT Transactions on Ecology and the Environment* **114**, 59–67 (2008).
5. Schmitt, O. Some interesting and useful biomimetic transforms, in *Proceedings of the 3rd International Biophysics Congress*, p. 297 (Boston, MA; 1969).
6. Bar-Cohen, Y. Biomimetics – using nature to inspire human innovation. *Bioinspiration & Biomimetics* **1**, P1–P12 (2006).
7. Bhushan, B. Biomimetics: lessons from nature – an overview. *Philosophical Transactions of the Royal Society A* **367**, 1445–1486 (2009).
8. Vincent, J.F.V., Bogatyreva, O.A., Bogatyrev, N.R., Bowyer, A. & Pahl, A.-K. Biomimetics: its practice and theory. *Journal of the Royal Society Interface* **3**, 471–482 (2006).

9. De Stefano, M., De Stefano, L. & Congestri, R. Functional morphology of micro- and nanostructures in two distinct diatom frustules. *Superlattices and Microstructures* **46**, 64–68 (2009).

10. Yu, K., Fan, T., Lou, S. & Zhang, D. Biomimetic optical materials: Integration of nature's design for manipulation of light. *Progress in Materials Science* **58**, 825–873 (2013).

11. Fratzl, P. & Weinkamer, R. Nature's hierarchical materials. *Progress in Materials Science* **52**, 1263–1334 (2007).

12. Fratzl, P. Biomimetic materials research: what can we really learn from nature's structural materials? *Journal of the Royal Society Interface* **4**, 637–642 (2007).

13. Reaser, L. Finance innovation informed by nature's strategies, in *Risk and Value Creation Forum*. NASA, Singularity University (Mountain View, CA; 2015).

14. Barthelat, F. Nacre from mollusk shells: a model for high-performance structural materials. *Bioinspiration & Biomimetics* **5**, 035001 (2010).

15. Barthelat, F., Tang, H., Zavattieri, P.D., Li, C.M. & Espinosa, H.D. On the mechanics of mother-of-pearl: A key feature in the material hierarchical structure. *Journal of Mechanical and Physical Solids* **55**, 306–337 (2007).

16. Launey, M.E., Chen, P.Y., McKittrick, J. & Ritchie, R.O. Mechanistic aspects of the fracture toughness of elk antler bone. *Acta Biomaterials* **6**, 1505–1514 (2010).

17. Vukusic, P. & Sambles, J.R. Photonic structures in biology. *Nature* **424**, 852–855 (2003).

18. Ashby, M.F., Bréchet, Y.J.M., Cebona, D. & Salvo, L. Selection strategies for materials and processes. *Materials & Design* **25**, 51–67 (2004).

19. Csete, M.E. & Doyle, J.C. Reverse engineering of biological complexity. *Science* **295**, 1664–1669 (2002).

20. Buehler, M.J. & Yung, Y.C. Deformation and failure of protein materials in physiologically extreme conditions and disease. *Nature Materials* **8**, 175–188 (2009).

21. Damborsky, J. & Brezovsky, J. Computational tools for designing and engineering enzymes. *Current Opinion in Chemical Biology* **19**, 8–16 (2014).

22. Gronau, G. *et al.* A review of combined experimental and computational procedures for assessing biopolymer structure–process–property relationships. *Biomaterials* **33**, 8240–8255 (2012).

Part I

Biomimetic Structural Materials and Processing

2

Strong, Tough, and Lightweight Materials

2.1 Introduction

Strength and toughness are the two most important mechanical properties for engineering structural materials. Strength (or hardness) measures the resistance of a material to failure, given by the applied stress (or load per unit area), whereas toughness (or damage tolerance) is the resistance of a material to fracture or the energy needed to cause fracture. The attainment of both strength and toughness is a vital requirement for most engineering structural materials, and stronger and tougher materials are demanded because of increasingly severe service conditions and environments. However, the strength and toughness of structural materials are two conflicting mechanical properties of materials. The traditional strategies to strengthen materials are to inhibit polymer chain motion in polymeric materials through controlling the second-phase and rigid particles, and to impede dislocation motion in metallic materials by decreasing microstructural length scales (e.g., the spacing between second-phase particles, grain size, etc.). It has been demonstrated empirically that monotonically changing micro-structural length scales often leads to an imbalance between the strength and toughness of materials, that is, high strength is achieved at the cost of the lowered toughness, or *vice versa*. How to make an effective tradeoff between strength and toughness, or find ways to increase material strength while maintain toughness has become one of critical issues for the design of advanced structural engineering materials.

Natural designs of materials can be extremely efficient in terms of fulfilling specific functions, and offer a vast reservoir of solved engineering problems.[1] Nature uses ingenious methods to activate various strengthening and toughening mechanisms at levels ranging from the nanoscale to the macroscale, thereby increasing both strength and toughness. Biological materials can exhibit remarkable combinations of stiffness, low weight, strength, and toughness, which are in some cases unmatched by synthetic materials. For this reason engineers and researchers are increasingly turning to nature for novel designs and inspiration. In particular,

Biomimetic Principles and Design of Advanced Engineering Materials, First Edition. Zhenhai Xia.
© 2016 John Wiley & Sons, Ltd. Published 2016 by John Wiley & Sons, Ltd.

high-performance structural materials produced by nature are appealing in their design principles for their high performance. Using materials commonly found in the environment (e.g., aragonite in abalone shell), nature is able to build biomaterials several orders of magnitude tougher than the minerals they are made from. While the mechanical properties of biomaterials cannot compare one-on-one with engineering materials (except for a few extreme cases such as spider silk), what makes them attractive is how they combine the building blocks and amplify properties.[2]

Biological materials can be classified into two broad groups: (1) soft (polymeric) materials, which consist of fibrous constituents (collagen, keratin, elastin, chitin, lignin, and other biopolymers) that exhibit widely varying mechanical properties, and (2) hard (mineralized) materials, composed of hierarchically assembled composites of minerals (e.g., hydroxyapatite, calcium carbonate, and amorphous silica). In the area of soft materials, spider silk has drawn attention from all sections of engineering on account of its superior properties compared to existing fibrous materials. Silk fibers have tensile strengths comparable to steel, and some silks are nearly as elastic as rubber on a weight-to-weight basis. In combining these two properties, silks display a toughness that is two to three times that of synthetic fibers like Nylon or Kevlar while keeping high strength. In addition, spider silk is antimicrobial, hypoallergenic, and completely biodegradable. In the area of hard tissues, abalone shells, teeth, and bone are known for their high stiffness and hardness; the most impressive property, however, is their fracture toughness, which is two to three orders of magnitude higher than the minerals they are made of. In particular, these biological ceramics are highly mineralized tissues (at least 95% mineral content); toughness is critical here, so that cracks emanating from initial defects in the shell can be resisted, effectively making the material damage tolerant.

Biological materials are quite diverse in nature. In addition to the soft and hard tissues mentioned above, many biological materials are composites with many components that are hierarchically structured and can have a broad variety of constitutive responses. Below, the structures of selected biological materials are discussed to reveal the design principles that nature uses to achieve exceptional mechanical properties and functionalities. Examples are given of how to utilize these bioinspired strategies to design and build engineering materials with the desired strength and toughness.

2.2 Strengthening and Toughening Principles in Soft Tissues

2.2.1 Overview of Spider Silk

Silk is an amazing natural fiber produced by various species such as silk months, silk worms, bees, wasps, ants, and spiders. Although silk from the silk moth is ideally suited for fashion textiles, and has been produced commercially and traded in China for at least 4000 years, because of its high yields, achieved via breeding the silk-producing larvae on large farms, spider silk stands out among the various silks owing to its higher toughness and tensile strength, as well as better chemical resistance.[3] In particular, spider silk is of practical interest in engineering applications for its unique combination of high strength and rupture elongation. It is incredibly tough and remains unbroken after being stretched to two to four times its original length. Whilst this is unlikely to be relevant in nature, dragline silks can hold their strength below −40 °C and up to 220 °C. Quantitatively, spider silk is five times stronger than steel of the same diameter. A spider silk fiber is finer than a human hair (most threads

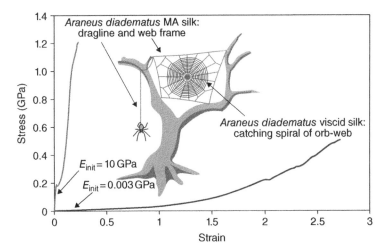

Figure 2.1 Schematic of a spider web showing major ampullate (MA) gland silk and viscid silks from the spider. Source: Gosline *et al.* (1999).[4] Reproduced with permission of the Company of Biologists Ltd.

are a few microns in diameter), designed by nature using subtle changes in chemical composition and, more importantly, morphological structure at a nanometer scale. Its outstanding mechanical properties, light weight, and biocompatibility render spider silk a very attractive target for material science, and an excellent model for the biomimetic design of high-performance fibers.

In nature, spiders produce various types of silk to construct their webs, wrap prey, and protect their offspring, and use silk as a lifeline to escape from predators. Among these silk fibers, frame and capture spiral are the most interesting. In an orb web (Figure 2.1), the frame and radii are made of strong and rather rigid silk, which is produced in the major ampullate glands and therefore named MA silk.[4] Spiders also use this particular silk as a lifeline (or roping thread), which has to always ready to escape predators – it is therefore always dragged, hence the nickname "dragline silk." The capture spiral of an orb web comprises fibers that are produced in the flagelliform (Flag) gland of spiders. Flag silk is highly elastic (up to 300%) and perfectly dissipates the impact energy of prey. For example, a typical honeybee with a body weight of 120 mg and a maximum flight velocity of about 3.1 m/s crashes into a spider's web with a kinetic energy of approximately 0.55 mJ.[5] Flag silk with a diameter of only 1–5 μm can withstand that massive impact. The enormous resilience of these threads is crucial for catching and holding prey that is sometimes bigger than the spider itself.

2.2.2 Microstructure of Spider Silk

Spider silk fibers, like many other biomaterials, are natural polymeric composites with a hierarchical structure, as schematically shown in Figure 2.2. At the macroscale, spider silk has a skin-core structure with a weak skin and a number of filaments. These micro-sized filaments with a circular cross-section are stuck together and aligned parallel to the filament axis (Figure 2.2b). On the nanoscale structure level (Figure 2.2c), each microfilament consists of

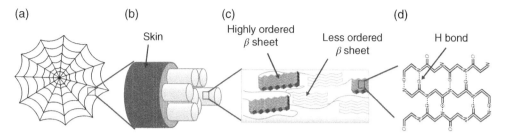

Figure 2.2   Schematic of hierarchical spider silk structure that ranges from macro to nano. (a) The web thread consists of (b) multiple filaments bonded together, while each filament is composed of (c) small crystalline β-sheet rich subunits (see close-ups), which are embedded into an amorphous structure. The crystalline and non-crystalline parts are covalently connected. (d) Inside the β-crystal peptide strands are held together by hydrogen bonds. Source: Adapted from Romer & Scheibel (2008) and Keten *et al.* (2010).[5,9]

strong nanometer-sized crystals and soft disordered chains filling up the remaining volume of the micron-sized fibrils. The crystals have a length/width ratio of approximately 5, with approximately 10–12% of the volume of the hydrated silk. In *B. mori* silk, these protein crystals occupy 40–50% of the total volume of the silk fiber. The remainder is occupied by protein chains having a much less ordered, and amorphous structure. While the crystals form the intermolecular connections between the fibroin molecules, they also act as reinforcing filler particles to stiffen and strengthen the network. Less-ordered but amorphous alanine-rich crystalline regions have also been identified, which connect the blanine-rich crystalline regions. The amorphous chains, which interconnect the crystals, are mostly 16–20 amino acid residues long. The rubber-like elasticity of the hydrated network arises from the large-scale extension of these coiled amorphous chains.[5]

At molecular level, the silk crystals consist of peptide strands held together by individually weak, but collectively strong, hydrogen bonds (Figure 2.2d). The crystalline regions are very hydrophobic, which aids the loss of water during solidification of spider silk. This also explains why the silk is so insoluble: water molecules are unable to penetrate the strongly hydrogen-bonded structures. The short side chain alanine is mainly found in the crystalline domains (β sheets) of the nanofibril, and glycine is mostly found in the amorphous matrix, which consists of helical and β-turn structures.[6,7] The glycine-rich spiral regions of spidroin aggregate to form amorphous areas. Other compounds are used in silks to enhance their properties. Pyrrolidine is found in especially high concentrations in glue threads. This substance has hygroscopic properties and helps to keep the thread moist. Potassium hydrogen phosphate releases protons in aqueous solution, resulting in a pH of about 4, making the silk acidic and thus protecting it from fungi and bacteria that would otherwise digest the protein. Potassium nitrate is believed to prevent the protein from denaturing in the acidic milieu.[7]

Different types of silk have different structural distributions (e.g., different compositions of crystalline and hydrogel parts). MA silk, which is used for constructing the frame of the web, contains a large number of crystalline (β-sheet) structures. In contrast, the much more flexible Flag silk consists almost exclusively of amorphous hydrogel-like regions. Thus, for spider silk function is closely related to structure.[5] Overall, the generalized structure of spider silk is a composite of crystalline regions in an amorphous matrix. It is the interplay between the hard crystalline segments and the strained elastic semi-amorphous regions that gives spider silk its extraordinary properties.[8]

2.2.3   Mechanical Properties of Spider Silk

Spider silk exhibits exceptional mechanical properties, such as high tensile strength and great extensibility. The most outstanding property of spider silk is its balance of high strength and high toughness. Among all natural and synthetic fibers, spider silk demonstrates maximal resilience – the ability of a material to absorb energy when it is deformed elastically, and release that energy upon unloading. The distinct spider silk threads are able to absorb three times more energy than, for example, Kevlar, one of the sturdiest man-made materials on a weight-to-weight basis (Figure 2.3). In contrast, most synthetic materials typically show a higher stiffness and strength compared to natural fibers, but they are much less elastic than natural fibers (Table 2.1).[4] As engineering fibers have an extremely rigid molecular structure that virtually excludes the possibility of large-scale molecular motion, their high stiffness and strength are achieved at the expense of extensibility. Synthetic carbon fibers, for example, have a yield point at approximate 4 GPa, which is more than three times higher than the best insect silk. The elasticity of carbon fibers, however, is only marginal. Spider silk shows a well-balanced combination of strength and elasticity, and therefore mechanically outperforms other natural fibers as well as synthetic threads under certain circumstances. In addition to its outstanding resilience, spider silk shows a torsional shape memory that prevents the spider from twisting and turning during its descent on a silk thread.[10] Because of its high damping coefficient, MA silk needs no extra stimulus for total recovery after being turned from its initial position; unlike synthetic fibers, it barely oscillates after twisting.

Spider silk also shows a high supercontraction rate. Absorbed water leads to significant shrinkage in an unrestrained dragline fiber and reversibly converts the material into a rubber. This process is known as supercontraction, a distinctive feature of MA and Flag silk tension

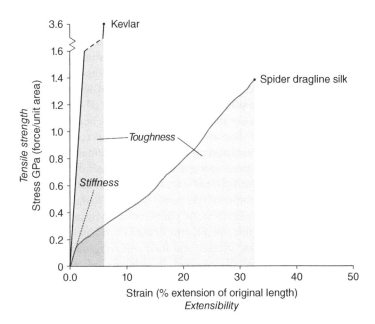

Figure 2.3   Tensile stress–strain behavior of *N. clavipes* spider silk compared to other textile fibers. Source: Shao & Vollrath (2002).[14] Reproduced with permission of Nature Publishing Group.

Table 2.1 Average values of the mechanical properties of silk fibers and high-performance synthetic fibers (adapted from Romer & Scheibel (2008)[5]).

Fiber type	Density (g/cm^3)	Young's modulus E (GPa)	Tensile strength σ_R (GPa)	Breaking strain ε_R (%)	Resilience (MJ/m^3)
Spider silk					
Argiope trifasciata	1.3	1–10	1.2	30	100
Nephila clavipes	1.3	1–10	1.8	30	130
Silkworm silk	1.3	5	0.6	12	50
Nylon 66 fiber	1.1	5	0.9	18	80
Kevlar 49 fiber	1.4	130	3.6	3	50
Carbon fiber	1.6	300	4	1.3	25
High-tensile steel	7.8	200	3	2	6

in a wet web. Supercontraction will occur in water or in air if the relative humidity reaches a level of approximately 90% or more,[11] therefore likely that the silk fibers in the webs of spiders may experience conditions for supercontraction in their normal use. At high humidity spider dragline silk "supercontracts" – unrestrained silk will shrink up to 50% in length whereas restrained silk generates stresses in excess of 50 MPa.[12] Absorption of water leads to shrinkage and tightens the thread, which is important to ensure the rigidity of the spider's web during its lifetime. This process is caused by the organization and arrangement of individual silk proteins.[13]

2.2.4 Strengthening and Toughening Mechanisms of Spider Silk

The incredible properties of spider silk originate from its unique hierarchical structure that spans multiple levels from molecular (amino acid sequence, proteins), to nanoscale (nanocomposite), microscale (fibrils), and macroscale (fibers) (Figure 2.2). These hierarchical levels are a manifestation of how the biochemical information that defines the protein sequence directly affects the behavior of the silk. Each level in the material contributes to the overall properties; the remarkable properties of the entire system emerge because of a series of synergistic interactions across the scales.[15] But how do these constituents contribute to the mechanics of silk fiber? What precisely are their functions? These fundamental questions are the key to the bioinspiration and rational design of strong and tough engineering materials.

As shown in the last section, at the 1–10 nm scale the molecular structure of dragline spider silk repeat units consists of semiamorphous and nanocrystalline β-sheet protein domains. The crystalline-semiamorphous domains in silkworm (*Bombyx mori*) respond to loading in a quite complex way.[9] At the low-stress region, the protein strands uncoil and straighten, followed by entropic unfolding of the amorphous strands, and then stiffening due to load transfer to the crystalline β sheets. The β-sheet crystals crosslink the fibroins into a polymer network with great stiffness, strength, and toughness. Upon tensile loading, the amorphous areas can partially deform, contributing to the elasticity and flexibility of the thread, whereas the nanocrystals provide fiber strengthening. Such an arrangement permits toughness while being strong.[9]

β-sheet nanocrystals that universally consist of highly conserved poly-(Gly-Ala) and poly-Ala domains, embedded in a rubbery protein matrix at the nanometer scale, are considered to be critical to the mechanical and biological performance of spider silk. This seems counter-intuitive because the key molecular interactions in β-sheet nanocrystals are hydrogen bonds, one of the weakest chemical bonds known. In fact, β-sheet nanocrystals confined to a few nanometers can achieve higher stiffness, strength, and mechanical toughness than larger nanocrystals. Large-scale molecular dynamics simulations revealed that an energy dissipative stick-slip shearing of the hydrogen bonds occurs during failure of the β sheets, as shown in Figure 2.4.[9] For a stack with height $L \leq 3$ nm (Figure 2.4c), the shear stresses are more substantial than the flexure stresses, and the hydrogen bonds contribute to the high strength obtained (1.5 GPa). Through nanoconfinement, a combination of uniform shear deformation that makes most efficient use of hydrogen bonds and the emergence of dissipative molecular stick-slip deformation leads to significantly enhanced mechanical properties. However, if the stack of

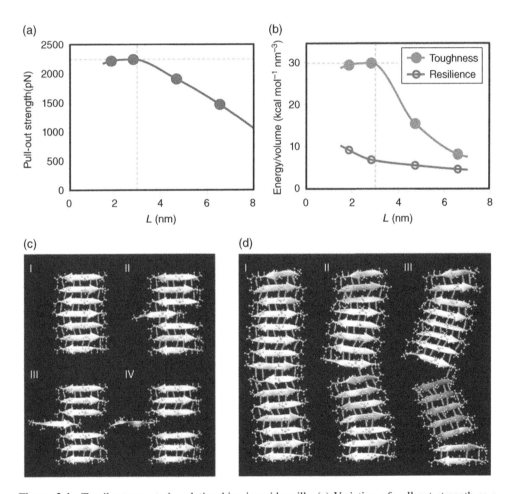

Figure 2.4 Tensile stress–strain relationships in spider silk. (a) Variation of pull-out strength as a function of β-sheet nanocrystal size L. (b) Toughness and resilience as a function of β-sheet nanocrystal size L. Molecular dynamics simulation of silk: (c) short stack and (d) long stack of β-sheet crystals. Source: Keten (2010).[9] Reproduced with permission of Nature Publishing Group.

β sheets is too high (Figure 2.4d), it undergoes bending with tensile separation between adjacent sheets. The nanoscale dimension of the β sheets allows for a ductile instead of brittle failure, resulting in high toughness values of silk. Thus, size affects the mechanical response considerably, changing the deformation characteristics of the weak hydrogen bonds. Bone and other biomaterials also demonstrated similar structures,[16] where sacrificial hydrogen bonds between mineralized collagen fibrils contribute to the excellent fracture resistance. This may explain how size effects can be exploited to create bioinspired materials with superior mechanical properties in spite of relying on mechanically inferior, weak hydrogen bonds. Thus, the geometric confinement of the silk fiber diameter to <100 nm is critical to exploit the full potential of the material behavior endowed from the molecular scale.

Apart from the typical nanocomposite structure – crystals embedded in an amorphous matrix – that makes up the fibrils, the fibril morphology at the hundreds of nanometers scale is also considered important to mechanical behavior and dissipation of energy. At this length scale, the core region of a spider silk fiber consists of braided bundles of 200 nm diameter fibrils (Figure 2.5) that align in the direction slightly off-axis (~10°) with respect to the overall fiber.[17] The fibrils are not smooth, but rather are characterized by globular protrusions approximately 100 nm wide and 150 nm long with respect to the fibril direction. Collagen and tendon show a similar interlocking structure. Under normal conditions, this morphology creates nonslip fibril kinematics, restricting shearing between fibrils, yet allowing controlled local slipping under high shear stress, dissipating energy. The local slipping dissipates energy without bulk fracturing. This mechanism provides a relatively simple target for biomimicry and, thus, can potentially be used to increase fracture resistance in synthetic materials.[17]

At fiber level, the deformation of the silk to tensile forces is very interesting because of its unique set of molecular and submicroscale configurational conformations. Silk fibers typically soften at the yield point to dramatically stiffen during large deformations until the point of failure. Viscid (catching spiral) spider silk, for example, has a stress–strain curve (so-called

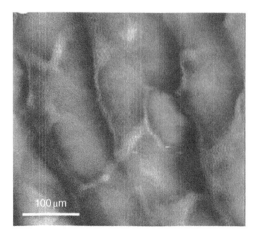

Figure 2.5 AFM image of fibril structure across a range of size scales. Fibrils in the core region of spider silk fibers. The clear globular/banding patterns can be seen in each fibril and the interlocking of the globules/bands between the fibrils. Source: Brown *et al.* (2012).[17] Reproduced with permission of ACS Nano.

J-shaped curve) where molecular uncoiling and unkinking occur with considerable deformation under low stress.[18] This stiffening as the chains unfurl, straighten, stretch, and slide past each other can be represented analytically in one, two, and three dimensions. Thus, the stress–strain curves of the silk can be divided into two regimes: (1) unfurling and straightening of polymer chains and (2) stretching of the polymer chain backbones. Such stress–strain behavior is what allows for localization of deformation upon loading, and is what makes spider webs robust and extremely resistant to defects, as compared to a hypothetical linear-elastic or elastic-plastic silk fiber. To achieve high strength and toughness, the initial work performed on extension should be small, to reduce energy expenditure, whereas the material should stiffen close to the breaking point, to resist failure.[18]

Other soft biological materials, for example wool, whelk eggs, and silks, have more complex responses, marked by discontinuities in *modulus*.[18] Several mechanisms are responsible for this change in slope, including the phase transition from α- to β-keratin and entropic changes with strain. When the materials are subject to tensile loading, the transformation from α-keratin helices to β sheets occurs. Upon unloading, the reverse occurs, and the total reversible strain is, therefore, extensive. This stress-induced phase transformation is similar to what occurs in shape-memory alloys, therefore this material can experience substantial reversible deformation (up to 80%) in a reversible fashion when the stress is raised from 2 to 5 MPa, ensuring the survival of whelk eggs, which are continually swept by waves.[19] With distinctive deformation characteristics at different length scales and synergetic effects between the length scales, these biopolymers show distinct properties that allow them to be strong and highly extensible.

2.3 Strong and Tough Engineering Materials and Processes Mimicking Spider Silk

2.3.1 Biomimetic Design Principles for Strong and Tough Materials

Spider silk represents an excellent prototype for the biomimetic design of super strong and tough materials. Considering its mechanical and physico-chemical properties, spider silk is perfectly suited to many industrial applications. One strategy for producing the silk is with recombinant protein using engineered host organisms.[20] Many prokaryotic and eukaryotic hosts have been employed but this strategy has been limited because few full gene sequences for spider proteins are available.

Alternatively, synthetic materials are used to produce silk-like fibers based on biological models. The key properties of web silks that emerge from the analysis of web function involve a balance between strength and extensibility, giving enormous toughness and a high level of internal molecular friction, and consequently high levels of hysteresis. At the 1–10 nm scale in spider silk, a combination of high strength and high toughness can be achieved by coupling crystalline with amorphous regions and manipulating the size of crystals, sacrificial bonding behavior, interaction between amorphous and crystalline domains, fraction of order, and strength of bonding between constituents. As alluded to above, spider silk has distinct microstructures: very stiff and strong crystals (hydrophobic) and more disorganized, labile, and very soft regions (hydrophilic). Using such a biological model advanced engineering materials could be designed by combining these building blocks to achieve very diverse mechanical signatures for different purposes.

The controlled sliding between fibrils and the spread of local dislocations over a large volume of material that result from this fibril morphology could potentially provide an additional toughening mechanism in synthetic materials. The repeated, interlocking globular morphology of spider silk fibrils in the range of hundreds of nanometers could be replicated in fibrous materials to combine high strength and stiffness with high toughness, in terms of energy to failure, which will act at high strains. Because of the larger scale (hundreds of nanometers) features on which this energy dissipating mechanism acts, control over this morphology will be a relatively simple target for the biomimetic design of tough materials. Rather than relying on the basic material properties of a fiber or fibril, it requires only morphological manipulation or design.[17]

2.3.2   Bioinspired Carbon Nanotube Yarns Mimicking Spider Silk Structure

Biomimetic approaches have been demonstrated in which lessons from nature can be incorporated into the design of spider silk-like super-tough composite yarns consisting of carbon nanotube and polymer.[21] In spider silk, an extensive network of hydrogen bonds and flexible proteins link stiff, "nanocrystalline" protein domains. As the material deforms, the hydrogen bonds break and reform, allowing the unfolding of the flexible protein regions.[22] In addition, fibrils are aligned along the fiber directions, and this controlled deformation allows fibrils to slide and dissipate energy. Carbon nanotubes (CNTs) have extremely high strength, very high stiffness, low density, and good chemical stability, and are an ideal one-dimensional nanomaterial building block as "nanocrystalline" in silk. To mimic the microstructure and toughening mechanism of spider silk, the nanotubes can be integrated into macroscopic composites with silk-like microstructures with a high degree of nanotube alignment. The soft mediates optimize the interactions between neighboring nanotubes to achieve hydrogen-bond breaking and reforming under the loading.

The load transfer mechanisms in natural materials can be reproduced by poly(vinyl alcohol) (PVA). As the polymer coating on the carbon nanotubes contains both hydrogen-bond donor and acceptor groups, nature's design strategy, in which stiff components can be linked together through a network of weak hydrogen bonds that can break and reform with deformation, can be implemented with the addition of a small amount of flexible polymer. The introduction of PVA, which mediates a dense network of hydrogen bonds between the CNT bundles, provides high stiffness, strength, and energy to failure to macroscopic nanotube-based yarns.[23] These hydrogen bonds work cooperatively to transfer loads, and break and reform with deformation, thus providing a mechanism for increased energy to failure.

Nanocomposite yarns are synthesized by spinning. After the chemical vapor deposition (CVD)-grown high-quality single-walled or double-walled carbon nanotubes have been functionalized with short polymer chains possessing carboxylic acid and ester functional groups, they are spun into composite fibers with PVA. Single-walled carbon nanotube (SWNT)–PVA composite fibers up to 100 m long have been fabricated, which are tougher than spider silk and any other natural or synthetic organic fibers reported previously. The superior toughness of SWNT–PVA composite fibers is attributed to the presence of PVA between SWNTs and the slippage between individual nanotubes within bundles, which results in the combination of high strength and high strain to failure. Through hydrogen bonding to this inherent coating, the flexible PVA chains link the adjacent stiff nanotube bundles, providing a structure that is

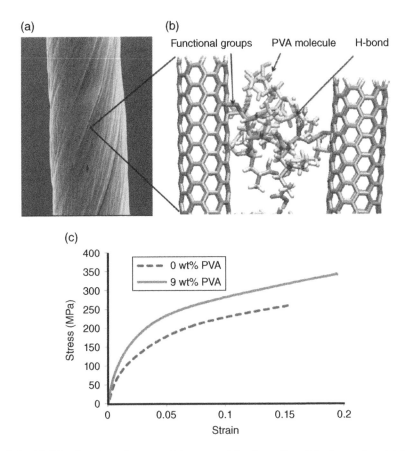

Figure 2.6 (a) Biomimetic carbon nanotube yarn, (b) schematics of hydrogen-bond interactions between the intrinsic organic coating containing carboxylic acid and ester functionalities and the polymer (PVA) containing donor–acceptor hydroxyl groups, and (c) representative stress–strain curves for DWNT-PVA composite yarns with 0 and 9 wt% PVA. Source: Beese (2013).[21] Reproduced with permission of ACS Nano.

functionally similar to spider silk and collagen (Figure 2.6). By varying the ratio of PVA to nanotubes, the density of hydrogen bonds can be adjusted to achieve the best mechanical performance. Multiwalled carbon nanotube (MWNT)–PVA composite yarns have also been prepared by introducing twist during spinning.[21] These yarns can retain their strength and flexibility after heating in air. High creep resistance and high electrical conductivity were retained after polymer infiltration, which substantially increases yarn strength.

2.4   Strengthening and Toughening Mechanisms in Hard Tissues

2.4.1   Nacre Microstructure

Nacre is one of the most appealing hard tissues in biomimicry because of its exceptional mechanical properties, especially toughness. In nature, nacre exists in the inner layer of the

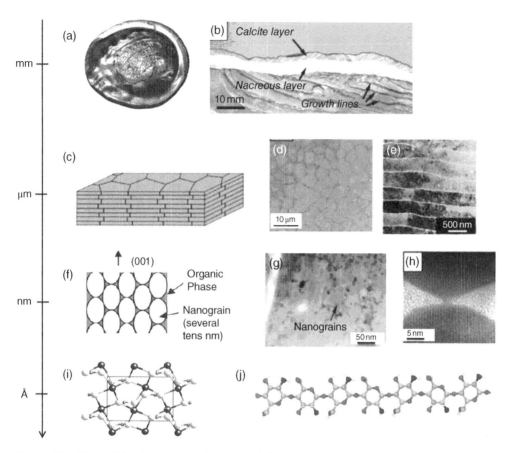

Figure 2.7   The multiscale structure of nacre (red abalone): (a) inside view of the shell, (b) cross-section of a red abalone shell, (c) schematic of the brick-wall-like microstructure, (d) optical micrograph showing the tiling of the tablets, (e) TEM showing tablet waviness, (f) schematic of the nanograin micro-structure, (g) TEM images showing single aragonite crystal with some nanograins, (h) high-resolution TEM of aragonite asperity and bridge, (i) calcium carbonate unit cell, and (j) chitin molecule structure. Source: (a–h) Espinosa *et al.* (2009)[31] and (i–j) Meyers *et al.* (2011).[24] Reproduced with permisson of Elsevier.

structure of some shells that protect the soft body of the mollusk against attack from predators, debris, and rocks moved by the current. Figure 2.7 shows the nacre structure over several length scales.[24] The characteristic structure of nacre at the microscale (Figure 2.7c–e) is a composite made of "brick and mortar", which is composed of 95% mineral calcium carbonate ($CaCO_3$) in the aragonite form. This mineral is in the form of polygonal microscopic platelets 5–15 μm in diameter and 0.5–1 μm in thickness. These platelets are tightly arranged into three-dimensional brick walls. The crystallographic *c* axis of the platelets points approximately perpendicular to the shell wall, but the direction of the other axes varies between groups. Adjacent platelets have dramatically different *c*-axis orientation, generally randomly oriented within ~20° of vertical. On the nanoscale level (Figure 2.7f–h) the platelets are composed of nanograins with a uniform crystal orientation and separated by a fine network of organic materials within the nacre layers. Platelets with such a consistent crystallographic texture act as a single crystal rather than a group of similarly oriented crystals. This unusual phenomenon has been described as the pseudo-single-crystal effect.[25] A significant waviness can be observed; its wavelength is in the order of one half to one tablet length, and its amplitude is about one quarter of the tablet thickness. Such waviness is important to the mechanical properties of nacre.

The remaining 5% of nacre is sheets of organic matrix composed of elastic biopolymers (Figure 2.7j) (such as chitin, lustrin, silk-like proteins, and polysaccharides) mostly located at the interface between platelets. This organic layer of nacre has a laminar structure consisting of a fibrous chitin core between layers of macromolecules. Chitin is identified underlying the interlamellar layers and is found to be present in high concentration in the intertabular matrix. The intertabular matrix is also rich in carboxylates and sulfates. The proteins have modular structures resembling the silks of spiders and silkworm cocoons, which bind to chitin or control the nucleation and growth of the aragonite on organic sheets.[26,27] As discussed in the previous section, the fundamental protein structures with modular repeating hydrophobic crystalline and hydrophilic domains adopt various conformations depending on the presence of water. Although having 5% of total volume, the soft organic phase plays an important role in enhancing strength and toughness by maintaining the cohesion of the platelets over large separation distances and forming tough ligaments. This high extensibility is the result of the uncoiling of modules along the chain of some proteins.[28]

There are other features at the interface between platelets. One of them is nanoasperities, which form a nanoscale roughness on the surface of the platelets, providing a frictional resistance to sliding. Mineral bridges are another feature that connect platelets across the interfaces to reinforce them. All these features, together with basic platelet structures and thin polymer layers, account for mechanical characteristics such as strength, elasticity, and toughness.[29] Because of the complexity of the structure of nacre, it is necessary to identify which features should be transferred to a successful biomimetic material.[30]

2.4.2   Deformation and Fracture Behavior of Nacre

Nacre has different mechanical responses depending on its hydration condition. While dry nacre exhibits a typical brittle behavior similar to bulk aragonite – linearly elastic up to brittle failure – hydrated nacre displays a totally ductile response (Figure 2.8a).[30,32] Nonelastic deformation or "yielding" occurs at tensile stresses of 60–70 MPa, and strain hardening develops up to failure at almost 1% strain and a tensile strength of ~100 MPa. Although this strain at failure is not impressive compared to many engineering metals, it is ten times the strain at failure of aragonite. This makes a significant difference in the area under the stress–strain curve, which is related to the energy required to break the material. Overall, the balance of strength and toughness in nacre is comparable to that of metals and polymers.

The non-elastic deformation and relatively large strains in hydrated nacre stem mainly from its unique platelet-layered structure and polymer interface. As schematically shown in Figure 2.8b, the applied tensile load is transmitted through the material by tensile stress in the platelets and shear stresses at the interfaces. At the initial loading, the platelets are interlocked firmly during the linear deformation. When the applied load is increased, these stresses increase accordingly, until one of several possible failure modes is activated, resulting in non-linear deformation or yielding. The first failure mode is interfacial debonding and sliding

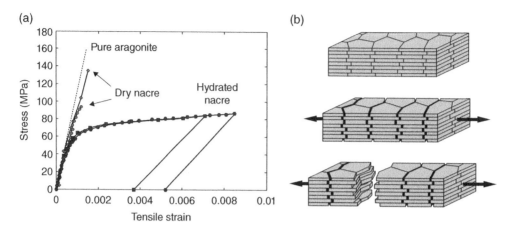

Figure 2.8 (a) Stress–strain curve in tension along the platelets for pure aragonite, dry nacre and hydrated nacre. (b) The relatively large strains measured in the hydrated case are the result of collective tablet sliding at the microscale. Source: Espinosa *et al.* (2009).[31] Reproduced with permisson of Elsevier.

along the interface between the platelets. As the sliding proceeds with loading, large strains with strain hardening are generated in the loading process. Interestingly, this sliding mechanism is also observed in bone and tendon in tension. Finally, the platelets are pulled out completely, causing a macroscopic fracture. Obviously, the organic phase between the platelets plays an important role in controlling the interface sliding behavior. As the gel-like organic phase is susceptible to water content, the interface properties are changed with nacre's hydrated condition, thereby influencing the macroscopic mechanical response of the nacre. The hydrated organic phase acts as a lubricant at the interface and controls interface shear sliding, but the dehydrated organic phase stiffens and loses flexibility and fluidity.[33] The dehydration results in strong bonding of the interface that brings about higher macroscopic strengthening but causes brittleness.

Dried nacre is quite brittle, but its fracture surface shows that the fracture is not the unstable crack propagation through the platelets from a defect, but rather the pullout of the platelets, as in hydrated nacre. Thus the catastrophic failure of the dried nacre is due to abrupt unlocking of the platelets at the maximum load after firm interlocking during the linear deformation (Figure 2.8a). There are several origins of the interlocking and frictional resistance, including mineral bridging, platelet waviness, and nanoasperity on the platelet surface.[34]

Sliding-induced platelet interfacial hardening is considered as the primary mechanism leading to an increase in both strength and ductility. In nacre platelets sliding spreads throughout the material so that all of the sliding sites are "activated" at failure. Each of the local extensions generated at the sliding site add up, contributing to the relatively high strains measured at the macro scale. Greater distribution of sliding across the material translates to an increase in dissipated energy and toughening, which contributes to the outstanding crack-arresting performance of nacre. In order to activate an increasing number of sliding sites, a hardening mechanism must take place at the interfaces to make it increasingly difficult to slide platelets further. Without hardening, sliding would lead to localization, that is, the platelets would only slide along a single band perpendicular to the loading direction, leading to small failure strains. The work hardening is caused by progressive platelet locking generated by the waviness of the platelets.[35] The waviness in some areas makes the platelets thicker at their periphery, effectively generating geometric "dovetails" (Figure 2.9c,d). Upon sliding, the end of the platelet is pulled out and the dovetail generates progressive locking and hardening. As a result, the spreading of these deformations over large volumes increases toughness.

The size of platelets or aspect ratio is well defined in nacre, which maximizes both composite strength and fracture toughness.[36,37] During platelet sliding, the forces are transferred through shear from a continuous polymeric matrix to a stiff discontinuous reinforcing element under tensile loading conditions. A simple shear-lag analysis on laminar structures with a platelet length L and thickness t can yield a critical aspect ratio, $s = t/L$, which is equal to the ratio of fracture strength of platelets (σ_f) to the shear strength (τ_y) of polymer between the platelets ($s_c = \sigma_f/\tau_y$).[37] Based on the shear-lag analysis, large s allows for the transfer of progressively higher stresses to the platelets and thus increases the strength of composite (Figure 2.9b). However, when $s > s_c$, the transferred stresses are high enough to fracture the stiff platelets and thereby cause catastrophic failure of the composite (Figure 2.9a).[38] Conversely, If $s < s_c$, several toughening mechanisms are enabled, including platelet pullout, inelastic and viscoelastic deformation of the polymer matrix, interfacial hardening, microcracking, and crack bridging. Composites combining both high strength and high toughness should therefore be achieved if the aspect ratio of the reinforcing platelets is selected to be slightly below the critical value.[34] Experiments have shown that the selection of platelets with aspect ratio slightly below the expected critical value for different biological composites is indeed a design principle used in nacre to achieve the highest strength and toughness levels possible with the initial set of building blocks available in the natural environment.[34] This quantitative design principle provides an effective approach for fabricating high-performance bioinspired composites using stronger synthetic building blocks, as is discussed in section 2.5.

The waviness of the platelets is also critical to strength and toughness, and the angle of waviness in nacre also seems optimal during its long evolution.[39] Figure 2.9e shows a parametric map of failure modes for natural and artificial nacres as a function of tablet geometry. With no dovetail (θ) there is almost no hardening after yielding and peak stress is relatively low because of the lack of an interfacial hardening mechanism. With a small dovetail angle ($\theta = 1°$, and maintaining the same dovetail overlap length), the degree of sliding, the failure stress, and strain are all improved significantly, and the energy dissipation per unit volume increases by more than 100% compared to the 0° sample. The interfacial hardening effect was clearly seen in the stress–strain curves. With shorter dovetail lengths, softening and platelet pullout occurred at lower stresses and strains, resulting in dramatically less energy dissipation because of shorter sliding distance of platelets over each other. For larger dovetail angles, the interfacial hardening strength was greater than that of the tablets themselves, and localized brittle tablet failure was observed. Interestingly, the platelet geometry of the artificial sample exhibiting the greatest energy dissipation falls directly within the range of tablet geometries found in natural nacre, indicating that the same design principle is applied to artificial nacre (Figure 2.9e).[39]

2.4.3 Strengthening Mechanism in Nacre

The basic function of shells is to protect the soft body of the mollusk against attack from predators, debris, and rocks moved by the current. A mollusk shell loaded from the exterior will

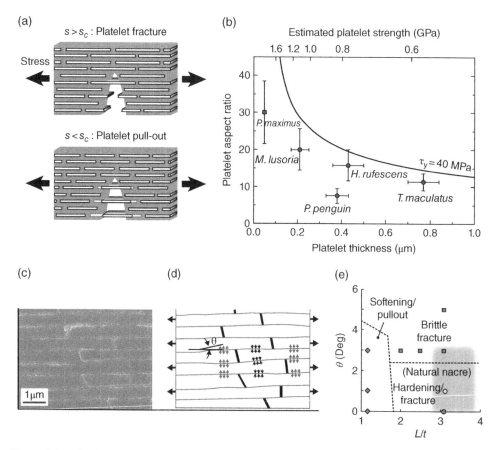

Figure 2.9 Platelet pullout and interfacial hardening in nacre. (a) Schematics showing the effect of the platelet aspect ratio on the failure mode of reinforced composites. (b) The aspect ratio of $CaCO_3$ platelets in the nacreous layer of different mollusk shell species, suggesting that the platelet aspect ratio (s) is optimized to be slightly lower than the expected theoretical critical value s_c (the solid line, τ_y, is the shear strength of the organic matrix around the platelets). Source: Studart (2012).[34] Reproduced with permission of John Wiley and Sons. (c) SEM image of a few dovetail like features at the periphery of the tablets. (d) Outline of the tablets contours, showing some of the stresses involved when nacre is stretched along the tablets. In addition to shear the interface is subjected to normal compression (black arrows), which generates resistance to tablet pullout. Equilibrium of forces at the interfaces requires tensile tractions at the core of the tablets. Source: Barthelet & Espinosa (2007).[40] Reproduced with the permission of Springer. (e) Failure mode diagram suggesting that the optimum dovetail geometry found in the synthetic composites matches with the shape of platelets in nacre. Source: Espinosa et al. (2011).[39] Reproduced with permission of Elsevier.

experience bending, which translates into tension in the inner nacreous layer. More severe loads may lead to the cracking of the outer brittle layer from contact stresses. In that case it is the nacreous layer that will ensure the integrity of the shell. For these two situations tensile stresses along the platelets are prominent in nacre, and nacre is probably constructed to maximize strength along that direction.[2] When platelets are subject to tensile stress, flaws in the platelets will cause stress concentration and crack propagation, which reduce the strength from the theoretical value ($\sim E/30$). According to fracture mechanics, the maximum stress (σ_{max}) that a material can sustain when a pre-existing crack of length a is present is given by the Griffith equation:

$$\sigma_{max} = \sqrt{\frac{2\gamma_s E}{\pi a}} = \frac{YK_{Ic}}{\sqrt{\pi a}} \tag{2.1}$$

where E is Young's modulus, γ_s is the surface (or damage) energy, Y is a geometric parameter, and K_{Ic} is the fracture toughness. The Griffith fracture criterion (Equation 2.1) can be applied to predict the flaw size (a_{cr}) at which the theoretical strength σ_{th} is achieved. For certain materials, material strength scales inversely with the flaw size a: *smaller is stronger*. This is a rational for pursuing nanomaterials. With typical values for the fracture toughness (K_{Ic}), σ_{th}, and E, the critical flaw size is in the range of tens of nanometers. Based on mechanics analysis on various biomaterials, Gao et al.[41] proposed that at sufficiently small dimensions (less than the critical flaw size), materials become insensitive to flaws, and the theoretical strength ($\sim E/30$) should be achieved at the nanoscale. This may explain why the size of platelets is small, reaching the nanoscale. However, the strength of the material will be determined by fracture mechanisms operating at all hierarchical levels.

2.4.4 Toughening Mechanisms in Nacre

In a composite material, the most successful toughening mechanism is crack deflection around reinforcements with *weak interfaces*, with "bridging" of the crack by the unbroken reinforcements restraining growth. Macroscopic toughness is set by the work required to pull apart these restraining ligaments and is thus enhanced at crack sizes larger than the microstructural scale. Fracture toughness K_{Ic} is a function of both microstructure and crack growth. For a fiber of radius r and embedded length l, the work per unit area to pull the fiber out against a interfacial friction τ is $W \propto l^2 \tau / r$,[42] increasing as the fiber sizes r and l increase together. Nanomaterials may therefore naturally have high strength but can only have high toughness if nanocomposite toughening mechanisms can be activated.

The toughening of shells is attributed to at least ten mechanisms,[43] and most of them are applicable to nacre. Based on the compositions, these toughening mechanisms can be organized into three categories: (1) submicrometer interactions of platelets (e.g., interfacial delamination, crack deflection, plate interlocking, mineral bridging, and plate pullout), (2) energy dissipation induced by the organic phase (e.g., organic bridging between plates, ligament formation, chain unfolding, and breaking of crosslinks), and (3) nanostructural toughening (e.g., rotation and sliding of nanograins, and organic bridging between nanograins). Among the mechanisms acting at the submicrometer scale within the brick and mortar structure, energy-dissipating sliding of the platelets at the submicrometer scale plays a predominant role in contributing to significant toughening of nacre.[39] Nano-order mechanisms are, however, important in controlling the platelet sliding behavior. Figure 2.10 illustrates the mechanisms classified according to the components involved and the dimensions at which the mechanism operates.

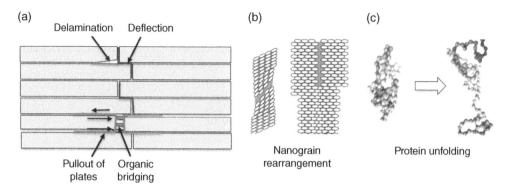

Figure 2.10   Hierarchical toughening mechanism of nacre. (a) An inter-platelet mechanism operating at the submicrometer scale, (b) an intra-platelet mechanism of the order of several tens of nanometers, and (c) individual organic molecules (nm scale).[44]

2.4.4.1   Toughening due to submicrometer interaction of platelets

There are three hallmarks of toughening mechanisms, interfacial debonding, plate bridging and pullout, which are well known in conventional laminar composites and fiber-reinforced composites, but these microfracture-accumulating mechanisms occur at two orders of magnitude lower in nacre than in synthetic composites. For example, in laminar composites, when the layer thickness is decreased 100 times, a hundredfold increase in the interface is introduced within the material.[44] Much more energy is dissipated in accumulating microcracks, delamination, and crack deflection during platelet sliding. The advantages of a segmented, that is, brick and mortar, structure over a continuous layer structure have been demonstrated in a large-scale model composite.[45]

In these mechanisms, interfacial interlocking plays an important role in toughening nacre. In conventional laminar composites, the sliding resistance decreases monotonically due to the decrease of contact area by pullout and the decrease of frictional coefficient by wear. However, in nacre, the platelet thickness is nonuniform and dovetail-shaped platelets are observed. During the sliding, the interface sliding resistance increases as the sliding proceeds. When the frictional resistance at a sliding interface becomes larger than the resistance for another interface, the sliding interface stops and the other interface starts to slide instead. This mechanism causes macroscopic work hardening, and more and more load is needed for further deformation. This mechanism is unique in nacre and significantly enhances the energy dissipation of platelet pullout and macroscopic work-hardening behavior.

2.4.4.2   Toughening contributed by the organic phase

The unique organic phase in nacre itself dissipates energy by viscoelastic deformation or redistributes the stress near the crack tip. The mechanical characteristics of the matrix at the molecular level are attributed to proteins with modular structures. When platelets are separated during tensile loading, the organic matrix ligaments between separating platelets are stretched and then subsequently fail and recoil. The organic phase shows nonlinear saw-tooth load–displacement curves due to unfolding of the domains, which increases energy dissipation during the deformation. Viscoelastic reversible behavior was observed and attributed to unfolding and refolding protein domains similar to those in silkworm and spider silks.[34]

2.4.4.3 Toughening via nanograin deformation

The platelet is composed of nanograins, as described in section 2.4.1. Although it is adequate to treat platelets as single crystals in macroscopic mechanical tests, the fracture behavior of platelets at the nanoscale may also contribute to the overall toughness of nacre. In platelets, organic matter is trapped inside the tablet in both an islet-like and sheet-like manner. More interestingly, the crystal structure of aragonite scaffold is continuous and homodromous throughout the whole tablet even in this case.[46] Nanoindentation showed that single platelets exhibit ductility and viscoelastic behavior, suggesting that the interaction between nanograins and organic material occurs at the nanoscale.[47] *In situ* tension and bending tests revealed rotation and deformation of nanograins.[48] It is now believed that these toughening mechanisms, such as nanograin rotation, separation, shearing, and bridging by organic material within the platelets (Figure 2.11), contribute to overall damage tolerance of nacre.

2.4.4.4 Toughening over hierarchical structure levels

Hierarchical structures responsible for mechanical protection of shells occur at various scales. The reason for this is that the origin of the intrinsic (plastic deformation) mechanisms tends to be on the smaller, submicrometer length scales, akin to the nanometer scale of dislocation Burgers vectors in metals, whereas the processes of extrinsic toughening and fracture occur on much coarser length scales typically well into the micrometer range.[50,51] On a macroscopic level (from millimeters down to tens of micrometers), the whole shell itself has a hybrid

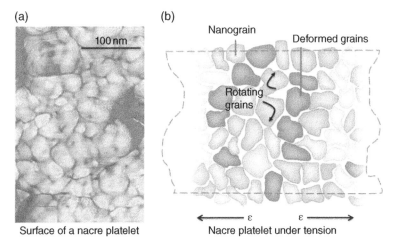

Figure 2.11   (a) Atomic force microscopy image of surface nanograins on an individual nacre platelet from California red abalone and (b) schematic of nanograin rotation under tension.[49] Source: Li (2006).[48] Reproduced with permission of the American Chemical Society.

structure consisting of several layers, such as prismatic, cross-lamellar, foliated and nacreous layers. Each layer has a different role against external mechanical actions, for example a columnar layer provides a hard protective layer on the outer side of the shell and nacre provides a tough structural component on the inner side.[51] On the smaller scale (from tens of micrometers down to nanometers), the cross-lamellar layer consists of three- or four-order lamella,[52,53] which provides damage tolerance behavior by accumulating multiscale fractures.

2.4.5 Strengthening/Toughening Mechanisms in Other Hard Tissues

Apart from seashells, there are many mineralized tissues that exhibit high strength and toughness. Typically these tissues form a protective shield or structural support, for example bone, deep sea sponge *Euplectella* species, the exoskeleton of an arthropod, radiolarians, diatoms, antler bone, tendon, cartilage, tooth enamel, and dentin. Most hard biological materials incorporate minerals in soft matrices, mostly to achieve the stiffness required for structural support or armored protection.[54] Universal building blocks, including the triple helical tropocollagen molecule or the occurrence of collagen fibrils in various types of collagenous tissues (e.g., bone, tendon, and cornea), can be found in these hard tissues although other features are highly specific to tissue types, such as particular filament assemblies, β-sheet nanocrystals in spider silk, or tendon fascicles (Figure 2.12).[55] Approximately 60 different minerals have been identified in biological processes, but the most common ones are calcium carbonate (found in mollusk shells) and hydroxyapatite (present in teeth and bones). The occurrence of

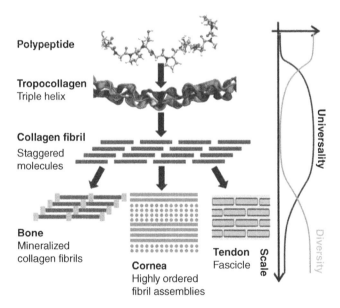

Figure 2.12 Universality and diversity in the structure of collagen-based biological materials.[31] Beyond the fibril scale, structural features vary significantly, here shown for bone, cornea, and tendon. This analysis reveals how biological materials with diverse properties are created through the use of universal features, a materials design paradigm that could be exploited also for the use synthetic materials.[56] Source: Espinosa (2009).[31] Reproduced with permission of Elsevier.

hierarchical structures is crucial to combine seemingly disparate properties such as high strength and high robustness.[55]

Bone is a typical biological material with an excellent balance of strength, stiffness, toughness, and light weight. Skeletal bone is composed of hierarchical assemblies of tropocollagen molecules, tiny hydroxyapatite crystals and water. On a volumetric basis, bone consists of ~33–43 vol% minerals, 32–44 vol% organics, and 15–25 vol% water. The strength and toughness of bone strongly depend on the interplay between different structural levels – from the molecular/nanoscale interaction between crystallites of calcium phosphate and an organic framework, through the micrometer-scale assembly of collagen fibrils, to the millimeter-level organization of lamellar bone.[57,58] In human cortical bone, there are a number of toughening mechanisms across different length scales, as shown in Figure 2.13.[59,60] Intrinsic toughening, that is, plasticity, derives from a fibrillar sliding mechanism on the scale of tens to hundreds of nanometers, the length scales associated with mineralized collagen fibrils. Molecular uncoiling and intermolecular sliding of collagen, fibrillar sliding of collagen bonds, and microcracking of the mineral matrix are identified at this length scale. At the microscale, the principal source of toughness in bone is extrinsic and arises from crack bridging and crack deflection as a growing crack encounters the more mineralized interfaces of the osteonal

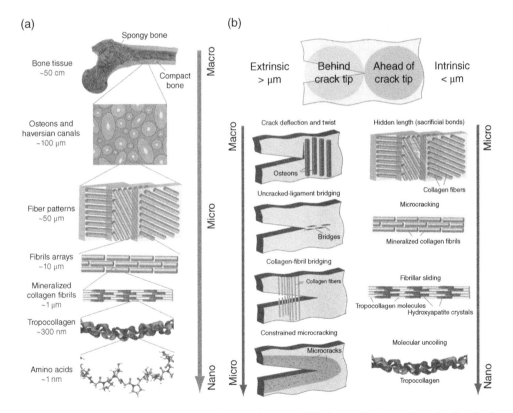

Figure 2.13 (a) Hierarchical structure of human bone and (b) the mechanisms of toughening (both intrinsic and extrinsic), which are directly related to a specific structural scale.[59,60] Source: Balasubramanian *et al.* (2013).[58] Reproduced with permission of Nature Publishing Group.

structures.[45] Collagen fibril bridging, uncracked ligament bridging, and crack deflection and twisting occur at this length scale.[50,61] As the size and spacing of the osteons are in the range of tens to hundreds of micrometers, the characteristic length scales for these events approach millimeter dimensions. The toughening effect in antlers has been estimated as crack deflection 60%, uncracked ligament bridges 35%, and collagen as well as fibril bridging 5%.[61] A particularly important feature in bone is that the fracture toughness increases as the crack propagates. This occurs by the activation of the extrinsic toughening mechanisms. In this manner, it becomes gradually more difficult to advance the crack.

Another example of tough hard tissue is the highly sophisticated, nearly purely mineral skeleton of glass sea sponges (*Euplectella aspergillum*) (Figure 2.14a). Glass is widely used as a building material in the biological world, despite its fragility. Sea sponges have evolved to effectively reinforce this inherently brittle material. Hexactinellid sponges can synthesize unusually long and highly flexible fibrous spicules for their skeletal systems. These spicules consist of a central core of monolithic hydrated silica, surrounded by alternating layers of hydrated silica and organic material (Figure 2.14c). This biological silica is amorphous and, within the spicules, consists of concentric layers, separated by an organic material, silicatein.[62,63]

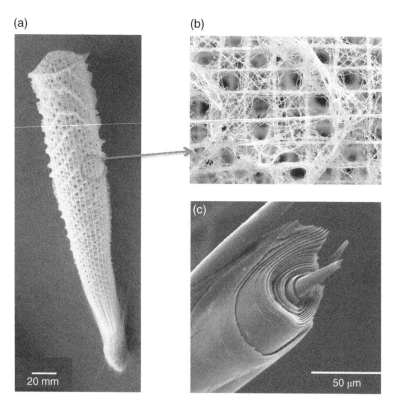

Figure 2.14 Silica sponge and the intricate scaffold of spicules. (a and b) The skeleton of *Euplectella aspergillum* and (c) fracture surface. Each spicule is a circumferentially layered rod: The interfaces between the layers assist in arresting crack propagation. Source: Woesz *et al.* (2006).[66] Reproduced with permission of Cambridge University Press.

This material exhibits exceptional flexibility and toughness compared with brittle synthetic glass rods of similar length scales.[64] The flexure strength of the spicule notably exceeds (by approximately fivefold) that of monolithic glass.[65] The principal reason for this is the presence of interfaces, which can arrest and/or deflect the crack. Under loading, fracture of this laminated structure involves cracking of the constituent silica and crack deflection through the intervening thin organic layers, leading to a distinctive stair step-like fracture pattern (Figure 2.14c). Crack deflection mitigates the high stress concentration that would otherwise be present at the crack tip, resulting in high spicule strength and toughness. A larger number of individual glass layers protect the spicule more effectively from this type of damage. Thicker inner layers help to enhance the mechanical rigidity of the spicule, whereas the thinner outer layers effectively limit the depth of crack penetration. This design strategy thus prevents the structure from failing catastrophically, as one would expect for a non-laminated glass rod.

Apart from the remarkable mechanical properties of their individual spicules, hexactinellid sponges are also known for their ability to form remarkably complex hierarchically ordered skeletal systems (Figure 2.14b). Seven hierarchical levels in the sponge skeleton are identified as major fundamental construction strategies, such as laminated structures, fiber-reinforced composites, bundled beams, and diagonally reinforced square-grid cells, etc. The resultant structure could thus be considered as a textbook example in mechanical engineering. Using the information gained from the study of these structures, it is possible to develop new design strategies for the synthesis of robust lightweight scaffolds for load-bearing applications and architectural designs of more cost-effective and energy-efficient buildings.[66]

2.5 Biomimetic Design and Processes for Strong and Tough Ceramic Composites

2.5.1 Biomimetic Design Principles for Strong and Tough Materials

As discussed above, there are various strengthening and toughening mechanisms in biological materials. Some of strengthening and toughening mechanisms found in nature and used in synthetic materials are listed in Table 2.2. In many biological materials such as mollusk shells, bones, and sea sponges, layered structures are common, consisting of strong inorganic platelets embedded in a soft, ductile organic matrix. The nano size of the platelets ensures the limited flaw size and thus high strength of the materials. Although the inorganic constituents (e.g., silica, calcium carbonates, and phosphates) are inherently weak, the high strength of the inorganic building blocks is ensured by limiting at least one of their dimensions to the nanoscale.[37] These tiny building blocks are usually organized into a hierarchical structure spanning over many length scales. Changes in the fraction of the inorganic phase (i.e., degree of mineralization) lead to hybrid materials ranging from soft tissues, such as calcified tendons, to strong, hard structures, such as bone and nacre. A good balance of strength and toughness is achieved by activating various toughening mechanisms at different length scales. These high-performance materials designed by nature provide a spectrum of materials design blueprints for advanced engineering materials.

Unusually long and highly flexible fibrous spicules are synthesized by hexactinellid sponges for their skeletal systems. These spicules consist of a framework made of a central core of monolithic hydrated silica, surrounded by alternating layers of hydrated silica and organic

Table 2.2 Some biomimetic design strategies for strengthening and toughening in hard materials.

Mechanisms	Strategies	Biological prototypes	Illustrations	Examples of possible application
Strengthening	Use nano-platelet/ nano-fibers	Nacres, bones		Reinforcement in composites
Stiffening/ lightweight	Hierarchically ordered skeletal framework	Silica sponge		Strong and ultra-lightweight materials
Stiffening/ toughening	Introduce two continuous interinfiltrated phases	Nacres, tooth enamel		Laminates, composites, wear-resisting
Toughening	Introduce soft polymer thin layers to generate weak but tough interface	Nacres, teeth, bone		Laminates, composites
	Dovetail: waviness of platelets	Nacres		Laminates, composites
	Micro/nano-asperities for interfacial interlocking	Nacres		Laminates, composites
	Reinforcing phase bridging	Nacres		Laminates, composites

material, which may inspire the design of stiff, tough, ultra-lightweight materials. A network of nearly isotropic microscale unit cells with high structural connectivity and nanoscale features, similar to sponge skeletal structures, has been designed and synthesized.[67] These microlattices can be produced with polymers, metals, or ceramics as constituent materials by projection microstereolithography (an additive micromanufacturing technique) combined with nanoscale coating and postprocessing. These materials exhibit ultrastiff properties across more than three orders of magnitude in density, regardless of the constituent material.

Interface structures, such as interfacial roughness, bridging by either soft polymer or ceramic, dovetails etc., are found to be effective in enhancing the toughness of biological structural systems. These surface nanoarchitectures on the aragonite tablets in nacre increase the contact area between organic and inorganic materials, resulting in higher magnitude of plastic strains after yield, while the morphology of the interdigitation of cranial bones was found to play an important role in determining the mechanical properties of the suture joints.[68] Introduction of these surface nanoarchitectures into the synthetic materials systems could also significantly increase the toughness of the materials. The following examples demonstrate how the principles found in natural composites are applied to fabricate high-performance biomimetic composites combining high tensile strength and ductile behavior.

2.5.2 Layered Ceramic/Polymer Composites

A high-performance nacre-like platelet-reinforced polymer composite has been developed by replicating the biological design principle of layered composites.[38] As described in section 2.4.2, one of the design principles of nacre being strong and tough is to ensure platelets with nanoscale thickness and an aspect ratio slightly smaller than the critical aspect ratio s_c, which is given by the ratio between the fracture strength of the platelet and the shear yield strength of the polymer matrix ($s_c = \sigma_f / \tau_y$).[37] Considering that synthetic alumina platelets are estimated to be fivefold stronger than the aragonite platelets in nacre, the critical aspect ratio in an alumina-reinforced synthetic system also increases by a factor of five in comparison to the biological material for polymeric matrices of comparable strengths. As a result, replication of nacre's design principle requires that the aspect ratio of synthetic alumina platelets is five times higher than that of the aragonite counterparts. Guided by these design principles, alumina platelets with an estimated σ_f of 2 GPa and a chitosan polymer with τ_y around 40 MPa were chosen for the artificial hybrid materials. With a high aspect ratio of 50, the combination of these materials should lead to strength values higher than that of nacre while ensuring that fracture occurs by platelet pullout.

The layered hybrid materials were fabricated by layer-by-layer assembly that sequentially deposits inorganic and organic layers at ambient conditions (Figure 2.15).[38] First, a smooth and perfectly oriented monolayer of platelets was formed at the surface of water on ultra-sonication. The two-dimensional assembled platelets were transferred to a glass substrate by dip-coating and were then spin-coated with an organic layer of chitosan solution. The thickness of the polymer layer was controlled by changing the chitosan concentration in the spin-coating solution. Repetition of these steps in a sequential manner led to multilayered inorganic–organic films with a total thickness typically less than a few tens of a micrometer. Free-standing films were obtained by peeling them off the substrate with a razor blade.

Similar to nacre, the mechanical behavior of the biomimetic layered alumina platelet/polymer composites strongly deviated from the linear elastic regime, when the yield tensile strength of the organic matrix was reached. At this stress condition, yielding of the polymer

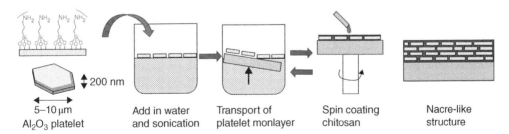

Figure 2.15 Surface modified platelets are assembled at the air–water interface to produce a highly oriented layer of platelets after ultrasonication. The two-dimensional assembled platelets are transferred to a flat substrate and afterwards covered with a polymer layer by conventional spin-coating. Source: adapted from Bonderer *et al.* (2011).[38]

phase between the inorganic platelets led to a pronounced plastic deformation of the composite film. Because of load transfer to the platelets, the tensile stresses required for plastic yielding increased from 50 MPa to values as high as 300 MPa when the volume fraction of platelets V_p was increased from 0 to 0.15. Most remarkably, films containing inorganic volume fractions up to 0.15 fractured at a total strain (ϵ_{rupt}) typically between 4 and 35% as a result of extensive plastic yielding of the polymeric matrix before rupture. Flexible composite films that are simultaneously strong (tensile strength $\sigma_c \sim 300$ MPa) and ductile ($\epsilon_{rupt} \sim 20\%$) were successfully produced at rather low V_p values ($V_p = 0.15$).[38]

Similar layer-by-layer assembly approaches were used to fabricate clay/polyelectrolytes nanocomposites[69] and alumina microplatelets–graphene oxide nanosheet–poly(vinyl alcohol) (Al_2O_3/GO–PVA).[70] The clay/polyelectrolyte nanocomposites with alternating organic and inorganic layers exhibited the unique saw-tooth pattern of differential stretching curves attributed to the gradual breakage of ionic crosslinks in polyelectrolyte chains. The tensile strength approached that of nacre, whereas their ultimate Young's modulus was similar to that of lamellar bones. Structural and functional resemblance makes clay/polyelectrolyte multilayers a close replica of natural biocomposites. In the (Al_2O_3/GO–PVA artificial nacre, Al_2O_3 and GO–PVA act as "bricks" and "mortar", respectively (Figure 2.16a–c). The artificial nacre has a hierarchical "brick-and-mortar" structure and exhibits excellent strength (143 ± 13 MPa) and toughness (9.2 ± 2.7 MJ/m³) (Figure 2.16d), which are superior to those of natural nacre (80–135 MPa, 1.8 MJ/m³). It was demonstrated that the multiscale hierarchical structure of ultrathin GO nanosheets and submicrometer-thick Al_2O_3 platelets can deal with the conflict between strength and toughness, thus leading to excellent mechanical properties that cannot be obtained using only one size of platelet.[70] This provides a strategy for the biomimetic design of new composites with excellent strength and toughness.

Bulk nacre-like composites were also fabricated by the so-called cold finger method.[71] Using controlled directional freezing of the suspension placed on a copper cold finger, large cylindrical porous ceramic scaffolds (50 mm diameter and 50 mm high) were produced with architectures that were templated by ice crystals.[71,72] Directional freezing was first employed to promote the formation of lamellar ice with prescribed dimensions; this then acted as the "negative" for creation of the layered ceramic scaffolds, which were subsequently freeze-dried and sintered (Figure 2.17). First, ceramic nanoparticles were dissolved in water. Under directional cooling conditions, the water started to crystallize directionally and nanoparticles stuck between crystals formed lamellar structures. After the solution had completely frozen, the structure was freeze dried, and then a heat treatment was used to bond the particles and the polymer resin was infiltrated into porous areas. In the last step, additional hot pressing is applied to change the lamellar structure to a brick-and-mortar structure and decrease resin content. Figure 2.18a shows the microstructure of the fabricated nanocomposites. In the process, cellulosic polymer was selected and dissolved in the water to form nacre-like interlaminar connections. The polymer makes interconnections between platelets.

The mechanical properties of such composites with different fabrication conditions are shown in Figure 2.18c,d. Whereas grafting to improve Al_2O_3/PMMA interface adhesion resulted in a mildly higher strength and initiation toughness for lamellar structures, a very significant increase was seen for the brick-and-mortar structure. However, the most notable feature of the synthetic composites is that they replicate the mechanical behavior of natural materials (Figure 2.18b). Specifically, the brick-and-mortar structures with ~80% alumina display a remarkable 1.4% strain to failure. There is exceptional toughness for crack growth

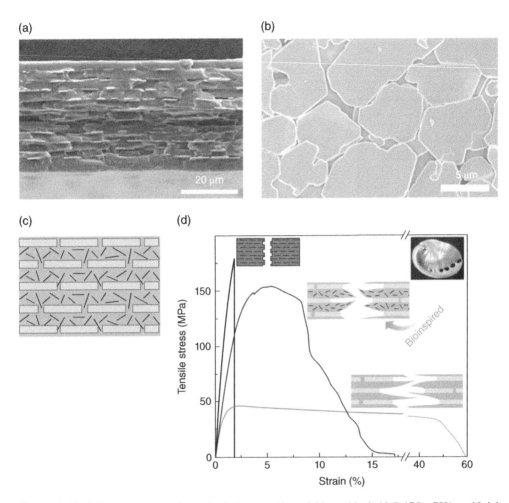

Figure 2.16 Microstructure and mechanical properties of hierarchical Al_2O_3/GO–PVA artificial nacre.[70] (a) SEM image of sample cross-section, showing the highly oriented layered structure, (b) the assembled Al_2O_3 monolayer onto GO–PVA layer, (c) schematic of layer-by-layer bottom-up assembly of multilayered Al_2O_3/GO–PVA artificial nacre, and (d) comparison of mechanical properties of hierarchical Al_2O_3/GO–PVA artificial nacre with those of layered GO/PVA and Al_2O_3/PVA with the same volume fraction of inorganic platelets, indicating that the nacre-like hierarchical structure helps to achieve a good balance of strength and toughness. Source: Wang *et al.* (2015).[70] Reproduced with permission of the American Chemical Society.

(Figure 2.18c). The bulk hybrid ceramic-based materials have high yield strength and fracture toughness (~200 MPa and ~30 MPa·m$^{1/2}$), comparable to those of aluminum alloys. The toughness values of the best hybrid materials are an order of magnitude higher than standard homogeneous nanocomposites consisting of 500 nm Al_2O_3 particles dispersed in PMMA with the same nominal composition. This unique processing and model materials can be used for the synthesis of bioinspired ceramic-based composites with balanced strength and toughness.

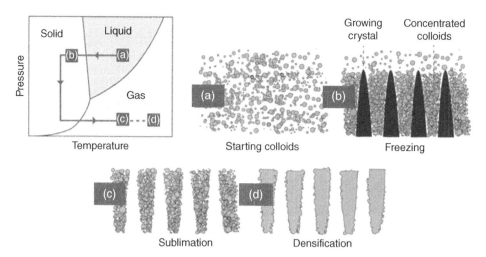

Figure 2.17 Schematic diagram of the ice-templating principles. While the ceramic slurry is freezing, the growing ice crystals expel the ceramic particles, creating a lamellar microstructure oriented parallel to the direction of the freezing front. A small fraction of particles are entrapped within the ice crystals by tip splitting, leading to the formation of inorganic bridges and roughness on the walls. Source: Deville (2013).[73] Reproduced with permission of Cambridge University Press.

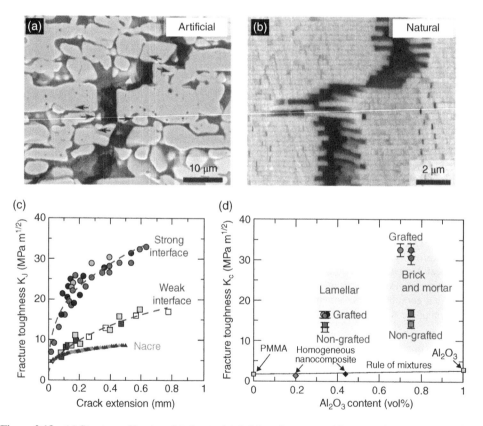

Figure 2.18 (a) Structure of ice-templated materials brick-and-mortar architectures, the structure and fracture of which is similar to (b) natural nacre. (c) Mechanical response and toughening mechanisms in the synthetic hybrid composites. These materials show exceptional toughness for crack growth, similar to natural composites, and display significant rising R curve behavior. (d) Fracture toughness of the lamella and brick-and-mortar Al_2O_3–PMMA composites, showing their exceptional fracture toughness as compared to that of their constituent phases. Source: Launey et al. (2009).[72] Reproduced with permission of Elsevier.

2.5.3   Layered Ceramic/Metal Composites

The cold finger freeze-casting method has been extended to fabricate nacre-like layered ceramic/metal composites.[74] After lamellar alumina scaffolds were fabricated in bulk form, the porous ceramics were subsequently infiltrated with molten Al–Si eutectic alloy by fitting the scaffold in an alumina crucible with pieces of Al–Si on top. This alloy was selected for the infiltration because of its low melting point, viscosity, and contact angle of Al–Si on alumina, which facilitate infiltration at relatively low pressures. The assembly was heated to 900 °C in a 10^{-4} Pa vacuum, and then gaseous argon gas was admitted into the furnace up to a pressure of approximately 70 kPa to force the molten alloy into the scaffold porosity. The final composites consisted of a fully dense Al–Si-infiltrated Al_2O_3 scaffold.

The nacre-like ceramic–metal hybrid materials have excellent combinations of strength and toughness, and can operate at elevated temperatures. In particular, the lamellar Al_2O_3/Al–Si composites with 36 vol% ceramic content display strengths of approximately 300 MPa and fracture toughness values that exceed 40 MPa.m$^{1/2}$.[74] The fracture surface of the ceramic–metal hybrids shows nacre-like fracture mode–platelet pullout. However, a refinement of the microstructure leads to a brittle behavior owing to a transition from a high-toughness "multiple-cracking" fracture mode to a low-toughness "single-cracking" mode. Nevertheless, the natural design concept of a hard ceramic phase with optimal volume fraction providing material strength, separated by a softer "lubricant" phase to relieve high stresses in order to enhance toughness, can be successfully used to further develop new structural materials with unprecedented damage-tolerance properties.

2.5.4   Ceramic/Ceramic Laminate Composites

A bioinspired approach was developed based on the cold finger processing techniques for the fabrication of bulk ceramics without a ductile phase (e.g., metals, polymers) and with a unique combination of high strength (470 MPa), high toughness (22 MPa m$^{1/2}$), and high stiffness (290 GPa).[45] Inspired by the nacre structure, researchers fabricated nacre-like ceramic materials with five structural features, spanning several length scales. As shown in Figure 2.19, the artificial nacre consists of long-range structural order at the macroscale; closely packed ceramic platelets of dimensions identical to that of nacre at the microscale, ceramic bonds

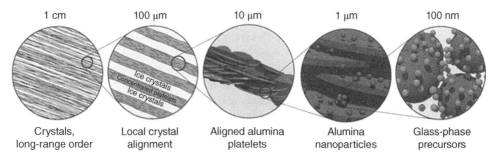

Figure 2.19   Design strategy describing the control at multiple scales of structural self-organization and densification strategy.[45] Self-organization of all the structural features occurs during the freezing stage. The growth of ordered ice crystals triggers the local alignment of platelets. Alumina nanoparticles and liquid-phase precursors are entrapped between the platelets. Source: Bouville *et al.* (2014).[45] Reproduced with permission of Nature Publishing Group.

(bridges) linking the platelets, nanoasperities at the surface of the platelets, and a secondary phase with lower stiffness ensuring load redistribution, crack deflection, and delamination.

The process is based on ice-templating, similar to that discussed in the previous section. The first step of the process is the preparation of an aqueous colloidal suspension containing all the required building blocks and processing additives, followed by ice templation of this suspension and a pressure-assisted sintering step at 1500 °C. The elementary building blocks of the structure are the platelets, with the same dimensions as nacre platelets (500 nm thickness, 7 µm diameter). When the suspension is directionally frozen, the metastable growth of the ice crystals repels and concentrates the particles present in the suspension.[75] The concentration of the particles occurs at a length scale where self-assembly of platelets can occur. Alumina nanoparticles (100 nm) incorporated in the initial suspension serve as a source of both inorganic bridges between the platelets and nanoasperities at the surface of the platelets, similar to what is observed in nacre. Finally, instead of polymer resin infiltrated as interlaminar connections, smaller nanoparticles (20 nm) of liquid-phase precursors (silica and calcia) are added to help fill the remaining gaps during the sintering stage. The material is thus composed of 98.5 vol% alumina, 1.3 vol% silica, and 0.2 vol% calcia. This simple strategy, where all the constituents are incorporated in the initial suspension and self-organized in one step, allows precise and easy tailoring of the final material composition. The long-range order of the ice crystals is obtained through freezing under a flow method. After the porous samples are preformed, they are simply pressed and sintered by field-assisted sintering. The preformed materials are 86% porous, but almost all of the macroporosity was removed after the pressing at 100 MPa prior to sintering.[76] During the sintering, the platelets are rearranged by a liquid phase formed by nanometric silica through capillary forces, facilitating platelet packing under the applied load by lubricating the contact points and filling the pore space between them. At the same time, the nanometric alumina particles form strong bridges by sintering to the platelets.[76]

The microstructure of the resulting synthetic material is very similar to that of natural nacre at several length scales; almost all the characteristic features of nacre can be found in the synthetic materials. The packing of platelets presents short-range order, compared to the microstructure of nacre, although the long-range order is not as perfect. The waviness of the stacking in the nacre-like ceramic, estimated at ±15° around the main orientation, comes from the organization of the ice crystals. The secondary phase between platelets mimics the organic layer in nacre. In the synthetic materials there are some alumina bridges between adjacent platelets, and nanoasperities on the platelets, analogous to those in nacre.

The nacre-like alumina is stiffer, with a flexural modulus of 290 GPa, compared to 40 GPa for nacre. Although the material lacks tough phases such as metals and polymers, more than 50% inelastic strain to failure is observed. This mechanical response is similar to that observed in ductile metallic or organic materials. During the loading nacre-like ceramic experiences ceramic platelet pullout, and crack deflection and delamination at the platelet interface (Figure 2.20a).[45] The reinforcement mechanisms are extrinsic, which means that no true ductility (plastic deformation without crack propagation) is observed. The nacre-like alumina exhibits a flexural strength of 470 MPa at room temperature and 420 MPa at 600 °C, a value essentially similar to that of the reference alumina. The properties are retained at relatively high temperature (600 °C, Figure 2.20c,d), with a stress intensity factor for crack initiation (K_{Ic}) of 4.7 MPa m$^{1/2}$ and a maximum toughness K_{JC} of 21 MPa m$^{1/2}$. The maximum toughness is extremely high, around 22 MPa m$^{1/2}$, if the local deflection as well as the other dissipation

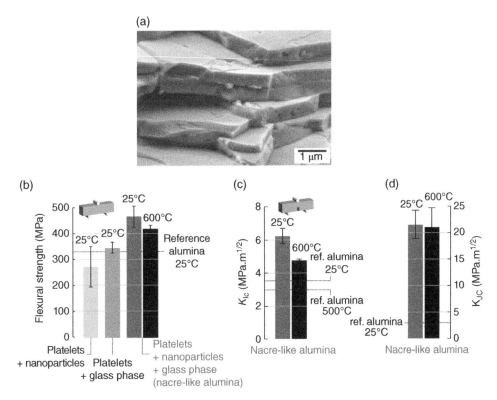

Figure 2.20 Mechanical properties of nacre-like alumina and nacre. (a) Detail of the fracture surface in nacre-like alumina, showing the crack deflection and delamination at the platelet interface. (b) Flexural strength of three different compositions: liquid phase (5 vol%) and platelets, nanoparticles (3 vol%) and platelets, and nacre-like alumina (1.5 vol% and 3 vol% of liquid phase and nanoparticles, respectively). The dotted line corresponds to an equiaxed fine-grain alumina. Error bars indicate standard deviation. (c, d) Comparison of toughness for crack initiation (K_{Ic}) and stable crack propagation (K_{Ic}) in nacre-like alumina and reference alumina at room temperature (25 °C) and high temperature (600 °C). Source: Bouville et al. (2014).[45] Reproduced with permission of Nature Publishing Group.

mechanisms with a J-integral are taken into account, and using the equivalence in the stress intensity factor. This corresponds to a 350% increase compared to the K_{Ic} toughness (600% increase with respect to the reference alumina). This far exceeds that of nacre and is equivalent to the best brick-and-mortar polymer/ceramic composites developed previously.[71] Because only mineral constituents are needed, these ceramics retain their mechanical properties at high temperatures (600 °C). This material and process verify the material-independent design principles from nacre and provide an excellent example of the bioinspired design of strong and tough ceramic materials for various applications.

Layered ceramic composites with shell architecture at the micrometer level were also fabricated on the macroscopic scale by different processes.[77,78] One example is the larger-scale segmented layered SiC composites with ceramic platelets, which make use of the toughening mechanism of nacre to overcome the brittleness of ceramics. Nacre-like ceramic composites were fabricated by using thin square tiles (50 mm × 50 mm × 200 µm) of SiC doped with boron.[77]

The tiles were coated with graphite to retain a weak interface after sintering. Under a three-point bending test, the crack is deflected along the weak interfaces, preventing catastrophic failure. The load-deflection curve shows the load continuing to rise after crack growth starts. The laminated composite exhibited a toughness and an increase in work of fracture by factors of 5 and 100, respectively, over monolithic SiC.

References

1. Vincent, J.F.V., Bogatyreva, O.A., Bogatyrev, N.R., Bowyer, A. & Pahl, A.K. Biomimetics: its practice and theory. *Journal of the Royal Society Interface* **3**, 471–482 (2006).
2. Barthelat, F. Nacre from mollusk shells: a model for high-performance structural materials. *Bioinspiration & Biomimetics* **5**, 035001 (2010).
3. Eadie, L. & Ghosh, T.K. Biomimicry in textiles: past, present and potential. An overview. *Journal of the Royal Society Interface* **8**, 761–775 (2011).
4. Gosline, J.M., Guerette, P.A., Ortlepp, C.S. & Savage, K.N. The mechanical design of spider silks: From fibroin sequence to mechanical function. *Journal of Experimental Biology* **202**, 3295–3303 (1999).
5. Romer, L. & Scheibel, T. The elaborate structure of spider silk Structure and function of a natural high performance fiber. *Prion* **2**, 154–161 (2008).
6. Simmons, A.H., Michal, C.A. & Jelinski, L.W. Molecular orientation and two-component nature of the crystalline fraction of spider dragline silk. *Science* **271**, 84–87 (1996).
7. Heimer, S. Wunderbare Welt der Spinnen. *Urania* **271**, 12 (1988).
8. Papadopoulos, P., Ene, R., Weidner, I. & Kremer, F. Similarities in the Structural Organization of Major and Minor Ampullate Spider Silk. *Macromolecular Rapid Communications* **30**, 851–857 (2009).
9. Keten, S., Xu, Z.P., Ihle, B. & Buehler, M.J. Nanoconfinement controls stiffness, strength and mechanical toughness of beta-sheet crystals in silk. *Nature Materials* **9**, 359–367 (2010).
10. Emile, O., Le Floch, A. & Vollrath, F. Biopolymers: Shape memory in spider draglines. *Nature* **440**, 621–621 (2006).
11. Liu, Y., Shao, Z.Z. & Vollrath, F. Relationships between supercontraction and mechanical properties of spider silk. *Nature Materials* **4**, 901–905 (2005).
12. Bell, F.I., McEwen, I.J. & Viney, C. Fibre science – Supercontraction stress in wet spider dragline. *Nature* **416**, 37–37 (2002).
13. Brown, C.P. *et al.* The critical role of water in spider silk and its consequence for protein mechanics. *Nanoscale* **3**, 3805–3811 (2011).
14. Shao, Z.Z. & Vollrath, F. Materials: Surprising strength of silkworm silk. *Nature* **418**, 741–741 (2002).
15. Nguyen, A.T. *et al.* Crystal networks in silk fibrous materials: from hierarchical structure to ultra performance. *Small* **11**, 1039–1054 (2015).
16. Fantner, G.E. *et al.* Sacrificial bonds and hidden length dissipate energy as mineralized fibrils separate during bone fracture. *Nature Materials* **4**, 612–616 (2005).
17. Brown, C.P. *et al.* Rough fibrils provide a toughening mechanism in biological fibers. *ACS Nano* **6**, 1961–1969 (2012).
18. Meyers, M.A., McKittrick, J. & Chen, P.Y. Structural biological materials: critical mechanics–materials connections. *Science* **339**, 773–779 (2013).
19. Miserez, A., ScottWasko, S., Carpenter, C.F. & Waite, J.H. Non-entropic and reversible long-range deformation of an encapsulating bioelastomer. *Nature Materials* **8**, 910–916 (2009).
20. Tjin, M.S., Low, P.L. & Fong, E. Recombinant elastomeric protein biopolymers: progress and prospects. *Polymer Journal* **46**, 444–451 (2014).
21. Beese, A.M. *et al.* Bio-Inspired Carbon Nanotube–Polymer Composite Yarns with Hydrogen Bond-Mediated Lateral Interactions. *Acs Nano* **7**, 3434–3446 (2013).
22. Keten, S. & Buehler, M.J. Nanostructure and molecular mechanics of spider dragline silk protein assemblies. *Journal of the Royal Society Interface* **7**, 1709–1721 (2010).
23. Beese, A.M. *et al.* Hydrogen-bond-mediated bio-inspired high performance double-walled carbon nanotube-polymer nanocomposite yarns. *Abstracts of Papers of the American Chemical Society* **246** (2013).

24. Meyers, M.A., Chen, P.-Y., Lopez, M.I., Seki, Y. & Lin, A.Y.M. Biological materials: A materials science approach. *Journal of the Mechanical Behavior of Biomedical Materials* **4**, 626–657 (2011).
25. Li, X.D. & Huang, Z.W. Unveiling the formation mechanism of pseudo-single-crystal aragonite platelets in nacre. *Physical Review Letters* **102** (2009).
26. Thompson, J.B. et al. Direct observation of the transition from calcite to aragonite growth as induced by abalone shell proteins. *Biophysical Journal* **79**, 3307–3312 (2000).
27. Suzuki, M. *et al.* An acidic matrix protein, Pif, is a key macromolecule for nacre formation. *Science* **325**, 1388–1390 (2009).
28. Oaki, Y., Kotachi, A., Miura, T. & Imai, H. Bridged nanocrystals in biominerals and their biomimetics: Classical yet modern crystal growth on the nanoscale. *Advanced Functional Materials* **16**, 1633–1639 (2006).
29. Wegst, U.G., Bai, H., Saiz, E., Tomsia, A.P. & Ritchie, R.O. Bioinspired structural materials. *Nature Materials* **14**, 23–36 (2015).
30. Knipprath, C., Bond, I.P. & Trask, R.S. Biologically inspired crack delocalization in a high strain-rate environment. *Journal of the Royal Society Interface* **9**, 665–676 (2012).
31. Espinosa, H.D., Rim, J.E., Barthelat, F. & Buehler, M.J. Merger of structure and material in nacre and bone – Perspectives on de novo biomimetic materials. *Progress in Materials Science* **54**, 1059–1100 (2009).
32. Barthelat, F., Tang, H., Zavattieri, P.D., Li, C.M. & Espinosa, H.D. On the mechanics of mother-of-pearl: A key feature in the material hierarchical structure. *Journal of Mechanical and Physical Solids* **55**, 306–337 (2007).
33. Lopez, M.I., Chen, P.Y., McKittrick, J. & Meyers, M.A. Growth of nacre in abalone: Seasonal and feeding effects. *Materials Science and Engineering C – Materials* **31**, 238–245 (2011).
34. Studart, A.R. Towards high-performance bioinspired composites. *Advanced Materials* **24**, 5024–5044 (2012).
35. Katti, K.S., Katti, D.R., Pradhan, S.M. & Bhosle, A. Platelet interlocks are the key to toughness and strength in nacre. *Journal of Materials Research* **20**, 1097–1100 (2005).
36. Fleischli, F.D., Dietiker, M., Borgia, C. & Spolenak, R. The influence of internal length scales on mechanical properties in natural nanocomposites: A comparative study on inner layers of seashells. *Acta Biomaterialia* **4**, 1694–1706 (2008).
37. Gao, H.J., Ji, B.H., Jager, I.L., Arzt, E. & Fratzl, P. Materials become insensitive to flaws at nanoscale: Lessons from nature. *Proceedings of the National Academy of Sciences of the United States of America* **100**, 5597–5600 (2003).
38. Bonderer, L.J., Studart, A.R. & Gauckler, L.J. Bioinspired design and assembly of platelet reinforced polymer films. *Science* **319**, 1069–1073 (2008).
39. Espinosa, H.D. *et al.* Tablet-level origin of toughening in abalone shells and translation to synthetic composite materials. *Nature Communications* **2**, 173 (2011).
40. Barthelat, F. & Espinosa, H.D. An experimental investigation of deformation and fracture of nacre-mother of pearl. *Experimental Mechanics* **47**, 311–324 (2007).
41. Gao, H.J. & Yao, H.M. Shape insensitive optimal adhesion of nanoscale fibrillar structures. *Proceedings of the National Academy of Sciences of the United States of America* **101**, 7851–7856 (2004).
42. Curtin, W.A. Stochastic damage evolution and failure in fiber-reinforced composites. *Advanced Materials Mechanics* **36**, 163–253 (1999).
43. Mayer, G. Rigid biological systems as models for synthetic composites. *Science* **310**, 1144–1147 (2005).
44. Kakisawa, H. & Sumitomo, T. The toughening mechanism of nacre and structural materials inspired by nacre. *Science and Technology of Advanced Materials* **12**, 064710 (2011).
45. Bouville, F. *et al.* Strong, tough and stiff bioinspired ceramics from brittle constituents. *Nature Materials* **13**, 508–514 (2014).
46. Wang, S.-N. et al. Nanostructured individual nacre tablet: a subtle designed organic–inorganic composite. *Crystal Engineering Communications* **17**, 2964–2968 (2015).
47. Li, X.D., Chang, W.C., Chao, Y.J., Wang, R.Z. & Chang, M. Nanoscale structural and mechanical characterization of a natural nanocomposite material: The shell of red abalone. *Nano Letters* **4**, 613–617 (2004).
48. Li, X.D., Xu, Z.H. & Wang, R.Z. In situ observation of nanograin rotation and deformation in nacre. *Nano Letters* **6**, 2301–2304 (2006).
49. Ortiz, C. & Boyce, M.C. Materials science – Bioinspired structural materials. *Science* **319**, 1053–1054 (2008).
50. Launey, M.E., Buehler, M.J. & Ritchie, R.O. On the mechanistic origins of toughness in bone. *Annual Review of Materials Research* **40**, 25–53 (2010).
51. Luz, G.M. & Mano, J.F. Biomimetic design of materials and biomaterials inspired by the structure of nacre. *Philosophical Transactions. Series A, Mathematical, Physical, and Engineering Sciences* **367**, 1587–1605 (2009).

52. Li, X.D. & Nardi, P. Micro/nanomechanical characterization of a natural nanocomposite material – the shell of Pectinidae. *Nanotechnology* **15**, 211–217 (2004).
53. Kamat, S., Su, X., Ballarini, R. & Heuer, A.H. Structural basis for the fracture toughness of the shell of the conch *Strombus gigas*. *Nature* **405**, 1036–1040 (2000).
54. Currey, J.D. The design of mineralised hard tissues for their mechanical functions. *Journal of Experimental Biology* **202**, 3285–3294 (1999).
55. Alberts, B. & May, R.M. Scientist support for biological weapons controls. *Science* **298**, 1135–1135 (2002).
56. Buehler, M.J. & Yung, Y.C. Deformation and failure of protein materials in physiologically extreme conditions and disease. *Nature Materials* **8**, 175–188 (2009).
57. Hassenkam, T. *et al*. High-resolution AFM imaging of intact and fractured trabecular bone. *Bone* **35**, 4–10 (2004).
58. Balasubramanian, P., Prabhakaran, M.P., Sireesha, M. & Ramakrishna, S. collagen in human tissues: structure, function, and biomedical implications from a tissue engineering perspective. *Advanced Polymer Science* **251**, 173–206 (2013).
59. Ritchie, R.O. The conflicts between strength and toughness. *Nature Materials* **10**, 817–822 (2011).
60. Ritchie, R.O., Buehler, M.J. & Hansma, P. Plasticity and toughness in bone. *Physics Today* **62**, 41–47 (2009).
61. Launey, M.E., Chen, P.Y., McKittrick, J. & Ritchie, R.O. Mechanistic aspects of the fracture toughness of elk antler bone. *Acta Biomaterialia* **6**, 1505–1514 (2010).
62. Cha, J.N. *et al*. Silicatein filaments and subunits from a marine sponge direct the polymerization of silica and silicones in vitro. *Proceedings of the National Academy of Sciences of the United States of America* **96**, 361–365 (1999).
63. Aizenberg, J. *et al*. Skeleton of Euplectella sp.: Structural hierarchy from the nanoscale to the macroscale. *Science* **309**, 275–278 (2005).
64. Sarikaya, M. *et al*. Biomimetic model of a sponge-spicular optical fiber – mechanical properties and structure. *Journal of Materials Research* **16**, 1420–1428 (2001).
65. Qiao, L., Feng, Q.L., Wang, X.H. & Wang, Y.M. Structure and mechanical properties of silica sponge spicule. *Journal of Inorganic Materials* **23**, 337–340 (2008).
66. Woesz, A. *et al*. Micromechanical properties of biological silica in skeletons of deep-sea sponges. *Journal of Materials Research* **21**, 2068–2078 (2006).
67. Zheng, X. *et al*. Ultralight, ultrastiff mechanical metamaterials. *Science* **344**, 1373–1377 (2014).
68. Zhang, Y., Yao, H., Ortiz, C., Xu, J. & Dao, M. Bio-inspired interfacial strengthening strategy through geometrically interlocking designs. *Journal of the Mechanical Behavior of Biomedical Materials* **15**, 70–77 (2012).
69. Tang, Z.Y., Kotov, N.A., Magonov, S. & Ozturk, B. Nanostructured artificial nacre. *Nature Materials* **2**, 413–U418 (2003).
70. Wang, J., Qiao, J., Wang, J., Zhu, Y. & Jiang, L. Bioinspired hierarchical alumina–graphene oxide–poly(vinyl alcohol) artificial nacre with optimized strength and toughness. *ACS Applied Materials & Interfaces* **7**, 9281–9286 (2015).
71. Munch, E. *et al*. Tough, bio-inspired hybrid materials. *Science* **322**, 1516–1520 (2008).
72. Launey, M.E. *et al*. Designing highly toughened hybrid composites through nature-inspired hierarchical complexity. *Acta Materialia* **57**, 2919–2932 (2009).
73. Deville, S. Ice-templating, freeze casting: Beyond materials processing. *Journal of Materials Research* **28**, 2202–2219 (2013).
74. Launey, M.E. *et al*. A novel biomimetic approach to the design of high-performance ceramic–metal composites. *Journal of the Royal Society Interface* **7**, 741–753 (2010).
75. Deville, S. *et al*. Metastable and unstable cellular solidification of colloidal suspensions. *Nature Materials* **8**, 966–972 (2009).
76. Hunger, P.M., Donius, A.E. & Wegst, U.G.K. Platelets self-assemble into porous nacre during freeze casting. *Journal of Mechanical Behaviors of Biomedical Materials* **19**, 87–93 (2013).
77. Clegg, W.J., Kendall, K., Alford, N.M., Button, T.W. & Birchall, J.D. A simple way to make tough ceramics. *Nature* **347**, 455–457 (1990).
78. Mayer, G. New classes of tough composite materials – Lessons from natural rigid biological systems. *Materials Science and Engineering: C* **26**, 1261–1268 (2006).

3

Wear-resistant and Impact-resistant Materials

3.1 Introduction

Surface contacts and relative motion are commonly seen in our daily lives and are essential to technological applications with moving parts. When two surfaces in contact are moving, friction and wear occur. While contact friction leads to the force resisting the relative motion, the progressive removal of material from a surface in sliding or rolling contact results in wear of the surfaces. As a serious consequence, wear may result in performance degradation and/or damage to components. Direct and consequential annual loss to industries in the United States due to wear is estimated to be approximately 1–2% of GDP.[1] Currently, various methodologies have been used to minimize wear. In addition to the traditional method of lubrication, various methods have been developed, including soft or hard film coating, multiphase alloying, and composite structuring on contact surfaces. Since the impact of wearing on the economy is significant, better solutions for protecting surfaces from wear are needed to enable further advances in materials surface engineering.

Biological systems also involve friction and wear. Teeth, jaws, and mandibles, for example, are subjected to intense cyclic mechanical loading during feeding and mastication, and must sustain a high tolerance to contact forces and wear in order to fulfill this critical function.[2] To protect these important organs against wear and abrasion for survival, many living organisms, after billion-year perfection, have developed unique materials and mechanisms to minimize the loss of material in their armor or protective structures from wear. From the engineering perspective, mimicking these hard tissues that have evolved to fulfill functions demanding a high wear tolerance many lead to the design of novel wear-resistant materials.

Among numerous mineralized tissues in the mammalian body, enamel stands out as the tissue with the most robust mechanical properties.[3] Mature enamel is the hardest material that vertebrates ever form and is the most highly mineralized skeletal tissue, comprising 95–97% carbonated hydroxyapatite (HA) by weight with less than 1% organic material.[4] Although

Biomimetic Principles and Design of Advanced Engineering Materials, First Edition. Zhenhai Xia.
© 2016 John Wiley & Sons, Ltd. Published 2016 by John Wiley & Sons, Ltd.

enamel is hard tissue like bone, it has distinct architecture, pathology, and biological mechanisms mediating its formation. In addition to enamel, there is a dentin–enamel junction (DEJ) between dentin and enamel, which provides a unique crack-arrest barrier for flaws formed in the brittle enamel.[4] Inspired by the unique enamel, DEJ, and dentin structures, synthetic materials have been fabricated to enhance wear resistance and damage tolerance. In this chapter, the structures and biomimetic design principles of enamel, dentin, and DEJ are discussed to understand how these materials achieve exceptional mechanical properties and functionalities, and how to utilize these bioinspired strategies to design and build engineering materials with the desired wear properties.

Mineralized tissue and non-mineralized biological composites with high impact and energy absorbance fulfill protection tasks or act as body armor. Some well-known examples of such composites are hooves, turtle shell, ram horn, and armadillo shell. All of these materials are composed of the same protein, keratin, which is also found in many other tough materials, for example skin, hair, fur, claws, and hooves.[5] The unique material structures and mechanical deformation behavior of the materials result in an excellent ability to resist penetration and absorb energy. These lessons from nature can be used as inspiration for the development of lightweight armor systems. Some remarkable materials in terms of energy absorption and impact resistance are discussed in this chapter, with examples of biomimetic materials with enhanced energy absorption and impact resistance.

3.2 Hard Tissues with High Wear Resistance

Dental friction and wear is an inevitable lifetime process due to normal oral function. As an important masticatory organ, human teeth are subjected to friction and wear every day. Thus, anti-wear properties are critical for both human teeth and dental restorative materials.[6] The human tooth is a hierarchical structure consisting of two major layers, enamel and dentin. Enamel, the outer layer, must retain its shape as well as resist fracture and wear during load-bearing functioning, ideally for the entire life of the individual. Unlike bone, damage to teeth is not repairable. Teeth are subject to a range of loading conditions: (1) direct contact with other objects and/or opposing teeth and (2) normal and sliding contact that results in wear. During mastication of food, teeth can generate masticatory forces ranging from 28 N to more than 1200 N, and the contact area can be as small as a few square millimetres.[7] The outer layer, enamel, is exposed to both normal and shear forces due to oblique contact with opposite teeth and external objects during mastication. It is amazing that the tooth, a multilayer structure with an outer coating of less than 2 mm, can sustain and survive such high cyclic stresses for a long period of time. The outstanding performances of teeth in wear and fatigue are closely related to their compositional and hierarchal microstructural characteristics, and they can serve as a role model for the biomimetic design of high wear-resistant materials.

3.2.1 Teeth: A Masterpiece of Biological Wear-resistance Materials

Teeth are among the most distinctive (and long-lasting) structures of mammal species. While their primary function is mastication of food, some species use them for attacking prey and for defense. Tusks are very similar to teeth in many aspects among various taxa. A tooth consists of four distinct tissues: pulp, dentin, cementum, and enamel.[8] As schematically shown in

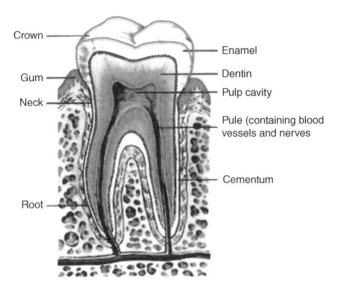

Figure 3.1 Structure of tooth, schematically illustrating enamel, dentin, cementum, pulp, and dentino–enamel junction. Source: Zhou & Zheng (2006).[10] Reproduced with permission of SAGE Publications Ltd.

Figure 3.1, the uppermost 1–2 mm of the tooth crown is enamel containing a high mineral content, whereas underneath the enamel is dentin, which tightly connects and supports enamel. The composition of dentin is different from enamel but similar to that of bone. The cementum, the mineralized layer, surrounds the root of the tooth covering the dentin layer and some of the enamel layer. The tooth is anchored to the alveolar bone (jawbone) through the periodontal ligament between the cementum and the alveolar bone. Within a tooth, tubules extend perpendicularly from the pulp and are surrounded by the mineral phase. These tubules are filled with fluid in the live animal. At the very center is the pulp and a vascular, nerve-containing core that connects to the body's main vascular and nervous systems.[8]

The macroscale architecture with specific zones of enamel and dentin, with transition layers between them, contributes to whole-tissue functioning for mastication. Enamel is the hardest material in our body while dentin is less mineralized and less hard than enamel. The DEJ transition region exhibits a gradual transition from dentin to enamel. This region is considered to be a primitive area of the tooth that serves as a toughening mechanism because of its flexible nature.[3,9] These components work together to give rise to a tough, damage-tolerant, and abrasion-resistant tissue through their unique architectures and mineral compositions.

3.2.2   Microstructures of Enamel, Dentin, and Dentin-enamel Junction

Dental enamel is a masterpiece of bioceramics, secreted by cells known as ameloblasts, which differentiate at the DEJ and migrate outward towards what becomes the surface of the crown. The prominent hierarchy of enamel structure is achieved through the highly precise regulation of the biomineralization procedure. As shown in Figure 3.2, the highly organized hierarchical structure can be divided into several different levels ranging from

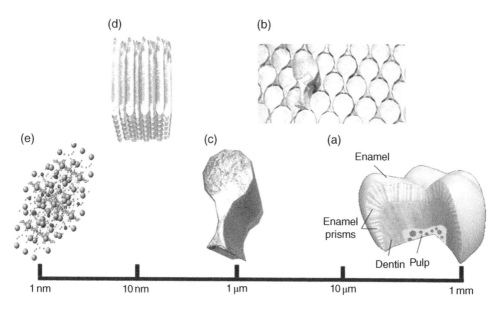

Figure 3.2 Computer-generated images of a posterior tooth showing the hierarchical structure of tooth enamel on the micro- to nanoscale levels. (a) Enamel is the external tooth layer protecting the softer underlying structures (dentin and pulp). (b) Enamel consists of prisms bands, "keyhole"-shaped structures, packed together. (c) A single "keyhole"-shaped structure is made up of enamel prisms. (d) Each enamel prism is composed of multiple CAP nanocrystals. (e) The atomic composition of a single CAP nanocrystal. Source: Eimar (2012).[12] Reproduced with permission of Elsevier.

nanoscale to microscale: HA nanocrystals, nanofibrils, fibril/fiber bundles, microscale prisms, keyhole shaped structures, and prism bands.[11,12] The smallest structural units are needle or plate-like HA crystallites (Figure 3.2e), which have a roughly rectangular shape in cross-section with a mean width of 68 nm and a mean thickness of 26 nm. These particles are bonded together by enamelin and assemble into nanofibrils (Figure 3.2d). These mineralized nanofibrils are further assembled into larger-scale fibers, which are further organized and bundled together by organic molecules into larger-scale structures known as "keyhole-shaped" enamel rods (Figure 3.2c). Within the rods, the orientation of the HA crystallites varies, depending on the location of the minerals. In the central part of the rod, the crystallite plates are parallel to the rod axis while those near the edge of the rods usually have an angle of nearly 45° to the longitudinal axis of the rods.[13] The next level of the teeth is the arrangement of rods that determines the enamel type. Human enamel consists of ~5 μm diameter rods encapsulated by thin protein-rich sheaths that are arranged parallel in a direction perpendicular to the DEJ from dentin to the outer enamel surface (Figure 3.2b). In some areas, rods may twist together or change their direction slightly to reinforce the whole structure.[14] The mechanical properties of these rod units and surrounding organic interrod sheath structures are different in the orientation of HA crystals: the rod contains aligned crystallites, whereas the mineral in the interrod is less ordered and rich in protein.[15] The hardness and elastic modulus of the sheaths are 74% and 53% lower, respectively, than for the rods.[15] In addition to rods and interrods, there is a structure called aprismatic enamel that contains HA crystals that show no mesoscale or macroscale alignment.

The enamel represents the highest mineralized biological material. Overall, human tooth enamel is composed of mineralized crystal with 92–96% inorganic substances, 1–2% organic materials, and 3–4% water by weight.[16,17] The enamel crystals are hexagonal and contain relatively large amounts of carbonate ions (~2–5 wt%) and small amounts of incorporated trace elements such as F, Cl, Mg, K, and Fe. Enamel carbonated apatite (CAP) is divided into two types, type A and type B. Type A (~11% of overall enamel CAP) is formed when the carbonate ion replaces the hydroxyl ion (OH⁻), whereas type B (~89% of overall enamel CAP) is formed when the carbonate ion replaces the phosphate ion within the crystal. The proteins and collagen lying between crystallites bond HA crystallites together level by level, forming the hierarchical structure of enamel. Moreover, with special nanomechanical properties, these proteins may introduce unique strengthening and toughening mechanisms to regulate in bulk the mechanical responses of enamel. These basic units are organized hierarchically by proteins to form the tough tissue of enamel, which can withstand high forces and resist wear.[3]

Similar to enamel, dentin is also a hierarchical structure with 45 vol% apatite, 30 vol% collagen, and 25 vol% fluid (Figure 3.3). At the nanoscale, dentin consists of a carbonated nanocrystalline apatite mineral phase (~50 vol%) and a felt-work of type I collagen fibrils. The collagen fibrils (~30 vol%, roughly 50–100 nm in diameter) are randomly oriented in a plane perpendicular to the direction of dentin formation. Within this collagen scaffold, the mineral locates at two sites: intrafibrillar (inside the periodically spaced gap zones in the collagen fibril) and extrafibrillar (in the interstices between the fibrils). The mineral crystallites, ~5 nm in thickness, are needle-like near the pulp; the shape continuously progresses to plate-like with proximity to the enamel.[18]

The distinctive feature of dentin is the distribution of cylindrical tubules (1–2 μm diameter) that run from the soft, interior pulp to the DEJ. The tubule density varies between 4900 and

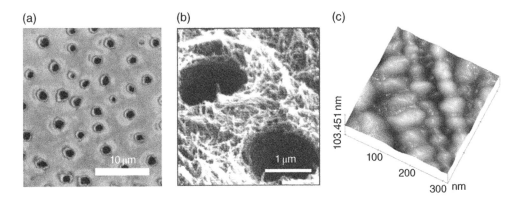

Figure 3.3 (a) SEM image of fully mineralized dentin specimens showing the tubule lumens with surrounding cuffs of peritubular dentin. Source: Kinney *et al.* (2001)[20]. Reproduced with permission of Springer. (b) SEM image of a fixed, demineralized dentin specimen showing the collagen fibrils that are randomly oriented in the plane perpendicular to the tubule lumens. Source: Marshall *et al.* (1997).[21] Reproduced with permission of Elsevier. (c) A higher-resolution AFM image of an unfixed specimen obtained in water. The AFM image shows the periodic 67-nm hole and overlap zones characteristic of the type I collagen fibrils found in dentin and bone. Source: Habelitz *et al.* (2002).[22] Reproduced with permission of Elsevier.

57,000 mm^{-2} increasing from the enamel region to the interior.[5] These tubules are surrounded by a collar of highly mineralized peritubular dentin (~1 µm thick) and are embedded within a matrix of mineralized collagen, called intertubular dentin.[19] At this length scale, dentin can be considered a continuous fiber-reinforced composite, with the intertubular dentin matrix and the cylindrical fiber reinforcement composed of the tubule lumens with their associated cuffs of peritubular dentin (Figure 3.3b). The tubules run continuously from the DEJ to the pulp in coronal dentin, and from the cementum-dentin junction to the pulp canal in the root. The regular, almost uniaxial, aligned tubules could play an important role in enhancing the mechanical properties of teeth.

The DEJ also plays an essential role in enhancing the toughening properties of the teeth. This component is a complex structure that unites the brittle overlying enamel with the dentin that forms the bulk of the tooth. The DEJ itself has a hierarchical microstructure with a three-dimensional scalloped appearance along the interface. Although very thin, the DEJ is a functionally graded zone where the structure and properties transfer gradually from the enamel to dentin, making it a truly functionally gradient material. This property gradient stems from the morphology of the collagen, in which type I fibrils emanate from the dentin and project fibrils (~100 nm in diameter) perpendicular to the DEJ, and directly into the enamel across the DEJ and porosity gradually increases from dentin to enamel. In contrast, collagen fibrils in bulk dentin are either parallel or at an angle of less than 90 Å to the plane of the junction.[23] With these complex structures, the DEJ possesses at least three levels of microstructure: scallops with varying size and location, microscallops housed within each scallop, and a finer nanolevel structure within each microscallop.[23] This unique structure appears to confer excellent toughness and crack-arresting properties to the tooth, and is considered to be an excellent biomimetic model of a structure uniting dissimilar materials.

3.2.3 Mechanical Properties of Dental Structures

Teeth have widely varying functions, from crushing food to fighting. Although their functions are quite different, all teeth have similar structures (e.g., hard outer sheath mitigated by a soft but tough interior) and mechanical properties (e.g., strong, tough, and hard). Since the porosity gradient starts low at the surface and increases into the interior, the elastic modulus inversely changes. The mechanical properties of enamel, dentin, and DEJ are summarized in Table 3.1. Basically, enamel has a hardness (~1.5 GPa) that is three times higher than that of dentin (Figure 3.4). The elastic modulus of enamel is characteristically 60–80 GPa,[24] whereas the Young's modulus of dentin is highest for the highly mineralized peritubular dentin and is lower for the less mineralized intertubular dentin. Human dentin is strong hard tissue (its

Table 3.1 Mechanical properties of enamel, dentin, and DEJ (adapted from Imbeni et al. (2005),[23] Xu et al. (1998),[28] Elmowafy & Watts (1986),[31] Chen et al. (2008),[32] and Chun et al. (2014).[33]

Tissue	Young's modulus (GPa)	Hardness (GPa)	Compressive strength (MPa)	Fracture toughness (MPa.m$^{1/2}$)
Enamel	60–90	2.8–3.7	95–140	0.44–1.55
Dentin	18–22	0.53–0.63	230–370	3.0
DEJ				3.38

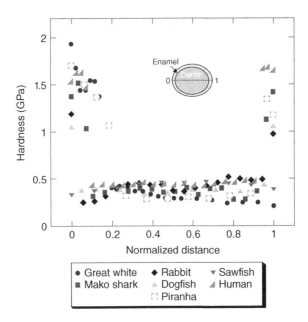

Figure 3.4 Vickers hardness number data across teeth for various taxa. The high values are for enamel and the lower values for dentin. Source: McKittrick (2010).[5] Reproduced with permission of Elsevier.

ultimate compressive strength is 250–350 MPa). Although elastic deformation prior to failure is relatively low (<2%), dentin does show some plasticity after elastic deformation under compression.[25,26] Under loading, the main crack grows by joining with satellite cracks, and with such a fracture mechanism dentin can demonstrate considerable plasticity, just like the plastic metals silver, nickel, and gold.[26,27]

The DEJ is a critical structure that combines hard and brittle enamel with tougher and softer dentin. The fracture toughness of human enamel typically ranges from ~0.7 MPa m$^{1/2}$ in the direction parallel to the enamel rods to ~1.3 MPa m$^{1/2}$ in the perpendicular direction.[28] Conversely, dentin, as a biological composite, is tougher than enamel and similar at the nano-structural level to bone. The anisotropic dentin has a K_c toughness ranging between 1.0 and 2.0 MPa m$^{1/2}$ in directions perpendicular and parallel to the tubules.[29,30] It was noticed that the work of fracture of the dentin increases near the DEJ. The measured fracture of DEJ is about 3.38 MPa m$^{1/2}$, very close to dentin. In spite of this, studies on the fracture process of the DEJ have suggested that the DEJ appears to have greater functional width than anatomic appearance; it probably undergoes plastic deformation during crack propagation. In other words, the DEJ most likely serves as a crack deflector and blunter to microstructurally distinct and mechanically tougher. Cracks tend to penetrate the (optical) DEJ and arrest when they enter the tougher mantle dentin adjacent to the interface due to the development of crack-tip shielding from uncracked-ligament bridging.[23]

Teeth are subject to intense cyclic mechanical loading during feeding and mastication. It is thus important to maintain the structural integrity of "hard" mineralized tissues since these tissues make up the primary load-bearing structures in the body. In general, these mineralized tissues display S/N (stress/number of cycles) curves similar to ductile metals, with N_f increasing

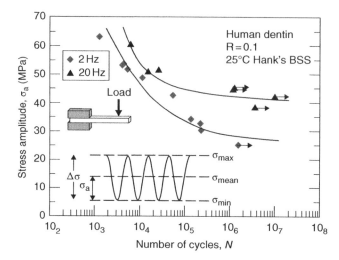

Figure 3.5   Traditional stress-life (*S/N*) fatigue data for human dentin at two different cyclic frequencies, tested in 25°C Hanks' Balanced Salt Solution (HBSS). Source: Nalla *et al.* (2003).[34] Reproduced with permission of John Wiley and Sons.

with decreasing stress, σ_a (Figure 3.5).[34] Among the microstructural features, tubule orientation can significantly affect fatigue life. When the tubules in human dentin are orientated parallel to the longitudinal beam axis, the fatigue life is significantly lower than when the tubules are oriented parallel to the loading direction, indicating that the tubules enhance mechanical properties, like traditional fiber-reinforced composites.[35] In addition to the microstructures, testing and environmental conditions also affect the fatigue life. *S/N* fatigue studies on human dentin have shown that by changing the frequency from 20 to 2 Hz, the fatigue limit and fatigue lifetimes (in terms of cycles) were both lowered, although when such S/N data were plotted on the basis of time, this frequency effect was reduced.[36] This suggests that the fatigue in dentin may be predominantly time, rather than cycle, that is, there may not be a true cyclic fatigue mechanism. When the teeth were exposed to lactic acid the fatigue strength significantly reduced, with nearly a 30% reduction in the apparent endurance limit (from 44 MPa to 32 MPa).[36] The reduction in pH also caused a significant decrease in the threshold stress intensity range required for the initiation of cyclic crack growth, and a significant increase in the incremental rate of crack extension. The reduction may be caused by demineralization of the tooth structure when exposed to lactic acid.[37]

3.2.4   Anti-wear Mechanisms of Enamel

Enamel is one of the best anti-wear and damage-tolerant biomaterials. Although cracks and craze lines are often observed in enamel, they rarely cause tooth fracture. The excellent wear-resistance of human tooth enamel stems from its unique microstructure. As shown in Figure 3.2, human tooth enamel has a hierarchical structure running from macroscale to nanoscale. The fundamental structure of enamel is the aligned microscale rods that have a unique arrangement (Figure 3.6a).[6] Between these rods is the tougher interrod enamel, which

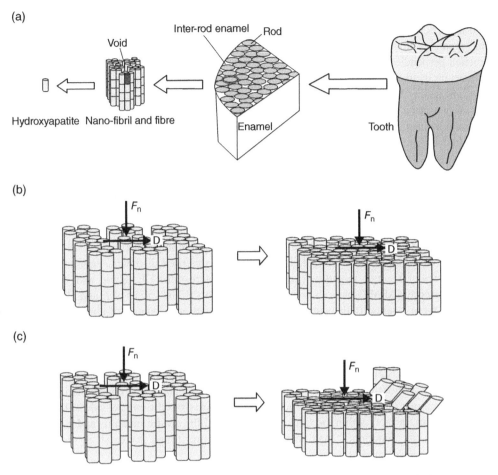

Figure 3.6 Schematic diagrams of enamel microstructure and wear process: (a) enamel microstructure, (b) wear process at low loads, and (c) wear process at high loads. Source: Zheng *et al.* (2003).[6] Reproduced with permission of Elsevier.

is richer in organic matter. The organic matter within the enamel can spread the damage laterally over a much larger volume and allow limited differential movement between adjacent rods to protect the enamel from catastrophic damage.[38] At low compressive loads the wear is mainly due to the collapse of nanofibrils and fibers, and plastic deformation (Figure 3.6b). During the process, the HA particles bonded together by enamelin are debonded by the external action of the normal load, causing a decrease in the size of the particles.[6] Under the high compressive load, HA particles would be detached from the surface of the enamel by the load, resulting in material removal and then particle packing, as shown in Figure 3.6c.[6] However, cracks are difficult to form even under high loads because of the voids between nanofibers within rods and the rich-in-protein interrod enamel between rods, therefore brittle delamination rarely occurs in enamel.

The importance of hierarchical structure on mechanical behaviors has also been verified by the fracture behavior of enamel. Because of its anisotropic microstructure, such as rod orientation, and organic components, cracks in the enamel axial section were significantly longer in the direction perpendicular to the occlusal surface rather than parallel. Moreover, the macroscale architecture includes specific zones of enamel that have unique characteristics that contribute to the whole tissue. Aprismatic regions of enamel have been proposed to be primitive areas of the tooth serving as a toughening mechanism due to their flexible nature.[3,9] While it is difficult to express resistance to mechanical wear in the same detail, it is clear that enamel hardness (compared to that being contacted) and toughness are the critical properties that resist damage.[39]

3.2.5 Toughening Mechanisms of the DEJ

The DEJ has a functionally graded microstructure that durably unites dissimilar hard brittle enamel and tough flexible dentin, and serves as a crack-arrest barrier for flaws formed in the brittle enamel. Unlike the interfaces in traditional ceramic composites, which are either too strong or too weak, the DEJ rarely fails other than in inherited disorders. Figure 3.7 shows typical values of Young's modulus E and hardness H across the DEJ along with the atomic force microscopy (AFM) image showing the indentations.[40] The values of E and H in enamel and dentin were essentially the same as in the bulk tissues at a distance of about 10 μm from the DEJ region ($E = 20$ GPa and $H = 0.7$ GPa for dentin, $E = 65$ GPa and $H = 3.5$ GPa for enamel). Within the DEJ region from dentin to enamel both E and H increase from bulk values of dentin to those of enamel. The toughness of the DEJ, measured by indentation, has a value between those of enamel and dentin (~5 to 10 times higher than enamel but ~75% lower than dentin),[41] indicating that a smooth transition occurs between the harder and stiffer enamel and the softer dentin. This transition appears to be a monotonic decrease from the enamel side of the DEJ to the bulk dentin. In fact, within the thickness of the DEJ the elastic modulus and hardness of enamel from its functional load-bearing direction are found to exist in exponential

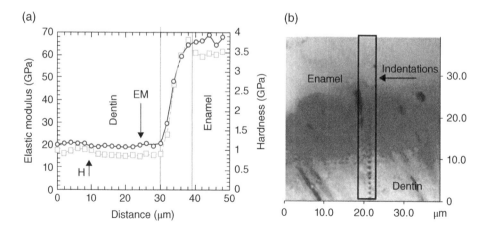

Figure 3.7 (a) Nanoindentation measurements of Young's moduli of dentin, dentino-enamel junction, and enamel. (b) Atomic force microscopy image of indented enamel, dentino-enamel junction, and dentin. Source: Marshall *et al.* (2001).[40] Reproduced with permission of John Wiley and Sons.

relationships with its normalized thickness.[42] The creep ability of enamel was also observed to increase towards the DEJ, therefore, in addition to porosity, there may also be a gradient in mineral volume fraction across the interface because the mineral fraction of the calcified tissue is closely related to its mechanical properties, such as modulus and hardness. The graded properties of the biocomposite are the result of both microstructural and compositional changes across the DEJ.

The DEJ zone is not a weak mechanical interface, but has a fracture toughness of 0.8 MPa.m$^{1/2}$.[23] Although some delamination occurs at the optical DEJ interface, damage is spread over a wide area and the cracks are arrested at the DEJ (Figure 3.8b). There are distinct toughening mechanisms at the DEJ interface.[23] The primary toughening mechanism involves the specialized aprismatic enamel close to the DEJ interface, which prevents catastrophic interfacial failure through multiple convoluted branching, cracks spread and diffuse damage over a wide area of adjacent enamel. A second toughening mechanism involves short microcracks in the DEJ adjacent to dentin with possible crack bridging. A third toughening mechanism involves plastic deformation of the DEJ without delamination.[23] These specific toughening mechanisms combine to control, diffuse, and spread damage over the DEJ zone, protecting the teeth from catastrophic failure.

The graded mechanical properties of DEJ are considered a major source of toughening. Finite element analysis showed that compared with no-graded interface, the smooth transition in mechanical properties of DEJ leads to the reduced development and distribution of stress at a lower level, along with lower peak stress at the DEJ under the loading point and smoother stress distribution along the DEJ. Because of smaller differences between the maximum and minimum principle stresses, the multilayer structure with the graded coating is more reliable, and these features ensure the integrity of the enamel and DEJ under functional cyclic loads.[42] Thus, the graded properties of enamel have an important role in maintaining the integrity of the multilayer tooth structure.

The concept of functionally graded materials (FGMs) derived from biomaterials such as DEJ provides a new approach for the improvement of dental post-material performance compared to traditional homogeneous and uniform materials. This technique allows the production of very different characteristics within the same material at various interfaces.[43] The functionally graded biomaterials for implants in medical and dental applications integrate dissimilar materials, without severe internal stress, by combining diverse properties in a single material. The toughening mechanisms related to the FGM can therefore be introduced in the materials. More broadly, the concept of FGMs can be extended to the biomimetic design of other engineering materials.

3.3   Biomimetic Designs and Processes of Materials for Wear-resistant Materials

3.3.1   Bioinspired Design Strategies for Wear-resistant Materials

Wearing is a complex, multifactorial process involving breaking, chipping, or cutting, and is strongly system dependent: the materials surface structures, contact stresses, contact time, wear environments (e.g., temperature and humidity), and the presence of a third body (lubricating layer) all strongly affect the wear response.[2] The importance of these various factors in contributing to wear damage may be different. From a materials science point of view, the

(a)

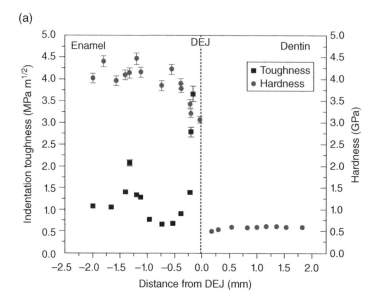

(b)

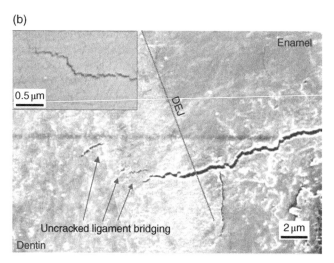

Figure 3.8　(a) Typical profiles of the Vickers hardness and the indentation toughness. Profiles taken normal to and across the DEJ region from the enamel to the dentin in a human molar. (b) EM examples of arrested cracks. Shown are cracks from the enamel that are normally incident on the optical DEJ and are arrested after propagating less than ~10 μm beyond the interface into the mantle dentin. Behind the arrested crack tip, numerous uncracked-ligament "bridges" can be seen; these are regions of uncracked material that oppose the opening of the crack and sustain load that would otherwise be used for crack growth. Such bridging, which is a form of cracktip shielding and is a prominent toughening mechanism in dentin and bone, acts to reduce the effective driving force for crack extension, thereby arresting the crack. Inset shows the lack of such bridging for cracks in the enamel. Cracking can also be seen near, and nominally parallel, to the optical DEJ. Source: Imbeni *et al.* (2005).[23] Reproduced with permission of Nature Publishing Group.

wear damage depends on a combination of materials properties, such as hardness, elastic modulus, or fracture toughness, and on the type of wear mechanisms. In fact, even for engineering materials, predicting the tribiological performance of a material remains rather elusive,[44] so "the harder the better" is not always true.[2] For example, a tool steel and chrome carbide iron with the same 600 BHN hardness are different in an abrasion application by as much as 5 times. Enamel materials have quite different structures and wear environments, and their wear response strongly depends on the biological function, for example food grasping, attacking prey and feeding, or shielding of a soft body, and whether the function occurs in a hydrating environment. These environments have a fundamental influence on the evolution of the chemistry and structural organization of these materials, so what can we learn from them and what characteristics contribute to the superb mechanical behavior of enamel coating?

One design principle that contributes to enamel's wear properties is its hierarchical structure. Unlike traditional ceramics, which usually consist of grains at micron level, enamel has a hierarchical microstructure and inelastic functions of organic components inside the structure. This unique structure is believed to be responsible for the long-lasting performance of teeth.[45] In teeth, high volume percent minerals are deposited in the biocomposites, which results in high hardness. In addition, reinforcement is achieved from nanocrystalline carbonated HA crystals, which have a platelet morphology with a large aspect ratio. This yields a higher Young's modulus than that of spherical particles. Finally, in mineralized tissues the mineral phase is nanocrystalline, therefore it does not fracture, but strengthens the matrix. There is a high degree of interaction between the mineral phase and the biopolymer, an interlocking that is chemical and mechanical in nature, which strongly enhances the mechanical properties.[42] As a result, biological composites have orders of magnitude higher toughness than single-phase minerals.

A second design principle that contributes to enamel's wear properties is its layered structure where the enamel as a hard coating deposits on soft dentin. In enamel, prism bands, "keyhole"-shaped structures, are packed together and interlocked with each other. Comparing the two-layer tooth structure (enamel and dentin) with other multilayer systems, such as dental porcelain-fused-to-metal or all-ceramic crowns, teeth seldom have chipping or cracking problems.[46]

The third design principle is the DEJ, which perfectly combines the enamel and dentin. Although very thin, the DEJ unites the two quite different tissues and serves as a key component of the crack-arrest barrier. The DEJ is an FGM: it can distribute loads and protect the substrate better than non-graded coatings.[47] If a crack is generated in the enamel, it propagates along the interface of the prism bands and stops at the junction, preventing the chipping or failure of entire tooth. Thus, the DEJ is one of the key designs in nature that contributes to the long lifespan of heavily loaded tissue.

3.3.2   Enamel-mimicking Wear-resistant Restorative Materials

In restorative medicine, wear damage is recognized as a major reason for implants loosening and failing,[48] for example metallic alloys give rise to large elastic incompatibilities at articular joints, whereas many synthetic polymers suffer from biocompatibility issues or from excessive abrasion damage, leading to rapid degradation of the prostheses.[2] It is therefore highly desirable that the Young's modulus of dental materials should be close to that of dentin, their hardness should be between that of enamel and dentin, and the wear of the material itself and the antagonistic tooth should be minimized. Most importantly the new material should be very thin to preserve sound tooth structure.

Emulating the structure and properties of natural teeth, an artificial enamel material, named polymer-infiltrated-ceramic-network (PICN), was developed.[49] This material consists of a hybrid structure with two interpenetrating networks of ceramic and polymer, a so-called double network hybrid (DNH), mimicking the interlocking of prism bands in natural teeth. In contrast to traditional composites, which have one continuous phase filled with inorganic particles, PICN consists of two continuous interpenetrating networks. One network is a ceramic material and the other is a polymer (commonly methacrylates in dental applications).[49] In the first step of the PICN application process a porous pre-sintered feldspar ceramic with adjustable densities is fabricated by compressing the ceramic powder into blocks and then sintering these to a porous network. Ceramics with different porosities can be produced by manipulating the initial ceramic particle size and utilizing different firing temperatures. Resin is then added to the porous ceramic network structure. The ceramic network is treated by a coupling agent prior to resin infiltration so that the polymer network is chemically cross-linked to the ceramic network to form an interpenetrating network system with strong bonding interfaces between the polymer and ceramic. The chemically conditioned porous inorganic network is infiltrated with a crosslinking polymer by capillary action. Finally, the PICN material is obtained by inducing heat polymerization to form a polymer network within the interstitials of ceramics particles.[50]

The PICN material combines the properties of ceramic and polymer. Due to the interpenetrating/interlocking structure of feldspar ceramic and the acrylate polymer network, this material has a similar abrasion, high flexural strength, and elasticity to dentin because of its continuous ceramic phase. Wear is comparable to common dental ceramic, while antagonistic tooth wear is also lower. The material can be milled very thinly (0.2–0.5 mm) to preserve the enamel structure. The Weibull modulus, which measures the reliability and strength of a material, is surprisingly high, which can probably be attributed to its continuous polymer phase. The Vickers hardness test on a single ceramic showed that a typical crack line is clearly visible, while in similar tests the crack line in the hybrid ceramic is stopped by the interpenetrating polymer within the hybrid network.[50] This behavior is an indication of the damage tolerance of PICN.

3.3.3   Biomimetic Cutting Tools Based on the Sharpening Mechanism of Rat Teeth

Some biological cutting systems, such as animal teeth, use abrasive wear in order to form sharp cutting edges. This approach is completely different to that of industrial cutting tools. Rat teeth, for example, have an extremely wear-resistant material, dental enamel, in the horseshoe-shaped outer zone. The cutting face, however, consists of soft, bonelike dentine. There is an extremely strong connection between the two materials, which is achieved by a three-dimensional interlocking structure, combined with an organic membrane. As the soft dentine surface wears during use, the underlying hard enamel is exposed at the cutting edge, sharpening the rat's tooth (Figure 3.9a).

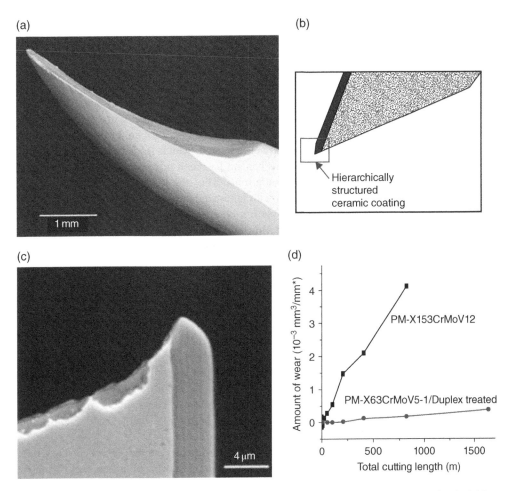

Figure 3.9 (a) Rodent incisor of a vole (Arvicolinae), colored SEM picture. (b) Rat tooth-mimicking cutting tool. (c) Cross-section of duplex-treated PM-X63CrMoV5-1, 52 HRC after 1638 m total cutting length. (d) Specific wear volume plotted against cutting length. Source: Rechberger *et al.* (2013).[51] Reproduced with permission of Elsevier.

Inspired by the unique structure and sharpening mechanism of rats' teeth, highly wear-resistant cutting tools were developed.[51] To emulate the dentin body of rodent incisors, a biomimetic cutting tool was made of tool steel with sufficient toughness that the rake surface of the tool is able to wear in a controlled way, like dentin. To mimicking the dental enamel, a hard ceramic layer is deposited, but only the flank surface is thinly coated. This layer provides high abrasive resistance by combining enough toughness and hardness to avoid brittle failure of the cutting edge. In this way the design principle of the rat tooth's hierarchic structure is transferred into the architecture by combining ductile and hard phases on the nanoscale (Figure 3.9b).[51]

The biomimetic cutting tools were fabricated using cold work tool steels by powder metallurgy (PM). To construct a hierarchically laminated coating that mimics enamel's hierarchic structure, a ternary Ti–B–N multilayer coating of thickness 4 µm was formed on the cutting tool by combining plasma diffusion treatment and the plasma-enhanced chemical vapor deposition (PECVD) coating process. This was generated by controlling the PECVD process atmosphere, which allows the assembly of a special multilayer design with repeated gradients in material composition. In this way, wear-resistant and high-wearing zones were formed at the edge of the tool flank (Figure 3.9c). Unlike conventional cutting tools, which show continuous wear at the cutting edge during use, after a short time of use the biomimetic blade exhibits a near-zero wear rate (Figure 3.9d). This can be attributed to the fact that the wear on the flank and on the rake faces is well adjusted, which leads to a self-sharpening effect. Since this bioinspired sharpening mechanism has been introduced, dissipated cutting work has reduced significantly for biomimetic tools. Thus the mechanical and tribiological properties of the steel body in combination with laminated hard coatings enable much lower wear rates compared to conventional hard-phase rich cold work tool steels. The biomimetic cutting tools have several benefits, including low energy consumption in the cutting process, stable characteristics of the cutting process over time, and an ultralow wear rate, resulting in higher endurance and efficiency.[51]

3.3.4 DEJ-mimicking Functionally Graded Materials

In dentistry, ceramic dental restorations are generally used to reconstruct an impaired tooth. In most cases, ceramic crowns are directly glued to the dentin layer using dental cement after the enamel and DEJ layers are removed. However, these flat-layered structures have relatively high failure rates within the first 5–10 years of service in the oral cavity. Radial cracking usually occurs in the top ceramic layer of ceramic crowns. These cracks initiate in the vicinity of the dental cement that attaches the dental restoration to the dentin/dentin-like ceramic filled polymers.

To improve dental restorations, lessons can be learnt from the natural tooth. As discussed in the previous sections, natural teeth consist of two distinct materials: hard enamel (Young's modulus $E \sim 65\,GPa$) and relatively soft dentin ($E \sim 20GPa$), which are bonded together at the DEJ. The Young's modulus across the DEJ gradually decreases from enamel to dentin. Because of the graded transition, the stress in the enamel is dramatically reduced.[42] Inspired by the DEJ structure, a new dental crown restoration structure has been designed and fabricated using an FGM layer at the bottom of the ceramic.[52] In this layer, E gradually decreases from that of the dental ceramic to a lower value. The structure is bonded with the dentin-like polymer by dental cement. If E for the FGM layer is high, the stress concentration will occur in the FGM layer at the interface between the FGM layer and the cement. Otherwise, high stress will present at the interface between the dental ceramic and the FGM layer. The optimal design is to ensure that the stress is uniformly distributed in the FGM layer and is continuous at the interface of the ceramic and FGM layer such that the overall stress concentration is reduced. Based on this design criterion, the Young's modulus in the FGM layer should decrease from that of the ceramic at the interface of the ceramic and FGM layer to a lower value at the interface of the FGM and cement layer.[53]

Finite element simulations were used to examine a range of dental ceramics (Figure 3.10a). In all of these cases the graded architecture reduced the maximum principal stresses by ~30%. Such a reduction in stress is likely to improve the durability of dental multilayers with graded

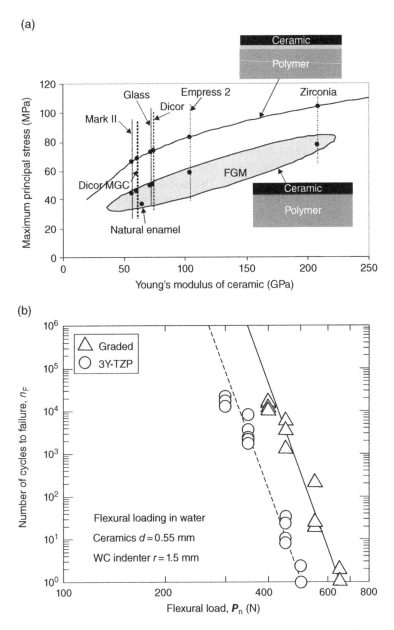

Figure 3.10 (a) Maximum principal stress in the ceramic and FGM layer predicted by finite element methods. Source: Huang *et al.* (2007).[52] Reproduced with permission of Elsevier. (b) Plot of cycles to failure as function of flexural load for graded and ungraded 3Y-TZP. Failure occurs by radial fracture from the cementation surface. Source: Zhang (2012).[53] Reproduced with permission of Elsevier.

interlayers between the dentin and crown layers, therefore functionally graded interfaces mimicking the DEJ structure offer a new approach that can be used to engineer reductions in the interfacial/layer stresses.

To validate the biomimetic design concept of FGMs, micron-scale bioinspired structures of dental restorations were fabricated.[52] To obtain gradient stiffness in the adhesive layer, the mass of ceramic (zirconia/alumina) was determined using a simple rule of mixtures to determine how much powder was needed to achieve the desired Young's modulus for the layer. The samples contained three layers: zirconia forming a hard surface, polymer matrix composite (PMC) as substrate, and FGM between them. The thicknesses of the zirconia, FGM, and polymer matrix layers were 1.0 mm, 100–150 μm, and 3.0 mm, respectively. The graded structure FGM consisted of 10 distinctive layers. The moduli of each of the layers in the FGM were varied linearly to produce an FGM layer with a modulus that varied between that of zirconia at the top to adhesive at the bottom. The samples were examined by Hertzian contact testing, in which the loads were increased continuously until failure occurred by the pop-in of radial cracks to the ceramic layer. It was observed that radial cracks propagated from the interface (between the ceramic and the dental cement) towards the contact surface between the glass and the indenter. The stresses significantly reduced with an improvement of ~32% in the pop-in loads. A three-point bend test was performed on graded glass–zirconia sandwich beams and their ungraded zirconia counterparts.[54] Specifically, graded sandwich and homogeneous zirconia beams (1.2×4×25 mm) were prepared and tested using a similar method. The graded beams have a higher fracture strength with an increase of ~28% compared to homogeneous zirconia beams. Modulus gradations also enhance the flexural fracture resistance of fine-grained alumina infiltrated with coefficients of thermal expansion (CTE)-matched silicate glass.[55]

Mimicking tooth contact and loading, various cyclic flexural tests have also been carried out to evaluate the long-term load-bearing capacity of graded and ungraded zirconia.[53] In one experiment, thin plates ($d = 0.55$ mm), representing thin restorations or connectors of bridges, were bonded to dentin-like composite supports with dental cement. To simulate the loading conditions on tooth wear, the specimens were mounted at an angle of 30° and flexural load applied at the ceramic top surface in water. Experimental results showed that radial cracks initiated at the cementation surface and this crack propagated, resulting in failure. *S/N* curves were obtained for graded glass–zirconia and homogeneous zirconia on compliant substructures (Figure 3.10b). The graded glass–zirconia showed a significantly higher number of cycles to failure than the pure zirconia plates, demonstrating an improved fatigue flexural damage resistance from surface grading. Similar beneficial effects of modulus gradients on fatigue fracture resistance have been demonstrated in fine-grained alumina infiltrated with CTE-matched silicate glass.[56] These results suggest that the bioinspired functionally graded architectures can significantly reduce stress concentrations and thus pop-in loads, and can therefore be used to improve the fatigue lifespan of the materials.

3.4   Biological Composites with High Impact and Energy Absorbance

Nature has developed some remarkable solutions using impact-resistant materials that show superior armor behavior against environmental threats. These biological materials can be divided into two categories: mineral-based and protein-based composites. Some examples of mineral-based composites are hard tissues such as antler, bone and seashells. Among them, the dactyl club of the smashing predator stomatopod (specifically, *Odontodactylus scyllarus*) is outstanding. This crustacean's club can withstand the thousands of high-velocity blows that it delivers to its prey. The endocuticle of this multiregional structure is characterized by a helicoidal arrangement of mineralized fiber layers, an architecture that results in impact resistance and energy absorbance.[57] Protein-based impact-resistant materials include hooves, turtle shell, ram horn, and armadillo shell, and comprise the same protein: keratin. Keratin is also found in many other tough materials, for example skin, hair, fur, claws, and hooves.[5] Keratin is classified as either α- or β-keratin, depending on its molecular structure. The unique material structures and mechanical deformation behavior of these materials give them excellent penetration resistance and energy absorption. These lessons from nature can be used as the inspiration for the development of lightweight armor systems. Some of the most remarkable materials in terms of energy absorption and impact resistance are discussed in the following sections.

3.4.1 Mineral-based Biocomposites: Dactyl Club

The brick-and-mortar structure of nacre in the inner layer of mollusk shells is considered to be an ideal model system for toughened biological composites, as described in Chapter 2. It was recently shown, however, that the nacreous structure is vulnerable to attack from a smashing predator known as the stomatopod or mantis shrimp. This aggressive marine crustacean uses a hammer-like strike to destroy the shells of mollusks to expose the soft body of the animal.[58] In captivity some larger species are capable of breaking through aquarium glass with a single strike. The dactyl club of one smashing species, *Odontodactylus scyllarus* (Figure 3.11a), for example, delivers high-velocity blows reaching accelerations up to $104\,\text{m/s}^2$ and speeds of $23\,\text{m/s}$, generating instantaneous forces of $1500\,\text{N}$ against the surface being attacked.[59] The striking is so rapid that it generates cavitation bubbles between the appendage and the surface. The collapse of these cavitation bubbles produces measurable forces on their prey in addition to the striking force, so the prey is hit twice by a single strike: first by the claw and then by the collapsing cavitation bubbles that immediately follow.[59]

The shells of the prey (e.g. mollusks), consisting of characteristically tough biological composites, are destroyed by a strike. The dactyl club itself, however, survives and can withstand thousands of repeated impacts without experiencing catastrophic failure.[58,59] To withstand these high impacts, the club must be hard and stiff, but also strong and tough enough to deliver momentum to its prey and resist damage from an equal and opposite impact force.

Microstructural analysis shows that the stomatopod dactyl club is a multiregional biological composite that consists of fibers of the long-chain polysaccharide α-chitin mineralized in the exocuticle by oriented crystalline HA.[58] The most heavily mineralized and hardest portion of the club is its impact region (outlined in dark line in Figure 3.11b) where the club impacts the prey. Underlying this hard outer layer is the endocuticle, which is mineralized with amorphous calcium carbonate and calcium phosphate. The endocuticle can be divided into two zones: striated and periodic regions. The striated region is located on either side of the endocuticle (bright lines in Figure 3.11b). This region prevents lateral expansion of the club during a strike through aligned mineralized fiber layers that form a circumferential band around the club. In the periodic region, the fibers exhibit a regular pattern of laminations with a characteristic nested arc pattern in each laminated region (Figure 3.11c). This nested arc appearance results

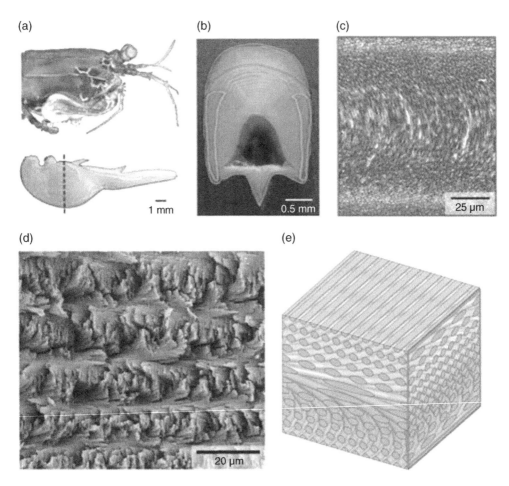

Figure 3.11   (a) Photograph of *Odontodactylus scyllarus*, along with an image of a dactyl club that has been removed and sectioned (as indicated by the dashed line marks). (b) Optical micrograph of a transverse section of the stomatopod dactyl club revealing the exocuticle (dark line), striated (bright line), and periodic region (remainder of club). (c) SEM image of a polished section through a single period, depicting a characteristic nested arc pattern. (d) SEM image of a fracture surface through multiple periods, revealing a helicoidal organization of fibers. (e) Schematic of Bouligand geometry, clarifying the origins of the nested arc pattern, a consequence of the helicoidal arrangement of aligned fiber layers. Source: Grunenfelder *et al.* (2014).[57] Reproduced with permission of Elsevier.

from a helicoidal stacking of aligned fiber layers (Figure 3.11d). The relationship between a rotated stacking of fiber layers and the nested arc pattern is schematically shown in Figure 3.11e. The sheets stack with each layer rotated by a small angle from the layer below, eventually completing a rotation of 180°. The helicoidally arranged α-chitin fibers are mineralized at the nanoscale with crystalline and/or amorphous mineral. Thus, the cuticle can be considered as a fiber-reinforced structure, with polymer fibers embedded in a ceramic matrix.[57]

The periodic region is important to energy dissipation and impact resistance.[57,58] A small rotation angle between consecutive layers can lead to large change in the properties of the helicoidal composite, especially the isotropic response to in-plane loading at the macroscale.[60] This rotated fiber architecture increases the toughness of the composites when loaded in tension, with the lamellae reorienting to adapt to the loading conditions and thus resist deformation.[61] In addition to rotated fiber stacking, each cuticle region has specific properties with variations of the degree of mineralization as well as the stacking density of the Bouligand layers. The combination of these regions with different composite structures of the cuticle provides the stiffness, strength, and hardness required to protect the animal and allow for movement and predation.[57] This structure of the dactyl club provides an excellent model for building strong and tough engineering composite materials for impact-protection applications.

3.4.2 *Protein-based Biocomposites: Horns and Hooves*

In nature, unmineralized biological materials are also designed to resist impact and can absorb a large amount of impact energy. Bighorn sheep (*Ovis canadensis*) horn is an example that shows excellent stiffness and strength under the impact loading that occurs during sparring between males. Rhinoceros horn also demonstrates high impact resistance. Similar to turtle and armadillo shell, ram horn is a hierarchical material, made up of a core of cancellous bone surrounded by a keratin sheath. Rhinoceros horn does not have a bony core.[62] The keratin sheath, a fibrous structural protein, is the primary impact load-bearing material in horns, hooves, hair, furs, claws, and fingernails.

Ram horn is composed of the hierarchical structure shown in Figure 3.12.[5] At the molecular level, a horn comprises helical, α-keratin protofibrils, which assemble into rope-like crystalline intermediate filaments, orienting along the growth direction and coiling up into hollow, elliptically shaped tubules (~20–100 μm in diameter). These tubules resemble hollow reinforcing fibers and are embedded in an amorphous keratin matrix. In the matrix keratin fibers are randomly distributed, forming a structure that is akin to a short fiber reinforced composite. There is also a porosity gradient across the thickness of the horn, with the porosity increasing from outer surface to the interior.[63] Like reinforcement in traditional composites, the keratin fibers strengthen and stiffen the structure by forming long, hollow, fiber-like tubules, but tubules, as discussed later, may provide an additional toughening mechanism for the horn. This dispersed tubule microstructure has been observed in other tough biological materials such as hoof, bone, antler, and dentin.[5] At the macroscale, a horn takes the shape of a logarithmic spiral. The porosity of the keratin sheath in ram horn was found to be approximately 6%. The keratin fibers are parallel to the growth direction and are stacked in a lamellar fashion through the thickness of the horn. This fiber distribution through the cross-section resembles that in a unidirectional fiber composite, and thus material behavior in the other two directions (transverse and radial) is nearly identical. Horn is therefore a transversely isotropic composite material, that is, it is isotropic in the transverse and radial directions.

The tubules are one of the key structural characteristics that determine the energy dissipation and impact resistance of materials. In horn and similar materials, keratin surrounds a closed-cell, foam-like core. The porous outer layer introduces toughening mechanisms such as crack deflection, crack arrest, and penetration resistance, whereas the central core functions to absorb large amounts of energy during collapse with a low cost in weight.[5]

Sheep horn materials were tested under high impact conditions. Figure 3.13 shows the compressive behavior of the horn at quasi-static rates and under high strain rates in the split

(a)

(c)

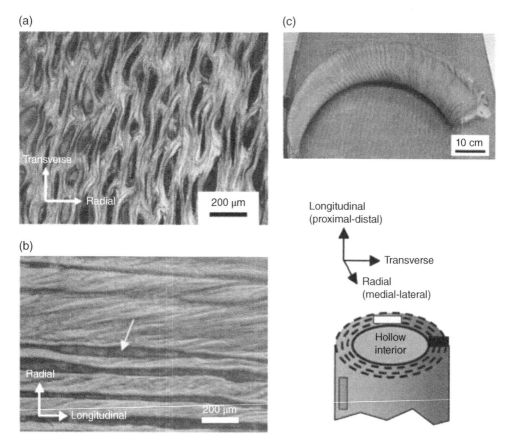

Figure 3.12 Optical micrographs of ambient dried horn. (a) Cross-section showing the dark elliptical-shaped tubules, (b) longitudinal section showing the outline of the parallel tubules (arrow points to a tubule), and (c) orientation of the samples in the horn. Source: McKittrick *et al.* (2010).[5] Reproduced with permission of Elsevier.

Hopkinson bar.[5] The horn loaded in the radial direction provides more energy absorption than in the longitudinal direction. Similar to the typical behaviors of polymers, the Young's modulus, the yield strength, and the toughness increase with increasing strain rate in both directions. In general, as the strain rate increases, the polymer chains do not have time to align and thus behave more as a polymer network. However, the strains to failure can reach up to 80%, which is an order of magnitude higher than in a typical network polymer. Such high strain is the result of dissipative microdeformation processes such as compression of the tubules (Figure 3.13c). The horn has a higher Young's modulus and yield strength in the longitudinal direction than in the radial direction. This is attributed to the alignment of the lamellae in the longitudinal direction. The toughness is lower in this direction because of microbuckling and delamination of the lamellae (Figure 3.13d).

Similar to horn, hooves are also composed of α-keratin fibers that are wound into circular lamellae that surround a hollow, empty channel. These tubules only serve mechanical functions and are not used to keep the hoof hydrated because all the tubules are empty.[64] The tubules have the

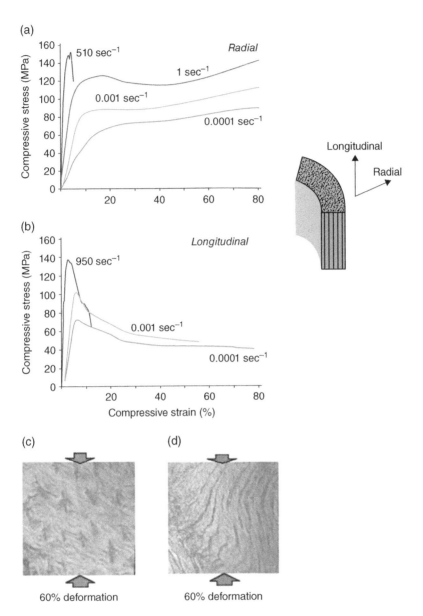

Figure 3.13   Effect of strain rate on the compression stress–strain curve for big horn sheep horn in the (a) radial and (b) longitudinal directions. (c) Micrograph showing tubule collapse after a compressive load and (d) micrograph showing microbuckling of the lamellae. Source: McKittrick *et al.* (2010).[5] Reproduced with permission of Elsevier.

highest density at the outer surface and a lower density towards the inner part, resulting in a gradient of porosity through the thickness of the hoof wall, with the highest porosity at the outer surface. However, the elastic modulus is highest at the outer surface because of the gradient distribution of tubules, resulting in an increase in the volume fraction of crystalline α-keratin filaments.[64]

3.4.3 Bioinspired Design Strategies for Highly Impact-resistant Materials

Both mineral-based and protein-based biocomposites have been designed by nature for resisting impact. For mineral-based biocomposites, such as enamel and dactyl club, the structures are intended to sustain the contact impact. These tissues contain as high a level of minerals as possible (e.g., over 98%) and a small amount of polymer to enhance the hardness and Young's modulus. At the same time, various toughening mechanisms are promoted by introducing unique structures. For dactyl club, which needs to deliver large momentum to its prey, the deposited minerals between the protein fibers in combination with the cuticle structure provide stiffness and hardness as well as strength and toughness, which are needed to protect the animal and allow for movement and predation. Thus, cuticle structure is an effective biological model that can be mimicked to fabricate strong and tough composite materials.

Protein-based biocomposites such as horn and hoof are composed of keratin fibers that strengthen and stiffen the structure. These materials have several design similarities in their microstructures:[5]

- the presence of long tubules (vascular channels in the case of skeletal bone) that extend in the longitudinal (growth) direction
- a density gradient and Young's modulus gradient in the transverse direction due to the presence of tubules and oriented structural protein fibers
- an amorphous polymer matrix reinforced by crystalline polymer tubules that are assembled from strong, tough high aspect ratio crystalline protein fibers (collagen or keratin)
- a multiscale, hierarchical material structure comprising a sandwich composite structure with a high-porosity, foam-like interior region and denser exterior region.

Although some mineral-based biomaterials such as dentin and bone also have tubule-like structures, some differences do exist. For example, the tubules within bones and teeth contain fluid, but no fluid is present in hooves or horns. The tubules in hooves and horns are believed to be important in increasing elastic hysteresis and energy absorption, and also serve to prevent microbuckling of the lamellae. Compared with mineralized components, the biological materials are tougher but more compliant.

The high-energy absorption of these protein-based materials can be attributed to microde-formation mechanisms such as delamination and microbuckling of the lamella.[5] Among many structural characteristics, the tubule distribution is a primary factor that determines the deformation mechanisms. Each lamella is composed of fibrous proteins that form the tough base material. This design for energy absorption can be considered as another excellent model for bioinspiration for amour and other energy-absorbing materials. In addition, the design of protein-based biocomposites is quite similar to spider silk structure, which consists of crystals embedded in an amorphous matrix, although the structures are quite different.

According to the biological models described above, several bioinspired design strategies can be considered based on the unique structures of the high-energy absorption materials. One possible approach is to encase a central porous core with an impact-resistant polymer containing tubules oriented perpendicular to the loading direction.[5] To resist high impact, ideally the materials for the central core should have a higher elastic modulus than those chosen for the outer sheath. In the outer sheath, the tubules and matrix materials should have similar properties and very good bonding between them. There should be a high degree of interaction and synergism between the protein skin and the foam-like core, which will strongly enhance the mechanical properties. For example, to develop a lightweight armor system, a bio-inspired design strategy involves encasing a metallic foam core with a fiber-polymer composite laminate. This sandwich structure mimics the structure found in natural energy-absorbent materials. The skin of this design resists penetration whilst the central core absorbs large amounts of energy at a low cost in weight. Good adhesion between the skin and the metallic core is crucial for this design to succeed and is an excellent opportunity for future research. This strategy has been shown to be very effective in resisting high-speed impact for drastic ballistic shock mitigation, weight savings, and significant reductions in penetration and load transmission under ballistic loading conditions.[65]

3.5 Biomimetic Impact-resistant Materials and Processes

3.5.1 Dactyl Club-Biomimicking Highly Impact-resistant Composites

The dactyl club of the smashing predator stomatopod (*Odontodactylus scyllarus*) is designed by nature to withstand the thousands of high-velocity blows that it delivers to its prey. The endocuticle of this multiregional structure is characterized by a helicoidal arrangement of mineralized fiber layers, an architecture that results in impact resistance and energy absorbance.[57] Mimicking the design of the dactyl club, a bioinspired corollary to the Bouligand structure has been achieved using fiber-reinforced composites. Traditionally in fiber-reinforced composites, fibers are unidirectionally aligned or cross-aligned. These layups form unidirectional and quasi-isotropic controls. Unidirectional composites consist of aligned fiber layers, similar to those observed in the crustacean exoskeleton, while quasi-isotropic layups are an aerospace industry standard design and a robust baseline architecture. Through the stacking of unidirectional layers (plies), a helicoidal composite can be readily obtained, as shown in Figure 3.14a,b.[57]

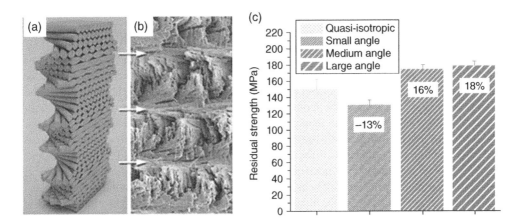

Figure 3.14 (a) Schematic representation and (b) SEM image of a helicoidal rotation of fiber layers (shown here for a medium-angle composite). (c) Residual strength calculated from compressive data, with percentage change listed for helicoidal samples. Source: Grunenfelder *et al.* (2014).[57] Reproduced with permission of Elsevier.

The bioinspired helicoidal architecture has been demonstrated to enhance the mechanical properties of the composites under static loading conditions. The toughness and strength of composite materials can be changed by changing the ply orientation, which alters the damage mechanisms and propagation, for example the flexural stiffness and shear strength of glass-fiber-reinforced helicoidal composites are better than those of a quasi-isotropic control sample.[66] The fracture toughness of a glass-fiber-reinforced helicoidal composite architecture is also better than that of a unidirectional control.[67]

The impact performance of bioinspired helicoidal carbon-fiber-reinforced polymer composites with different helicoidal rotation angles was examined and compared with unidirectional as well as quasi-isotropic controls.[57] Samples of size 100×150 mm were subjected to high-energy (100 J) impact using a drop tower. Following impact testing, the residual strength of the samples was determined through compression testing. The high-energy impact produced catastrophic failure in the unidirectional control samples, resulting in a splitting of the panels, whereas for the quasi-isotropic controls the impact event resulted in puncture through the backside of the panels, accompanied by fiber breakage. The helicoidal samples, in contrast, did not show indentation puncture, but rather splits on the back surface, depending on the helicoidal angle. According to dent depth measurements, the medium angle composites showed the greatest reduction in external indent damage, with an average depth reduction of 49%.[57] The internal damage field caused by impact was also investigated using a non-destructive pulse–echo ultrasound technique. The ultrasound data reveal that the quasi-isotropic control samples have the smallest internal damage field. The damage spread in the control samples is also symmetric. The helicoidal samples, however, show a more widespread and asymmetric damage field. The extent of the lateral spread of damage is most pronounced in small-angle composites and decreases with increasing ply rotation angle.[57] Finally, the measurements of compressive residual strength show that small-angle panels have the lowest residual strength while medium- and large-angle panels show a marked increase in residual strength when compared to the quasi-isotropic controls (16% and 18%, respectively) (Figure 3.14c).[57]

3.5.2 Damage-tolerant CNT-reinforced Nanocomposites Mimicking Hooves

Mineral-based (e.g., human tooth, bovine femur) and keratin-based (e.g., elk antler, sheep horn) impact-resistant biocomposites show striking similarities in their cross-sections: circular or elliptically shaped pores that arise from the tubules aligning in the longitudinal direction. The size of the tubular structures in the femur and antler are the same: osteons are $\sim 200 \mu$m with vascular channels around 30 μm in diameter, whereas hooves have smaller channels of 24 μm diameter and a much lower porosity of ~ 3%. The antler has a higher density of osteons due to its relatively young age compared with the femur. The tubules in horn are elliptical ($\sim 40 \times 100 \mu$m). The density of tubules in dentin is much higher than that of bone and the diameters are much less than in bone, around 1 μm, resulting in an areal porosity of 12%. The similarities in microstructure provide direct evidence of the importance of tubules in an energy-absorbent materials design. The tubules provide toughening mechanisms such as crack deflection and energy to collapse tubules, and prevent extended regions of microbuckling of the laminates.[68]

Similar toughening mechanisms were observed in synthetic Al_2O_3 ceramic nanocomposites reinforced with aligned carbon nanotubes (CNTs). As shown in Figure 3.15, this material has ordered CNTs of about 55 nm diameter, with similar morphology with the biomaterials.[69]

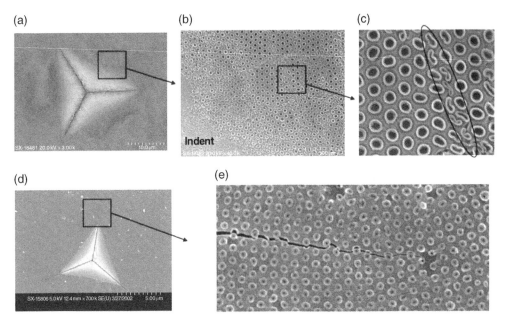

Figure 3.15 SEM photographs of deformation around indent in thin-wall CNT nanocomposites show-ing (a) indent marker without cracks generated, (b) array of shear bands formed, and (c) close-up of lateral buckling or collapse of the nanotubes in one shear band. SEM photographs of deformation around indent in thick-wall CNT nanocomposites showing (d) crack induced by top indentation and (e) the detail of the crack intersection with the successive CNT/matrix interfaces and deflection around the CNTs along the interface. Source: Xia et al. (2004).[69] Reproduced with permission of Elsevier.

Two nanocomposites with thin wall (4.5 nm) and thick wall (12.5 nm) CNTs were fabricated by chemical vapor deposition in anodic alumina with nanopores. For the thin wall CNT composites, top indentation onto the porous material did not cause cracking. Insted, in addition to the indentation mark, the deformation was accommodated by shear collapse of rows of CNTs. Figure 3.15a–c show the deformation at various scales. At large scales, as shown in Figure 3.15a, a deformation ring is observed around the indentation mark. Upon closer obser-vation, this ring is caused by the scattering from short bands of collapsed CNTs, as seen in Figure 3.15b. Further inspection shows the occurrence of shear cracks between opposite sides of neighboring collapsed CNTs, as seen in Figure 3.15c. However, when the wall of CNTs is thick, no shear band or CNT collapse was observed. Instead, a crack was formed at the corner of the indenter, indicating that the nanocomposites are brittle (Figure 3.15d,e).

The CNT collapse deformation mode is similar to shear band formation and to deformation modes observed in large-scale porous metals and polymers.[70] In this case, however, the matrix material is an amorphous alumina with a hardness of about 5 GPa and modulus 140 GPa. The deformed shapes of the sheared pores indicate that the shear cracks have formed only after substantial CNT deformation. To accommodate the shear deformation some flow of the alumina matrix is required. Although the anodized alumina is not well characterized, such flow is not normally expected in similar materials at macroscopic scales, room temperature, and under the expected loads around the indent mark. These results demonstrate similar design principles in that tubular structures enhance the toughness of high-impact biomaterials. The deformation behavior for loading in transverse directions, which is normally very weak and brittle in a larger-scale composite, can be controlled by the composite microstructure, i.e. the diameter and wall thickness of the CNTs, and possibly their organization. In particular, non-cracking deformation modes can be activated to accommodate aggressive contact loading. Although the nanocomposite itself was not fabricated based on the biological model, the observed toughening phenomena have indicated that the biomimetic design principles can be applied to various synthetic materials with tubule structures. Coupled with the existence of CNT crack-bridging toughening mechanisms for cracking perpendicular to the axial direction, these results indicate that nanotube ceramic composites can be engineered to exhibit multiaxial toughness or damage tolerance by adjusting the composite geometry and constituent properties.

References

1. Zum Gahr, K.-H. *Microstructure and Wear of Materials* (Elsevier, Amsterdam, New York; 1987).
2. Amini, S. & Miserez, A. Wear and abrasion resistance selection maps of biological materials. *Acta Biomaterialia* **9**, 7895–7907 (2013).
3. Palmer, L.C., Newcomb, C.J., Kaltz, S.R., Spoerke, E.D. & Stupp, S.I. Biomimetic systems for hydroxyapatite mineralization inspired by bone and enamel. *Chemical Reviews* **108**, 4754–4783 (2008).
4. Glauche, V. *et al.* Analysis of tooth surface elements by ion beam analysis. *Journal of Hard Tissue Biology* **20**, 99–105 (2011).
5. McKittrick, J. *et al.* Energy absorbent natural materials and bioinspired design strategies: A review. *Materials Science and Engineering: C* **30**, 331–342 (2010).
6. Zheng, J. *et al.* Microtribological behaviour of human tooth enamel and artificial hydroxyapatite. *Tribology International* **63**, 177–185 (2013).
7. Koc, D., Dogan, A. & Bek, B. Bite force and influential factors on bite force measurements: A literature review. *European Journal of Dentistry* **4**, 223–232 (2001).
8. Clemente, C. *Anatomy, a Regional Atlas of the Human Body* (Urban & Schwarzenberg, Baltimore; 1987).
9. White, S.N. *et al.* The dentino-enamel junction is a broad transitional zone uniting dissimilar bioceramic composites. *Journal of the American Ceramic Society* **83**, 238–240 (2000).
10. Zhou, Z.R. & Zheng, J. Oral tribology. *Proceedings of the Institution of Mechanical Engineers Part J – Journal of Engineering Tribology* **220**, 739–754 (2006).
11. Cui, F.Z. & Ge, J. New observations of the hierarchical structure of human enamel, from nanoscale to microscale. *Journal of Tissue Engineering and Regenerative Medicine* **1**, 185–191 (2007).
12. Eimar, H. *et al.* Regulation of enamel hardness by its crystallographic dimensions. *Acta Biomaterialia* **8**, 3400–3410 (2012).
13. Cevc, G., Cevc, P., Schara, M. & Skaleric, U. The caries resistance of human-teeth is determined by the spatial arrangement of hydroxyapatite micro-crystals in the enamel. *Nature* **286**, 425–426 (1980).
14. Macho, G.A., Jiang, Y. & Spears, I.R. Enamel micro structure – a truly three-dimensional structure. *Journal of Human Evolution* **45**, 81–90 (2003).
15. Ge, J., Cui, F.Z., Wang, X.M. & Feng, H.L. Property variations in the prism and the organic sheath within enamel by nanoindentation. *Biomaterials* **26**, 3333–3339 (2005).
16. Gwinnett, A.J. Structure and composition of enamel. *Operative Dentistry* 10–17 (1992).
17. Hsu, C.C., Chung, H.Y., Yang, J.M., Shi, W. & Wu, B. Influence of 8DSS peptide on nano-mechanical behavior of human enamel. *Journal of Dental Research* **90**, 88–92 (2011).
18. Kinney, J.H., Pople, J.A., Marshall, G.W. & Marshall, S.J. Collagen orientation and crystallite size in human dentin: A small angle X-ray scattering study. *Calcified Tissue International* **69**, 31–37 (2001).
19. Kinney, J.H., Habelitz, S., Marshall, S.J. & Marshall, G.W. The importance of intrafibrillar mineralization of collagen on the mechanical properties of dentin. *Journal of Dental Research* **82**, 957–961 (2003).
20. Kinney, J.H., Oliveira, J., Haupt, D.L., Marshall, G.W. & Marshall, S.J. The spatial arrangement of tubules in human dentin. *Journal of Materials Science – Materials in Medicine* **12**, 743–751 (2001).

21. Marshall, G.W., Marshall, S.J., Kinney, J.H. & Balooch, M. The dentin substrate: structure and properties related to bonding. *Journal of Dentistry* **25**, 441–458 (1997).
22. Habelitz, S., Balooch, M., Marshall, S.J., Balooch, G. & Marshall, G.W. In situ atomic force microscopy of partially demineralized human dentin collagen fibrils. *Journal of Structural Biology* **138**, 227–236 (2002).
23. Imbeni, V., Kruzic, J.J., Marshall, G.W., Marshall, S.J. & Ritchie, R.O. The dentin-enamel junction and the fracture of human teeth. *Nature Materials* **4**, 229–232 (2005).
24. Xie, Z.H., Mahoney, E.K., Kilpatrick, N.M., Swain, M.V. & Hoffman, M. On the structure–property relationship of sound and hypomineralized enamel. *Acta Biomaterialia* **3**, 865–872 (2007).
25. Bechtle, S. *et al.* Crack arrest within teeth at the dentinoenamel junction caused by elastic modulus mismatch. *Biomaterials* **31**, 4238–4247 (2010).
26. Zaytsev, D., Grigoriev, S. & Panfilov, P. Deformation behavior of human dentin under uniaxial compression. *International Journal of Biomaterials* **2012**, 854539 (2012).
27. Kruzic, J.J., Nalla, R.K., Kinney, J.H. & Ritchie, R.O. Mechanistic aspects of in vitro fatigue-crack growth in dentin. *Biomaterials* **26**, 1195–1204 (2005).
28. Xu, H.H.K., Liao, H. & Eichmiller, F.C. Indentation creep behavior of a direct-filling silver alternative to amalgam. *Journal of Dental Research* **77**, 1991–1998 (1998).
29. Iwamoto, N. & Ruse, N.D. Fracture toughness of human dentin. *Journal of Biomedical Materials Research A* **66a**, 507–512 (2003).
30. Imbeni, V., Nalla, R.K., Bosi, C., Kinney, J.H. & Ritchie, R.O. In vitro fracture toughness of human dentin. *Journal of Biomedical Materials Research A* **66a**, 1–9 (2003).
31. Elmowafy, O.M. & Watts, D.C. Fracture-toughness of human-dentin. *Journal of Dental Research* **65**, 677–681 (1986).
32. Chen, P.Y. *et al.* Structure and mechanical properties of selected biological materials. *Journal of the Mechanical Behavior of Biomedical Materials* **1**, 208–226 (2008).
33. Chun, K., Choi, H. & Lee, J. Comparison of mechanical property and role between enamel and dentin in the human teeth. *Journal of Dental Biomechanics* **5**, 1758736014520809 (2014).
34. Nalla, R.K., Imbeni, V., Kinney, J.H., Marshall, S.J. & Ritchie, R.O. On the development of life prediction methodologies for the failure of human teeth. *Materials Lifetime Science & Engineering*, 137–145 (2003).
35. Arola, D.D. & Reprogel, R.K. Tubule orientation and the fatigue strength of human dentin. *Biomaterials* **27**, 2131–2140 (2006).
36. Nalla, R.K., Kinney, J.H., Marshall, S.J. & Ritchie, R.O. On the in vitro fatigue behavior of human dentin: Effect of mean stress. *Journal of Dental Research* **83**, 211–215 (2004).
37. Do, D. *et al.* Accelerated fatigue of dentin with exposure to lactic acid. *Biomaterials* **34**, 8650–8659 (2013).
38. White, S.N. *et al.* Biological organization of hydroxyapatite crystallites into a fibrous continuum toughens and controls anisotropy in human enamel. *Journal of Dental Research* **80**, 321–326 (2001).
39. Lucas, P.W. & van Casteren, A. The wear and tear of teeth. *Medical Principles and Practice* **24** (Suppl 1), 3–13 (2015).
40. Marshall, G.W., Balooch, M., Gallagher, R.R., Gansky, S.A. & Marshall, S.J. Mechanical properties of the dentinoenamel junction: AFM studies of nanohardness, elastic modulus, and fracture. *Journal of Biomedical Materials Research* **54**, 87–95 (2001).
41. White, S.N. *et al.* Controlled failure mechanisms toughen the dentino-enamel junction zone. *Journal of Prosthetic Dentistry* **94**, 330–335 (2005).
42. He, L.H., Yin, Z.H., van Vuuren, L.J., Carter, E.A. & Liang, X.W. A natural functionally graded biocomposite coating – human enamel. *Acta Biomaterialia* **9**, 6330–6337 (2013).
43. Abu Kasim, N.H., Madfa, A.A., Hamdi, M. & Rahbari, G.R. 3D-FE analysis of functionally graded structured dental posts. *Dental Materials Journal* **30**, 869–880 (2011).
44. Zok, F.W. & Miserez, A. Property maps for abrasion resistance of materials. *Acta Materialia* **55**, 6365–6371 (2007).
45. Bechtle, S., Ang, S.F. & Schneider, G.A. On the mechanical properties of hierarchically structured biological materials. *Biomaterials* **31**, 6378–6385 (2010).
46. Anusavice, K.J. Standardizing failure, success, and survival decisions in clinical studies of ceramic and metal–ceramic fixed dental prostheses. *Dental Materials* **28**, 102–111 (2012).
47. Singh, R.K. *et al.* Design of functionally graded carbon coatings against contact damage. *Thin Solid Films* **518**, 5769–5776 (2010).
48. Mattei, L., Di Puccio, F., Piccigallo, B. & Ciulli, E. Lubrication and wear modelling of artificial hip joints: A review. *Tribology International* **44**, 532–549 (2011).

49. Coldea, A., Swain, M.V. & Thiel, N. Mechanical properties of polymer-infiltrated ceramic-network materials. *Dental Materials* **29**, 419–426 (2013).
50. Dirxen, C., Blunck, U. & Preissner, S. Clinical performance of a new biomimetic double network material. *The Open Dentistry Journal* **7**, 118–122 (2013).
51. Rechberger, M., Paschke, H., Fischer, A. & Bertling, J. New tribological strategies for cutting tools following nature. *Tribology International* **63**, 243–249 (2013).
52. Huang, M., Wang, R., Thompson, V., Rekow, D. & Soboyejo, W.O. Bioinspired design of dental multilayers. *Journal of Materials Science: Materials in Medicine* **18**, 57–64 (2007).
53. Zhang, Y. Overview: Damage resistance of graded ceramic restorative materials. *Journal of the European Ceramic Society* **32**, 2623–2632 (2012).
54. Zhang, Y. & Ma, L. Optimization of ceramic strength using elastic gradients. *Acta Materialia* **57**, 2721–2729 (2009).
55. Zhang, Y., Sun, M.J. & Zhang, D.Z. Designing functionally graded materials with superior load–bearing properties. *Acta Biomaterialia* **8**, 1101–1108 (2012).
56. Ren, L. *et al.* Improving fatigue damage resistance of alumina through surface grading. *Journal of Dental Research* **90**, 1026–1030 (2011).
57. Grunenfelder, L.K. *et al.* Bio-inspired impact-resistant composites. *Acta Biomaterialia* **10**, 3997–4008 (2014).
58. Weaver, J.C. *et al.* The stomatopod dactyl club: A formidable damage-tolerant biological hammer. *Science* **336**, 1275–1280 (2012).
59. Patek, S.N. & Caldwell, R.L. Extreme impact and cavitation forces of a biological hammer: strike forces of the peacock mantis shrimp *Odontodactylus scyllarus*. *Journal of Experimental Biology* **208**, 3655–3664 (2005).
60. Nikolov, S. *et al.* Revealing the design principles of high-performance biological composites using ab initio and multiscale simulations: The example of lobster cuticle. *Advanced Materials* **22**, 519–+(2010).
61. Zimmermann, E.A. *et al.* Mechanical adaptability of the Bouligand-type structure in natural dermal armour. *Nature Communications* **4** (2013).
62. Hieronymus, T.L., Witmer, L.M. & Ridgely, R.C. Structure of white rhinoceros (*Ceratotherium simum*) horn investigated by X-ray histology with implications computed tomography and for growth and external form. *Journal of Morphology* **267**, 1172–1176 (2006).
63. Tombolato, L., Novitskaya, E.E., Chen, P.Y., Sheppard, F.A. & McKittrick, J. Microstructure, elastic properties and deformation mechanisms of horn keratin. *Acta Biomaterialia* **6**, 319–330 (2010).
64. Kasapi, M.A. & Gosline, J.M. Micromechanics of the equine hoof wall: Optimizing crack control and material stiffness through modulation of the properties of keratin. *Journal of Experimental Biology* **202**, 377–391 (1999).
65. Huang, J., Durden, H. & Chowdhury, M. Bio-inspired armor protective material systems for ballistic shock mitigation. *Materials & Design* **32**, 3702–3710 (2011).
66. Cheng, L.A., Thomas, A., Glancey, J.L. & Karlsson, A.M. Mechanical behavior of bio-inspired laminated composites. *Composites Part A – Applied Science and Manufacturing* **42**, 211–220 (2011).
67. Chen, B., Peng, X., Cai, C., Niu, H. & Wu, X. Helicoidal microstructure of Scarabaei cuticle and biomimetic research. *Materials Science and Engineering: A* **423**, 237–242 (2006).
68. McKittrick, J. *et al.* Energy absorbent natural materials and bioinspired design strategies: A review. *Materials Science and Engineering: C* **30**, 331–342 (2010).
69. Xia, Z. *et al.* Direct observation of toughening mechanisms in carbon nanotube ceramic matrix composites. *Acta Materialia* **52**, 931–944 (2004).
70. Papka, S.D. & Kyriakides, S. Experiments and full-scale numerical simulations of in-plane crushing of a honeycomb. *Acta Materialia* **46**, 2765–2776 (1998).

4

Adaptive and Self-shaping Materials

4.1 Introduction

Adaptive materials are those able to spontaneously modify their physical properties (shape, color, conductivity, humidity, pH, etc.) under natural or provoked stimuli. The self-shaping or mechanical property change that adaptive materials exhibit on exposure to a predefined stimulus in a highly selective and reversible manner is useful in numerous engineering applications, including deployable space structures, drug-delivery systems, and morphing wings. In engineering materials for self-shaping, shape-memory alloys are a class of materials that have the ability to "memorize" or retain their previous form when subjected to certain stimulus such as thermomechanical or magnetic variations, and represent a design approach that relies on phase transitions at the molecular and/or atomic scale.[1] Similarly, piezoelectric and electrostrictive ceramics undergo dimensional change under external stimuli (voltage).[2] Other classes of widely investigated mechanically adaptable materials include temperature-, photo-, electro- and chemo-responsive polymers.[3] In addition to these adaptive materials with inherent phase transitions, composite structures, in which the components display different responses to certain external stimuli, may also exhibit a stimulated shape change. Such a concept has been extensively exploited to produce micro-scale multilayered objects with self-folding properties using mostly photolithographic techniques.[4]

In contrast to the aforementioned adaptive material systems, nature utilizes completely different strategies to accomplish property/shape changes on external stimuli. Instead of molecular phase transitions or combinations of dissimilar materials, many biological materials contain certain architectures with controlled local volume fraction, distribution, and orientation of stiff fibers embedded in a soft matrix. Such fiber architectures result in differential swelling/shrinkage or growth of the surrounding matrix to generate mechanical deformations that lead to well-defined macroscopic changes in shape.[4] For example, sea cucumbers and other echinoderms have the fascinating ability to rapidly and reversibly alter the stiffness of

Biomimetic Principles and Design of Advanced Engineering Materials, First Edition. Zhenhai Xia.
© 2016 John Wiley & Sons, Ltd. Published 2016 by John Wiley & Sons, Ltd.

their inner dermis when threatened. These animals can reversibly switch the modulus of their skin between "soft" and "rigid" within microseconds. This dynamic mechanical behavior is achieved through a nanocomposite architecture in which rigid, high-aspect ratio collagen fibrils reinforce a viscoelastic matrix.[5]

In natural adaptive materials, the deformation is controlled at different levels of tissue hierarchy by geometrical constraints at the micrometer level (e.g., cell shape and size) and cell wall polymer composition at the nanoscale (e.g., cellulose fibril orientation). This type of natural shape-changing effect does not stem from phase transitions at the atomic/molecular level, therefore bioinspired self-shaping materials can be potentially made with a wider variety of components by properly designing their constituent building blocks at the microscale.[4] These material design principles may provide new routes to create synthetic materials with unique adaptive capabilities. Inspired by nature, various mechanically adaptive materials have been developed, including nanocomposites with modulus changes of several orders of magnitude and self-shaped composites with shape changes of several hundred percent. The bioinspired design of adaptive microstructures provides a new pathway to control shape changes in synthetic materials.

In this chapter, typical self-shaping/reversible property change mechanisms in animals and plants are discussed, and biomimetic design strategies and synthetic materials with controllable property- and/or shape-changing capability are examined. Particular attention is paid to microstructural mimics and the design principles of fiber architectures and reversible interactions. The design principles may be particularly useful for a biomimetic translation into active composite materials and moving devices.

4.2   The Biological Models for Adapting and Morphing Materials

4.2.1   Reversible Stiffness Change of Sea Cucumber via Switchable Fiber Interactions

Many echinoderms (e.g., sea cucumbers, starfish, etc.) are highly distinctive, with a number of features found nowhere else in the natural world. Most prominent is a skeleton made of calcite crystals. It is well known that sea cucumbers can alter, rapidly and reversibly, the stiffness of their collagenous connective tissues, commonly referred to as mutable connective tissues (Figure 4.1a). This stiffness change occurs within seconds and creates significant survival advantages. The dermis of the *Cucumaria frondosa* (and other sea cucumber species) represents a compelling model of a chemo-responsive material in which a 10-fold modulus contrast (ca. 5–50 MPa) is possible.[6]

The structure of the inner dermis of sea cucumbers is complex but unique. In the collageneous inner dermis of these invertebrates there is a nanocomposite structure consisting of rigid, high-aspect ratio collagen fibrils organized within a viscoelastic matrix of fibrillin microfibrils (Figure 4.1b).[7] These collagen fibrils are isolated from each other in the relaxed state but can be linked into a strong network. These interactions between the collagen fibrils are regulated by soluble molecules that are secreted locally by neurally controlled effector cells. The aggregation of the isolated fibrils is induced by a constitutive glycoprotein of the extracellular matrix, stiparin, which was initially identified as a tissue-stiffening factor. There is another glycoprotein, stiparin inhibitor, which can bind to stiparin, thereby inhibiting its ability to induce fibril aggregation. A third component, tensilin, not only induces

(a)

(b)

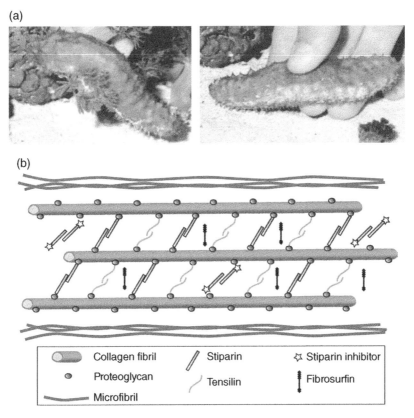

Collagen fibril	Stiparin	Stiparin inhibitor
Proteoglycan	Tensilin	Fibrosurfin
Microfibril		

Figure 4.1   The stiffening response of sea cucumbers triggered by chemical stimuli. (a) A sea cucumber in its soft (top) and stiff (bottom) states, and (b) hypothetical architecture of collagen fibers and other biomolecules in the dermal tissue of sea cucumbers, highlighting the reversible cross-linking molecules presumably involved in the stiffening process. Source: Studart & Erb (2014).[4] Reproduced with permission of the Royal Society of Chemistry.

collagen–fibril aggregation *in vitro*, but also increases tissue stiffness and appears to play an important role in stiffness change.[7,8] However, a detailed molecular mechanism of how the collagen–fibrils are regulated in sea cucumber dermis has yet to be developed.

The stiffness of the sea cucumber tissue depends on the ability of adjacent collagen fibrils to transfer stress via transiently established temporary bonding.[9] The stress transfer among the adjacent collagen fibrils is regulated by locally secreted proteins through either non-covalent or covalent bonds. The reversible tuning of fiber interactions in the dermal tissue of sea cucumbers is promoted by chemicals that are released from cells present in the outer and inner dermis of the animal. When glycoprotein stiparin inhibitor is secreted by the cells and appears on the surface of adjacent reinforcing fibrils in dermal tissue, weak bonds are formed between the fibrils, leading to bond breaking and consequently low stiffness. The crosslinking is on when the covalent bonds are established between the collagen fibrils by glycoprotein stiparin (Figure 4.2).[5] This leads to the rapid formation of a percolating network of collagen fibrils that significantly increases the elastic modulus of the dermal tissue. However, the crosslinking

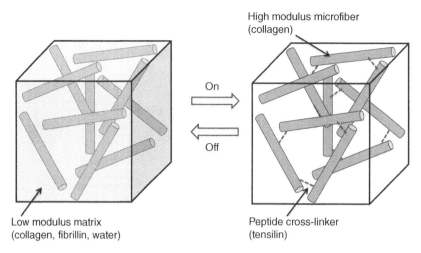

Figure 4.2 Mechanism of the soft-to-hard change in the dermal tissue of sea cucumbers. Source: adapted from White *et al.* (2013).[5]

bonds are revisable, and they can be disrupted again in the presence of a stiparin inhibitor, eventually resulting in softening of the tissue.

The rapid stiffness change mechanism of sea cucumbers provides an intriguing biological model for the biomimetic design of artificial polymer nanocomposites that exhibit similar architecture and chemo-mechanical behavior. To mimic the natural adaptive materials, synthetic microfibers could be used to form the frame in a soft matrix. In the soft-to-hard cycles, the elastic fibers need to link up, making a tough, resilient network that runs through the entire composite, stiffening it.[5] In softening process, the bonding between the fibers is removed through a mediator. Thus, the key point for the biomimetic design is how to establish revisable bonds between the fibers, like those in the inner dermis of sea cucumbers.

4.2.2 Gradient Stiffness of Squid Beak via Gradient Fiber Interactions

The squid beak is another interesting biological model for the biomimetic design of adaptive materials with gradient stiffness change. The beak of the Humboldt squid *Dosidicus gigas* represents one of the hardest and stiffest wholly organic (i.e., non-mineralized) materials known.[10] Although the tip of a squid's beak is hard, the base is as soft as the animal's jelly-like body. To bridge these two mechanically dissimilar parts of the squid, there is a mechanical gradient material between them. This stiffness gradient runs from the relatively compliant wing edge (elastic modulus ca. 50 MPa) to the razor-sharp rostrum (elastic modulus ca. 5 GPa) in the natural wet state.[11] This connector acts as a shock absorber so the bird can bite a fish with bone-crushing force yet suffer no wear and tear on its fleshy mouth. The squid's beak (the wing) performs the remarkable function of insulating the soft buccal envelope from the high interfacial stresses generated at the rostrum during feeding.

In fact, gradients are found frequently in nature at the interface of two mechanically dissimilar materials such as crustacean exoskeletons, polychelate jaws, teeth, and in the mussel byssus. It is now known that mechanical gradients in materials can lead to increased distribution of interfacial mechanical and thermal stress, improvements in the bonding of dissimilar mechanical components, reduced contact deformation and damage, elimination of stress singularities, and improved fracture toughness.[12]

Like the inner dermis of sea cucumbers, the beak is also a natural nanocomposite (Figure 4.3.a).[10] The nanocomposite is composed of a fibrous (ca. 30 nm fiber diameter) chitin network (Figure 4.3b)[10] embedded within a biopolymer matrix. The mechanical gradient correlates with a change in the crosslinking density along the length of the material (Figure 4.3c). Crosslinking is established in the composite between the imidazole functionality of peptidyl-histidine residues and both low molecular weight and peptidyl (via L-DOPA residues) catechol moieties. Hence, the crosslinking occurs predominantly within the biomatrix phase.[13] Because of the existence of di-, tri-, and tetra-histidine-catecholic adducts in the composite, there is a high degree of crosslinking in the mature squid beak. However, the mechanical gradient disappears on dehydration of the beak. In its natural hydrated state, the *Dosidicus* beak biocomposite possesses a stiffness (elastic modulus, E) of 5 GPa at the distal tip that decreases incrementally to 50 MPa in the proximal wing, which is tightly embedded within the muscular buccal mass.[14] In this biocomposite, high covalent crosslink densities at the rostrum correspond with a high stiffness and, not coincidentally, the least water of

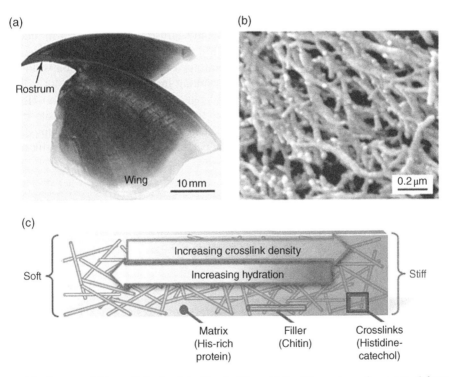

Figure 4.3 Images of (a) a split beak of the Humboldt squid *Dosidicus gigas* after removal from the buccal mass showing the relation of the wing to the rostrum and (b) a high-magnification scanning electron image of the chitin fiber network in the rostrum after alkaline peroxidation of the beak. (c) Schematic representation of water-enhanced mechanical gradient nanocomposite in the squid beak biomodel. Source: Fox *et al.* (2013).[15] Reproduced with permission of the American Chemical Society.

hydration (ca. 15 wt%). Conversely, the wing of the beak (i.e., the base) contains fewer cross-links, more chitin fibers, and consequently significantly more water (ca. 70 wt%). The stiffness gradient in the nanocomposite is therefore induced by the composition gradients of water content in combination with the chemical bonding gradient of crosslinking.[15] It is known that the composition gradients cause stiffness, as seen in other biological tissue and engineering materials, but it is quite unusual that the chemical bonding is used to generate stiffness. The microstructure of gradient nanocomposites provides design principles for attaching mechanically mismatched materials in engineering and biological applications.

4.2.3 Shape Change in Plant Growth via Controlled Reinforcement Redistribution

Many climbing plants, such as cucumbers, grapes, passion flowers, pumpkins, and gourds, have moving tendrils that search for stable support by changing their shapes. Among many shapes, chiral twisting is commonly seen in tendrils (Figure 4.4). A tendril of the cucumber plant, for example, grows straight until it contacts and attaches to a support. Once the tendril has contacted the support, one side of the coil exhibiting a ribbon of gelatinous fiber cells undergoes tissue morphosis in the form of lignification, water removal, and development of oriented cellulose microfibrils.[16] This leads to an asymmetric distribution of reinforcement (e.g., lignin) within the fiber ribbon, and consequently a frustrated geometry forms due to the differential local growth of the cell layers with different reinforcement levels. As the tendril grows, it minimizes the length of the more reinforced portion of the ribbon while maximizing the length of the less reinforced side, resulting in morphing by chirally twisting.[16] Sometimes this twisting can begin before a tendril contacts a support (Figure 4.4a). However, if the twisting conformation initiates only after the tendril contacts an opposite point of support, there must be zero net twist in the whole length of the tendril. To accommodate this fixed boundary condition, the tendril develops a perversion that is a point of helix reversal

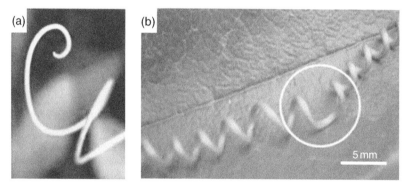

Figure 4.4 Tendrils of climbing plants. (a) Unbound tendrils. Source: Godinho *et al.* (2010).[18] Reproduced with permission of the Royal Society of Chemistry. (b) Tendrils that are initially bound between two points prior to twisting must maintain zero net twist and develop perversions (region enclosed by the circle). Source: Godinho *et al.* (2009).[17] Reproduced with permission of the Royal Society of Chemistry.

where the tendril changes from left-handed to right-handed chirality (Figure 4.4b).[17] This formation of twisting cucumber tendrils in a series of right- and left-handed helices draws the stem of the plant closer to the support. The spring-like tendril also provides a cushion between the support and stem.

Similar asymmetric distribution of reinforcement occurs during the growth of tree branches. As a tree grows in response to applied stresses induced by gravity, wind, or erosion, to gain exposure to more sunlight cellulose microfbril (CMF) orientation is actively regulated in plant stems and tree branches to achieve shape change. When the stem is programmed to restrict expansion in all directions, the plants generate a random fibril orientation.[19] To promote one-dimensional expansion (growth), coordinated restructuring of fibril architectures is produced to align the reinforcement perpendicular to the stem (Figure 4.5a,b).[4] For example, the conifer pine branch transitions its CMF orientation along the length of a branch from low angles to higher angles. This heterogeneous architecture leads to the natural bending of conifer tree branches to overcome gravity and regulate the stress in the branch (Figure 4.5c,d).[20] In tree growth, active responses are necessarily complex, requiring signal pathways between biomolecules to enable the creation or redesign of the CMF architecture. Understanding the rich mechanisms that drive these CMF architectures in plant systems may allow us to develop novel biomimetic designs and processes for actively controlling material shape changes in synthetic analogues.

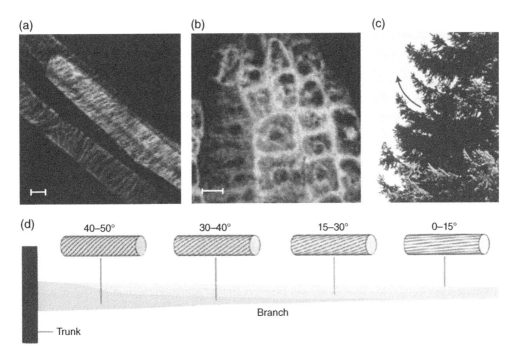

Figure 4.5 Biological systems that use fiber orientational control to enable structural bending and twisting include the plant stem (a, b) and the tree branch (c, d). Source (a, b): Foster *et al.* (2003).[19] Reproduced with permission of Elsevier. Source (c, d): Fratzl and Barth (2009).[20] Reproduced with permission of Nature Publishing Group.

4.2.4 Self-shaping by Pre-programed Reinforcement Architectures

While animals and live plants actively regulate property/shape changes through either neural systems or chemical signals, there are adaptive materials that control this process by local orientation of the reinforcement in composites. Plant cell walls are composites of reinforcement – stiff cellulose fibrils and matrix – pliant and highly swellable organics consisting of hemicelluloses, pectin, structural proteins and/or lignin. Apart from reinforcement to increase the mechanical strength of the composites, one of the major functions of cellulose microfbrils is to build specific architectures to restrict swelling or growth in certain directions, thus achieving pre-programed macroscopic self-shaping.[21] This mechanism has been observed in many plants, for example wheat awns that can dig their way into soil, and pinecones and other seedpods that cleverly open to release seeds under optimum conditions. While live plants actively regulate the fiber orientation and volume fraction during growth in response to stresses and other growth restriction, their seed dispersal units, such as pinecones, wheat awns, and seedpods, must have pre-designed fibril architecture to achieve pre-programmed shape change in their dead state before disconnection from their nutrient source. Although these seed dispersal units are passive systems, self-shaping systems predominantly rely on the reinforcement architecture – the local orientation, volume fraction, and distribution – rather than the intrinsic molecular/atomic scale properties of the plant constituent materials.[4]

The principle of self-shaping is simple in a composite: the (partially crystalline) cellulose fibrils do not swell, while the amorous matrix experiences significant expansion on taking up moisture. As a result, the extension on water uptake becomes extremely anisotropic; swelling will occur preferentially in the direction perpendicular to the fibrils due to the restriction of the fibrils in their axis direction. This pre-programed macroscopic morphing can be achieved by controlling the type, orientation, volume friction, and distributions of cellulose fibrils in cellulose nanocomposites. Because of the high swellable nature of matrix, the influx and efflux of water in the cell walls is considered to be the external stimuli that cause changes in cell and tissue geometry. Different cellulose alignments in subsequent tissue layers result in a substantial and directed bending of organs.[4]

Among many passive actuation systems with reversible shape change capability, pinecones possess an elegant pre-programmed fibril architecture in their scales. This architecture comprises two unique regions of different predominant fibril alignments in each pinecone scale. In the upper region of the scale, sclerenchyma fibers predominantly run along the length of the scale to form a laterally unidirectional composite, whereas sclereids are oriented perpendicular to the plane of the scale in the lower region (Figure 4.6a,b).[22] On drying, the lower region with perpendicular fibril distribution will shrink much more than the upper region with lateral fibrils. Since the two regions are tightly connected, the entire scale will bend downward to minimize internal stresses.[23] This fibril architecture allows the cones to close under wet conditions and open and release seeds upon drying.[24] A similar actuation mechanism was observed in the spore capsules of some mosses, but these systems are even more moisture sensitive than the conifer cones. In analogy, these systems can be considered as a kind of bimetal, but they have two substantial differences: (1) the actuation is mediated by humidity not temperature and (2) instead of different elements in bimetals, pinecones use the same chemical composition to achieve the pre-programmed deformation.[4]

A more interesting example of fibril architecture predesigned for plant actuation is the locomotion of wheat dispersal units – wheat awns that drive ripe grains into the soil. This action is

(a)   (b)

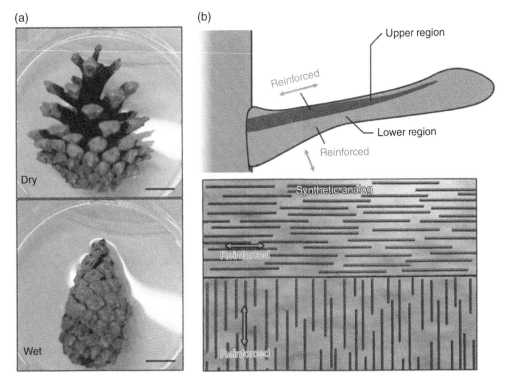

Figure 4.6 Biological systems can use fiber orientational control to enable structural bending and twisting in the pine cone. Source: Erb *et al.* (2013).[22] Reproduced with permission of Nature Publishing Group.

powered solely by the daily change in humidity, which induces a curvature of the awns depending on moisture (Figure 4.7).[21] Similar to pinecones, awns have tissues with different fibril architectures on the inner side, facing towards the other awn (called the cap), and the outer side, facing outwards (called the ridge). In the cap, the cellulose fibrils are oriented almost parallel to the awn axis, which makes the axial direction less sensitive to moisture changes. In this part, unidirectionally aligned fibrils mostly along the awn resist the axial contraction of the matrix with drying and a stable awn structure. In the lower section of the ridge, cellulose fibrils are distributed randomly and are more sensitive to moisture than the cap. These different orientations lead to different preferred axes of expansion and contraction. There is an intermediate unit between the active (ridge) and the resistance (cap) parts, which is composed of soft tissues of chlorenchyma and parenchyma. This part probably optimizes the movement and the connection between the two parts. Such a structure leads to cyclic movements during the day and night. In the dry daytime air, the wheat awn will bend its two stalks apart from each other while at night, dampened by dew, they move towards each other, driving the awn to move forward during the movement of two stalks. When humidity is high during the night, the awns of the spikelet become erect and draw together, and this action pushes the grain into the soil. During the daytime when the humidity drops, the awns slacken back again, but fine silica hairs on the awns act as ratchet hooks in the soil and prevent the spikelets from reversing back out

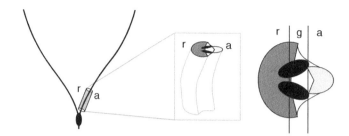

Figure 4.7 A schematic illustration of wheat awns. On the left is the general structure of the dispersal unit, showing the awns and the seed case. The actuating part of the awn is shown in the box. The outer active part (a) shrinks on drying. The resistance part (r) faces inwards. The chlorenchyma is represented in black. On the right is a map of the cross-section, showing the three components required for hygroscopic movement: the active part (a), the resistance part (r), and an intermediate gap (g) separating the two.[25] Source: Fratzl *et al.* (2008).[26] Reproduced with permission of the Royal Society of Chemistry.

again. Hence, over several days, this unidirectional movement, similar to the swimming strokes of a frog, drills the spikelet as much as an inch into the soil.[25]

4.2.5   Biomimetic Design Strategies for Morphing and Adapting

As discussed above, there are strategies in nature that can actively and reversibly change the properties or generate movement in the materials in response to external stimuli, even for dead tissues. The movement in the materials can be pre-programmed through adjusting fiber orientation and/or volume fraction. Table 4.1 lists some of extensively studied adaptive mechanisms based on nanofiber distribution and connection. These strategies could be applied to many fields, including transient implants, drug-delivery carriers, tissue-engineering scaffolds, thermo-responsive hydrogels, self-healing materials, cell cultures, bioseparation membranes, sensors, and actuators.

The intriguing feature exhibited by the dermis of sea cucumbers, that is, their ability to change mechanical properties "on command", provides an approach for the design of adaptive materials relying on the interactions between fibers in polymer nanocomposites. In these composites reinforcing nanofiber interactions (interfiber bonding) are stimuli-responsive and the mechanical properties of the bulk material are regulated by forming/breaking the bonding between the fibers. The high stiffness in the dry state and the ability to tailor the mechanical contrast via composition and processing makes the materials particularly useful as a basis for adaptive biomedical implants.[27] Unlike shape-memory polymers and metallic alloys, the shape-changing capabilities in these biological materials originate at the microstructural level rather than the molecular scale. This enables the creation of predefinable shape changes using building blocks that would otherwise not display the intrinsic molecular/atomic phase transitions required in conventional shape-memory materials. This provides an excellent biological prototype for the biomimetic design of adaptive materials using conversational fibers.

Squid beak provides another interesting model for creating gradient fiber reinforced composites. The mechanical gradient in this tissue is established by simply introducing different amounts of crosslinking along a direction. This provides new approach to fabricating mechanical gradient fiber-reinforced composites. Mechanical gradients in materials have been seen in many biological tissues such as tooth, and they can lead to better distribution of

Table 4.1 Biomimetic design strategies for morphing and adapting.

Functions	Adaptive mechanisms	Biological prototypes	Illustration	Examples of possible applications
Stiffness change	Nanofiber linking via switchable interactions (hydrogen bonding)	Sea cucumbers		Shape-memory, actuators, sensors
Gradient stiffness	Gradually changing crosslinking	Squid beak		Impact resistor
Shape change in growth	Controlled nanofiber distribution	Pine tree, cucumbers, grapes		Actuators, sensors
Self-shaping	Controlled nanofiber orientation	Pinecones, wheat awns, seedpods		Actuators, sensors

interfacial mechanical and thermal stress, enhanced bonding of dissimilar mechanical components, reduced contact deformation and damage, and improved fracture toughness.[12] Squid
beak is unique in its formation of the mechanical gradient by fibers with different amounts of
crosslinking, which is relatively easy to implement in biomimetic applications.

Swelling by adsorbing moisture in cell walls is one of the major mechanisms for plants to
change their shape. These configuration changes are controlled by the architecture of stiff
cellulose fibrils embedded in a swellable matrix. Thus, the main biomimetic strategy for an
adaptive/morphing composite mimicking plant cells is to control the distributions of stiff
fibrils in a swelling matrix (e.g., a gel). When the bonding between the matrix and the fibrils
is strong (e.g., by covalent bonding), the fibrils, essentially undeformable elements, can
restrict the matrix swelling, resulting in an anisotropic deformation of composites. There are
two ways to regulate the local deformation of the composites. One way is to make the distribution of the stiff fibrils uneven in the matrix, and the other is to control the orientation of the
fibrils. This reveals a general design principle that active gels can be directed in their swelling
behavior by stiff fibers distributed in a suitable way. From a materials science viewpoint, this
actuating mechanism by controlling microstructures provides new approaches for materials
design because the actuating movements occur without the need for an active metabolism,
unlike molecular motors in the human muscle, for example. This makes humidity-based
actuation systems in plants particularly interesting for biomimetic materials research.[4]

4.3   Biomimetic Synthetic Adaptive Materials and Processes

4.3.1   Adaptive Nanocomposites with Reversible Stiffness
Change Capability

Sea cucumbers can actively and reversibly change the stiffness of their dermis using switchable interactions between reinforcing building blocks. This adaptive behavior of materials has
been successfully mimicked in a new family of stimuli-responsive polymer nanocomposites
made of cellulose nanowhiskers as reinforcing elements in combination with different
polymeric matrices.[5,7] In these nanocomposites, the nanowhiskers are connected to each other
through hydrogen bonding to form a stiff percolating network throughout the composite.
When water is added to the nanocomposites, the materials are softened via the formation of
competitive hydrogen bonds that disrupt the initial percolating network. The elastic modulus
of the composites can thus be changed by switching the hydrogen bonding between reinforcing whiskers on and off. The variations in modulus can be achieved by a factor of ~ 10 if the
volume fraction of reinforcement is sufficiently high to ensure percolation of the fibers
(Figure 4.8). The elastic modulus of water-free samples can be reasonably described by a
percolation model (Figure 4.8b), indicating that the stiffening effect is indeed a result of the
onset of crosslinking interactions between the cellulose nanowhiskers. When the crosslinks
are removed by adding water to the composites, the elastic modulus reduces to values close to
the level expected from the Halpin–Kardos model for noninteracting whiskers (Figure 4.8).

While the stiffness of the nanocomposites is regulated by water, nanocomposites based on
poly(vinyl acetate) (PVAc) and cellulose whiskers display a "dual" responsive behavior, that
is, it is triggered by both water and cerebral spinal fluid.[28] On exposure to physiological
conditions (e.g., water or artificial cerebral spinal fluid), the materials slowly take up aqueous
fluid, which plasticizes the PVAc and reduces the glass transition temperature (T_g) from above
to below physiological temperature. In addition, the whisker–whisker interactions are switched

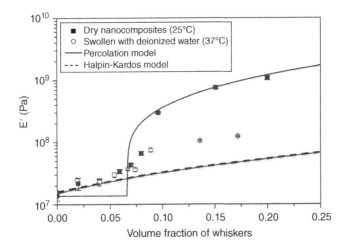

Figure 4.8 Tensile storage moduli E_0 of neat polyurethane and polyurethane/cellulose nanowhisker nanocomposites as a function of volume fraction of cellulose nanowhiskers in the dry state at 25°C and water-swollen state (after immersion in deionized water at 37°C for 5 days). Data points represent averages ($N=3$–7)±standard deviation. Lines represent values predicted by the percolation (solid) and Halpin–Kardos (dashed) models. Source: Mendez et al. (2011).[28] Reproduced with permission of the American Chemical Society.

off due to competitive hydrogen bonding with the water. This leads to a dramatic mechanical contrast between the dry state at room temperature and the water or artificial cerebrospinal fluid swollen state. These composites are sensitive to two stimuli and two different effects, plasticization of the PVAc matrix and disruption of the whisker network to manifest a dynamic change in modulus, and the mechanical contrast changes from ca. 5 GPa in the stiff state to ca. 5 MPa in the soft state, compared to the 50–5 MPa observed for the sea cucumber.

Built on the success of the above biomimetic nanocomposites with stiffness change on demand, the reversible formation of an elastic network of percolating fibers was further used to create composites exhibiting shape-memory effects.[28] These new biomimetic, stimuli-responsive mechanically adaptive nanocomposites are triggered by water stimulus, and the percolating network of reinforcing fibers was used to temporarily store a large amount of elastic energy that is initially applied to the surrounding polymer matrix of the material. On-demand rupture of the stiff percolating network activated by exposing it to water eventually releases the elastic energy stored within the system, causing a significant shape change in the entire composite.

Shape-memory materials are produced by introducing rigid cotton cellulose nanowhiskers (CNWs) into a rubbery polyurethane (PU) matrix. Figure 4.9a,b schematically illustrates the steps required to obtain the initial temporary shape and to eventually change it into the final stress-free state for the composites. In the first step, the composite is exposed to a wet environment to switch off the interactions between the reinforcing whiskers. The composite is then stretched on losing the crosslinking between cellulose whiskers. In this stage, the composite could be uniaxially stretched up to 200% strain due to the strong elastic nature of the polyurethane matrix. During the stretching the whiskers are aligned in the stretching direction because of a large deformation. In the third step, the composite is dried while still staying in the stretched state, leading to the formation of a stiff percolating network of cellulose whiskers that are

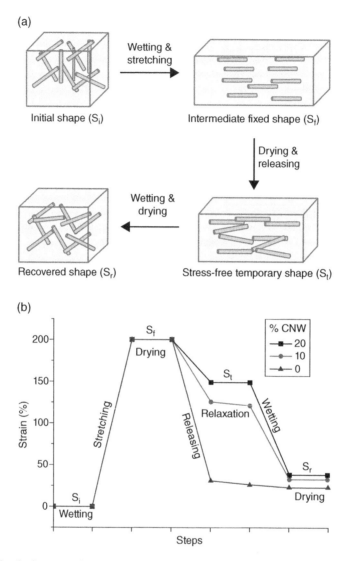

Figure 4.9 Synthetic composites undergo reversible shape changes through switchable interactions between reinforcing fibers. (a) The steps required to fix a temporary shape on drying and to recover the original shape on wetting. (b) Mechanical strains applied and recovered along the shape-changing process of polyurethanes containing 0, 10, and 20% cellulose nanowhiskers (CNWs). Source: Mendez *et al.* (2011).[28] Reproduced with permission of the American Chemical Society.

crosslinked through hydrogen bonding. When the applied stress is released after complete drying, the stiff cellulose network is able to retain a large fraction of the initial mechanical deformation imposed during stretching. Thus, a consolidated temporary shape (S_t) is obtained. To examine the whisker state and verify the process, polarized Raman spectroscopy was used. The composites before and after stretching and drying show that the high degree of alignment of whiskers is achieved during the initial stretching step. Indeed, the temporary shape is maintained by the formation of crosslinking interactions between whiskers, which do not let the elastomeric matrix relax back to its original configuration.[28]

On exposing the composite to water, it recovers from its temporary shape (S_t) to its unstressed shape (S_r). In this last step, the initially stretched polyurethane matrix is fully relaxed again when the stiff network is eventually broken by the competitive hydrogen bonding of the water molecules between adjacent cellulose whiskers. The entire cycle can be repeated multiple times, starting from the initial mechanical deformation of the composite into a temporary shape followed by its full relaxation in the presence of water molecules. This approach is appealing because it provides a general way of making shape-memory materials. In general, such concepts can be applied to advanced composites with reinforcing fibers whose interactions are sensitive to a specific external stimulus such as electric or magnetic signals, as is the case for cellulose whiskers in the presence of water. Once an interaction is established, bulk composites can be mechanically shaped into any arbitrary geometry and eventually relaxed to their original shape using a specific stimulus as an external trigger.[28]

4.3.2 Squid-beak-inspired Mechanical Gradient Nanocomposites and Fabrication

Inspired by the mechanical gradient composites found in squid beak, mechanical gradient synthetic materials were also fabricated using allyl-functionalized CNC/PVAc nanocomposites imbibed with a tetrathiol crosslinker and a UV initiator.[15] The amount of UV exposure can be used to control the degree of crosslinking, with longer exposure times resulting in stiffer wet materials (Figure 4.10a), for example the wet modulus of the CNC PVAc film can be increased from ca. 60 MPa to ca. 300 MPa after 20 min of UV crosslinking. This level of reinforcement in the nanocomposites matches up well with the predicted modulus by the percolation theoretical model using the modulus of the reinforcing phase as the wet modulus of the crosslinked CNC sheet (Figure 4.10a).

With the efficacy of the photocrosslinking process, mechanical gradient nanocomposites, which exhibit water-enhanced mechanical contrasts, were created by simply controlling the exposure time of different parts of the film. This photoinduced crosslinking technique could induce complex mechanical gradients that are programmed into a film by using specific photomasks. In this experiment, repeating the mask procedure three times gave a film which had sections along its length that were exposed to different amounts of UV irradiation (e.g., 0, 4, 8, 12, 16, and 20 minutes). As can be seen in Figure 4.10b, a dramatic increase in mechanical contrast was observed for wet film ($E'_{stiff}/E'_{soft} > 5$) over dry film either below or above T_g (E'_{stiff}/E'_{soft} of ca. 1.1 and 1.5, respectively). A similar water-enhanced mechanical contrast is observed in squid beak, which exhibits a mechanical contrast ratio of ca. 2 in the dry state (10–5 GPa) to ca. 100 in its natural wet state (5 GPa to 50 MPa).

These mechanical gradient materials could have broad applications across a number of biomedical applications wherever there is a need to interface stiff therapeutic interventions with soft biological tissue.[29] For example, a modulus buffer between a stiff implant and soft tissue would be advantageous in a range of percutaneous technologies, such as glucose sensors for diabetics, osseointegrated prosthetic limbs for amputees, long-term intravascular interfaces such as central venous port systems for chemotherapy, and chronically implanted intracortical microelectrodes.[15,29]

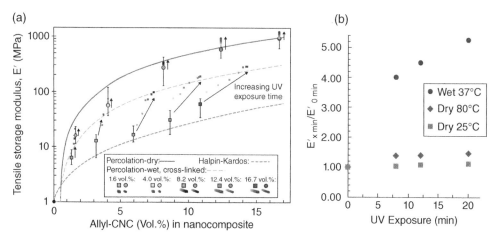

Figure 4.10 (a) Tensile storage modulus (as measured by DMA) of uncrosslinked (black-outlined symbols) and photo-crosslinked (non-outlined symbols) allyl-CNC/PVAc nanocomposites in both dry at 80°C (circles) and wet at 37°C (squares) states. Also shown for comparison are the curves for the percolation model (solid line), the Halpin–Kardos model (short-dashed line), and the wet, cross-linked percolation model (long-dashed line) used to model the nanocomposites. (b) Mechanical testing results plotted to highlight the contrast in gradient between the samples dry at 25°C (squares), dry at 80°C (diamonds), and wet at 37°C (circles). Source: Gerbode et al. (2012).[16] Reproduced with permission of the American Chemical Society.

4.3.3 Biomimetic Helical Fibers and Fabrication

Inspired by the chiral growth of cucumber tendrils, synthetic systems that exhibit identical cylindrical twisting to cucumber tendrils have been demonstrated. Canejo et al. synthesized materials with a helically twisting fiber architecture by electrospinning acetoxypropylcellulose (APC) fibers at high volume fractions in an anhydrous dimethylacetamide (DMAc) solvent.[30] If the volume fractions are high enough, the APC self-assembles into a liquid crystalline cholesteric phase and the solution is sufficiently fluid to be pumped through a capillary for electrospinning (Figure 4.11a). If there is no shear force, the assembled APC fiber axis and a disclination line will coincide with the capillary axis (Figure 4.11b).[31] However, as this liquid crystalline solution is forced through the capillary during the electrospinning process (Figure 4.11a), the APC solution is subjected to a parabolic flow, leading to gradually higher shear stresses. This causes the disclination line to shift from the central axis to the edge of the capillary, ultimately forming a chiral topological defect along the capillary (Figure 4.11b).

The disclination line forms a "hard" helix, as identified by the magnetic resonance (Figure 4.11b). This "hard" disclination line mimics the stiffened fibrous section of the cucumber plant tendril and acts as the guiding microstructural element that leads to a shape change. The asymmetric reinforcement in the synthetic system occurs between the disclination line and the cholesteric APC bulk phase (Figure 4.11b). As the solvent is evaporated from the APC solution during spinning and drying, the fiber undergoes a shrinking that is preferentially reduced along the "hard" disclination line. This microstructure produces a frustrated geometry that responds by twisting to minimize the system's energy. Thus, the helix in the macroscopic material is achieved using shear fields to generate a self-shaping microstructure that is similar to the

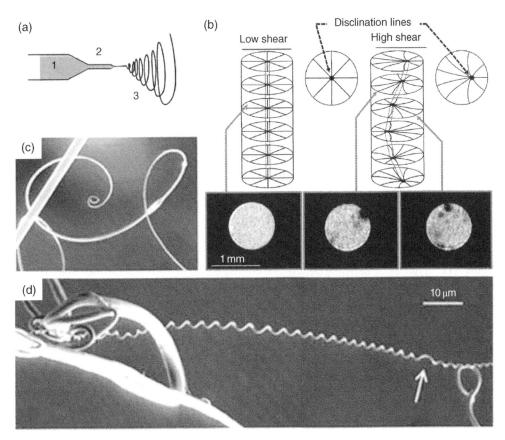

Figure 4.11 Synthetic route for making fibers that mimic the twisting shape change found in climbing plant tendrils. (a) Electrospinning of a solution of acetoxypropylcellulose (1) (APC) through a capillary (2) to generate a twisting macroscopic fiber (3). (b) Disclination line which parallels the stiff fiber of the cucumber plant tendril is seen as the dark (hard) regions in the inset MRI cross-sectional scans of the APC capillary flow. (c and d) On electrospinning at high shear, the macro fibers twist to reduce the high elastic energy created by the disclination line. Similar to plant tendrils, APC macro-fibers connected only at one point can form spirals (c), while APC macro-fibers that are connected at two points form helical twists (d) often with perversions (white arrow). Source: Studart & Erb (2014).[4] Reproduced with permission of the Royal Society of Chemistry.

reinforcement distribution of twisting climbing tendrils (Figure 4.11c,d). This is an interesting example in which the topological defect arising from the constrained packing of chiral molecules rather than the chirality of the molecule itself is responsible for the shape change.[4]

4.3.4 Water-activated Self-shaping Materials and Fabrication

Many plants utilize cellulose microfbril to construct specific architectures to control swelling in certain directions and pre-programed macroscopic actuation.[21] On swelling by water/moisture, these composites can exhibit complex shape changes depending on pre-designed architectures.

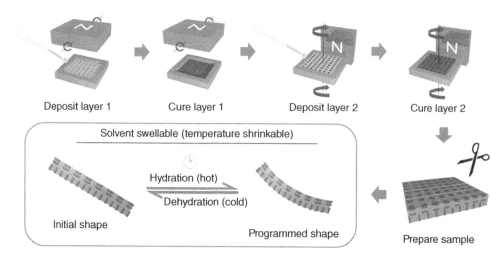

Deposit layer 1   Cure layer 1   Deposit layer 2   Cure layer 2

Solvent swellable (temperature shrinkable)

Hydration (hot)

Dehydration (cold)

Initial shape

Programmed shape

Prepare sample

Figure 4.12   Typical manufacturing method used to create bioinspired self-shaping reinforced composites. Source: Erb *et al.* (2013).[22] Reproduced with permission of Nature Publishing Group.

Inspired by these new self-shaping strategies, composites with such self-shaping capability have been fabricated by controlling fiber orientation during the assembly process. In fact, in commercial continuous-fiber reinforced composites, fibers are usually layered in different orientations to control mechanical behavior in different directions. In unidirectional composites, fibers are piled in one direction; the mechanical properties are strong in that direction. When the fibers are layered in different directions, quasisotropic composites are obtained. Although this design principle is very similar to biological materials design, for example pinecone architecture for commercial fiber-reinforced composites, the swelling effect of the matrix (usually resin) is usually too small to make large macroscopic shape change. In this regard, hydrogels (e.g., gelatin) have been used as the matrix that undergoes dramatic volumetric growth upon hydration. However, is it difficult to control the orientation of the short fibers when mixed with hydrogen matrix. To form the desirable fiber architectures in hydrogel matrix, micron-sized reinforcing elements are labeled with superparamagnetic iron oxide nanoparticles so that they are able to respond to external magnetic fields.[4] For the proper size of reinforcing particles this labeling technique leads to an ultrahigh magnetic response (UHMR) in which the reinforcing elements respond to very low magnetic fields in the order of 1 mT (Figure 4.12). Using this UHMR technique, 7.5 μm alumina microplatelets have been assembled in different architectures within hydrogels.[22]

A fiber architecture mimicking conifer pine branches (Figure 4.13a) has recently been synthesized using poly(vinyl alcohol) matrices with embedded reinforcing UHMR microplatelets exhibiting gradually changing orientations.[4] In conifer pine branches, the fibrils are oriented along the branches in the top half and perpendicular to the branch axis in the bottom half. To build such architectures, a magnetic field was applied to an aqueous precursor suspension containing the polymer and microparticles to create ordered particle distributions along the plate. In addition to static magnets, rotating magnetic fields were utilized to enable complete orientation of the microplatelets to control fiber architectures.[32] With these processes, a pinecone-mimicking gelatin bilayer structure was fabricated, in which the microplatelets were oriented along the plate in the top half and perpendicular to the plate axis

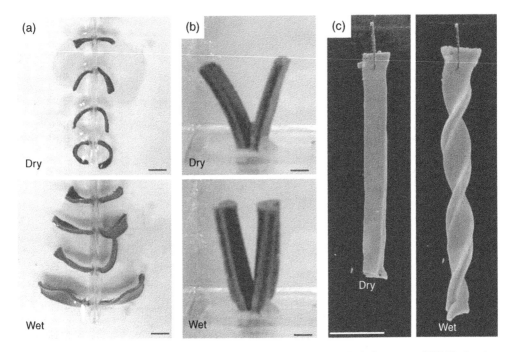

Figure 4.13 Bioinspired self-shaping synthetic systems with oriented reinforcement. UHMR alumina microplatelets can be positioned with magnetic fields in gelatin matrices to recreate CMF architectures of (a) pinecones, (b) wheat awns, and (c) orchid tree seed pods. Hydrating and dehydrating these systems allows macroscopic shape changes that mimic their natural counterparts. Scale bars in (a), (b), and (c) are 1 cm. Source: Studart & Erb (2014).[4] Reproduced with permission of the Royal Society of Chemistry.

in the bottom half. When these biomimetic bilayers are cycled between wet and dry states, they bend reversibly, in a similar manner to the natural pinecone, as shown in Figure 4.13a.

The natural wheat awn architecture has also been mimicked by synthetic composite structures fabricated by producing a gelatin bilayer with in-plane oriented micro-reinforcement in the top half and randomized micro-reinforcement in the bottom half (Figure 4.13b). The randomized reinforcement is formed simply by not applying a magnetic field during suspension casting. Since the randomly reinforced layer has slightly higher lateral swellability than the top layer with aligned reinforcements, it drives the bilayer to bend. As well as bending, the chiral twisting movements of the orchid tree seed pod were also recreated by building platelet-reinforced gelatin bilayers that exhibit orthogonal reinforcement of the two layers both angled at $\pi/4$ radians relative to the long axis of the material (Figure 4.13c). Such fiber architectures caused the hydrogels to undergo chiral twisting.[4]

The reinforced swellable hydrogels have also been exploited extensively to create nano- and microstructured surfaces that exhibit unique dynamic functionalities and adaptive properties.[33] In such hybrid systems, material surfaces consisting of environmentally sensitive hydrogels integrate within arrays of high-aspect-ratio nano- or microstructures (Figure 4.14). Unlike the bulk reinforcement architectures described above, which impose an anisotropic swelling of the matrix in a bulk bilayer structure, the hybrid materials rely on the actuation of surface-anchored posts or fins by the triggered contraction of the

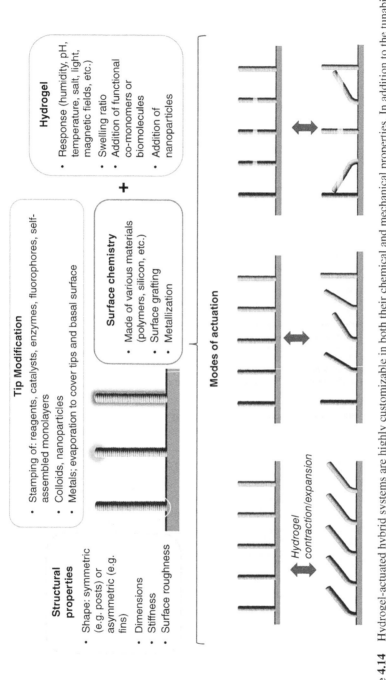

Figure 4.14 Hydrogel-actuated hybrid systems are highly customizable in both their chemical and mechanical properties. In addition to the tunability of each material (embedded structures or hydrogel) there are also various ways to combine them to create a range of modes of actuation, each with different chemomechanical behavior. Bottom left to right: large-area homogeneous actuation, localized actuation within a well-defined region, and localized radial actuation from tethered hydrogel pads. Source: Zarzar & Aizenberg (2014).[33] Reproduced with permission of the American Chemical Society.

surrounding confined hydrogel. With this actuation principle, the hydrogel-hybrid structures can exhibit three basic modes of actuation: large area coordinated bending, localized bending in specific areas, and localized radial bending (Figure 4.14). The coordinated bending of asymmetric fin structures has been demonstrated upon homogeneous chemical stimulus by using hydrogels that are chemically tethered to both the high-aspect ratio elements and the basal surface.[34] The actuation can be trigged by various stimuli, such as hydration (e.g., polyacrylamide), pH shifts (e.g., poly(acrylic acid-co-acrylamide)), or temperature changes (e.g., poly(N-isopropylacrylamide). In the design of this bioinspired hydrogen hybrid actuator, it is necessary to ensure that enough force is generated during hydrogel contraction, therefore metal elements (e.g., Si) with very high aspect ratios or soft flexible polymers (e.g., epoxies, polyurethanes, polydimethylsiloxane) have to be used to make high-aspect ratio elements sufficiently compliant for bending during actuation. This mechanical actuation of high-aspect ratio elements through the chemically triggered contraction of hydrogels is useful for the creation of functional surfaces with switchable wetting, optical, and chemo-mechanical responses.[33]

References

1. Jani, J.M., Leary, M., Subic, A. & Gibson, M.A. A review of shape memory alloy research, applications and opportunities. *Materials & Design* **56**, 1078–1113 (2014).
2. Li, F., Jin, L., Xu, Z. & Zhang, S. Electrostrictive effect in ferroelectrics: An alternative approach to improve piezoelectricity. *Applied Physics Reviews* **1**, 011103 (2014).
3. White, E.M., Yatvin, J., Grubbs, J.B., Bilbrey, J.A. & Locklin, J. Advances in smart materials: Stimuli-responsive hydrogel thin films. *Journal of Polymer Science: Polymer Physics* **51**, 1084–1099 (2013).
4. Studart, A.R. & Erb, R.M. Bioinspired materials that self-shape through programmed microstructures. *Soft Matter* **10**, 1284–1294 (2014).
5. Capadona, J.R., Shanmuganathan, K., Tyler, D.J., Rowan, S.J. & Weder, C. Stimuli-responsive polymer nanocomposites inspired by the sea cucumber dermis. *Science* **319**, 1370–1374 (2008).
6. Trotter, J.A. *et al.* Towards a fibrous composite with dynamically controlled stiffness: lessons from echinoderms. *Biochemical Society Transactions* **28**, 357–362 (2000).
7. Shanmuganathan, K., Capadona, J.R., Rowan, S.J. & Weder, C. Biomimetic mechanically adaptive nanocomposites. *Progress in Polymer Science* **35**, 212–222 (2010).
8. Koob, T.J., Koob-Emunds, M.M. & Trotter, J.A. Cell-derived stiffening and plasticizing factors in sea cucumber (Cucumaris frondosa) dermis. *Journal of Experimental Biology* **202**, 2291–2301 (1999).
9. Szulgit, G.K. & Shadwick, R.E. Dynamic mechanical characterization of a mutable collagenous tissue: Response of sea cucumber dermis to cell lysis and dermal extracts. *Journal of Experimental Biology* **203**, 1539–1550 (2000).
10. Miserez, A., Schneberk, T., Sun, C., Zok, F.W. & Waite, J.H. The transition from stiff to compliant materials in squid beaks. *Science* **319**, 1816–1819 (2008).
11. Waite, J.H. & Broomell, C.C. Changing environments and structure-property relationships in marine biomaterials. *Journal of Experimental Biology* **215**, 873–883 (2012).
12. Suresh, S. Graded materials for resistance to contact deformation and damage. *Science* **292**, 2447–2451 (2001).
13. Miserez, A., Rubin, D. & Waite, J.H. Cross-linking chemistry of squid beak. *Journal of Biological Chemistry* **285**, 38115–38124 (2010).
14. Uyeno, T.A. & Kier, W.M. Functional morphology of the cephalopod buccal mass: A novel joint type. *Journal of Morphology* **264**, 211–222 (2005).
15. Fox, J.D., Capadona, J.R., Marasco, P.D. & Rowan, S.J. Bioinspired water-enhanced mechanical gradient nanocomposite films that mimic the architecture and properties of the squid beak. *Journal of the American Chemical Society* **135**, 5167–5174 (2013).
16. Gerbode, S.J., Puzey, J.R., McCormick, A.G. & Mahadevan, L. How the cucumber tendril coils and overwinds. *Science* **337**, 1087–1091 (2012).

17. Godinho, M.H., Canejo, J.P., Pinto, L.F.V., Borges, J.P. & Teixeira, P.I.C. How to mimic the shapes of plant tendrils on the nano and microscale: spirals and helices of electrospun liquid crystalline cellulose derivatives. *Soft Matter* **5**, 2772–2776 (2009).
18. Godinho, M.H., Canejo, J.P., Feio, G. & Terentjev, E.M. Self-winding of helices in plant tendrils and cellulose liquid crystal fibers. *Soft Matter* **6**, 5965–5970 (2010).
19. Foster, R., Mattsson, O. & Mundy, J. Plants flex their skeletons. *Trends in Plant Science* **8**, 202–204 (2003).
20. Fratzl, P. & Barth, F.G. Biomaterial systems for mechanosensing and actuation. *Nature* **462**, 442–448 (2009).
21. Burgert, I. & Fratzl, P. Actuation systems in plants as prototypes for bioinspired devices. *Philosophical Transactions Series A: Mathematical, Physical, and Engineering Sciences* **367**, 1541–1557 (2009).
22. Erb, R.M., Sander, J.S., Grisch, R. & Studart, A.R. Self-shaping composites with programmable bioinspired microstructures. *Nature Communications* **4** (2013).
23. Reyssat, E. & Mahadevan, L. Hygromorphs: from pine cones to biomimetic bilayers. *Journal of the Royal Society Interface* **6**, 951–957 (2009).
24. Dawson, J., Vincent, J.F.V. & Rocca, A.M. How pine cones open. *Nature* **390**, 668–668 (1997).
25. Elbaum, R., Zaltzman, L., Burgert, I. & Fratzl, P. The role of wheat awns in the seed dispersal unit. *Science* **316**, 884–886 (2007).
26. Fratzl, P., Elbaum, R. & Burgert, I. Cellulose fibrils direct plant organ movements. *Faraday Discuss* **139**, 275–282 (2008).
27. Jorfi, M., Roberts, M.N., Foster, E.J. & Weder, C. Physiologically responsive, mechanically adaptive bio-nanocomposites for biomedical applications. *ACS Applied Materials and Interfaces* **5**, 1517–1526 (2013).
28. Mendez, J. *et al.* Bioinspired mechanically adaptive polymer nanocomposites with water-activated shape-memory effect. *Macromolecules* **44**, 6827–6835 (2011).
29. Seidi, A., Ramalingam, M., Elloumi-Hannachi, I., Ostrovidov, S. & Khademhosseini, A. Gradient biomaterials for soft-to-hard interface tissue engineering. *Acta Biomaterialia* **7**, 1441–1451 (2011).
30. Canejo, J.P. *et al.* Helical twisting of electrospun liquid crystalline cellulose micro- and nanofibers. *Advanced Materials* **20**, 4821–4825 (2008).
31. Cronin, D.W., Terentjev, E.M., Sones, R.A. & Petschek, R.G. Twisting transition in a capillary filled with chiral smectic-c liquid-crystal. *Molecular Crystals and Liquid Crystals Science and Technology Section A – Molecular Crystals and Liquid Crystals* **238**, 167–177 (1994).
32. Erb, R.M., Segmehl, J., Charilaou, M., Loffler, J.F. & Studart, A.R. Non-linear alignment dynamics in suspensions of platelets under rotating magnetic fields. *Soft Matter* **8**, 7604–7609 (2012).
33. Zarzar, L.D. & Aizenberg, J. Stimuli-responsive chemomechanical actuation: A hybrid materials approach. *Accounts of Chemical Research* **47**, 530–539 (2014).
34. Zarzar, L.D., Kim, P. & Aizenberg, J. Bio-inspired design of submerged hydrogel-actuated polymer microstructures operating in response to pH. *Advanced Materials* **23**, 1442–1446 (2011).

5

Materials with Controllable Friction and Reversible Adhesion

5.1 Introduction

Many living beings, from microscopic bacteria and fungi, to larger marine algae and invertebrates, insects, frogs and even terrestrial vertebrates (i.e., gecko), use specialized adhesive organs and secretions to attach to surfaces, temporally or permanently. These biological adhesives can vary widely in structure and capabilities, and often perform in ways that differ greatly from conventional man-made adhesives. The diversity and remarkable performance of these adhesives suggests the potential for developing artificial adhesive materials that are markedly different from those currently available. These biological adhesives provide elegant solutions to biomimetic design concepts for advanced engineering materials with high adhesion strength and reliability, reversible attachment, or attachment to any surfaces, including aquatic and fluid environments.

Biological adhesives can be classified into reversible and irreversible adhesion based on their bonding types. In the animal world, irreversible bonding is generally generated by covalent and hydrogen bonds. In most cases animals secret surface-reactive species and use these as a bioglue to stick to a surface, for example catechol units in 3,4-dihydroxyphenylalanine (DOPA) containing adhesive proteins secreted by mussels. For reversible adhesion, for example adhesive pads used for locomotion, nature utilizes physical interactions enhanced by complex topographical designs. This chapter focuses on reversible adhesion and the biomimetic design of advanced engineering materials with reversible adhesion.

In nature there are two main types of reversible adhesive designs found in the attachment pads of animals: "fibrillar" (in geckos, insects, and spiders) and "smooth" (in crickets, and tree and torrent frogs). While the fibrillar adhesion design usually works in dry environments (so-called dry adhesives), the smooth thin film pads adept to wet surfaces (wet adhesion). Both systems allow conformal contact to the substrate independent of the substrate roughness, using either flexible hairs or highly deformable materials.

Biomimetic Principles and Design of Advanced Engineering Materials, First Edition. Zhenhai Xia.
© 2016 John Wiley & Sons, Ltd. Published 2016 by John Wiley & Sons, Ltd.

The ability to climb is a significant advantage for animals as it makes available habitats not accessible to non-climbers or non-flyers. Indeed, many animal phyla and groups are represented in arboreal habitats. These animals are characterized by mechanisms that facilitate climbing and reduce the risk of falling. On rough surfaces, friction pads and claws can be effective, but on smooth surfaces and significant overhangs some mechanism of adhesion is essential.[1] Adhesion allows an organism to remain attached to an inclined, vertical, or even upside down surface whilst resisting falling or slipping. In addition to adhesion, controllable friction is also important in climbing. It has been shown that adhesion could be enhanced by friction, a phenomenon called frictional adhesion.[2]

While strong adhesion and friction are necessary for climbing propose, in many cases reduction of adhesion is useful and becomes a strategy for living in many organisms. Nature also provides many designs of materials and/or surfaces that have reduced adhesion and friction. An outstanding example is *Nepenthes* pitcher plants that use structures to lock in an intermediary liquid to form a slippery surface to capture their preys. Other examples include shark skin, which relies on an ingenious antidrag design to reduce drag, and lotus leaf, which can significantly reduce adhesion and drag to water droplets, and thus has self-cleaning capability. Mimicking these biological structures could create new materials with controllable friction and reversible adhesion for engineering applications.

5.2 Dry Adhesion: Biological Reversible Adhesive Systems Based on Fibrillar Structures

5.2.1 Gecko and Insect Adhesive Systems

The attachment/detachment and reversible adhesion strategies found in nature are sometimes quite different from our daily experiences. Unlike chemical glues and conventional pressure sensitive tapes, biological adhesive pads offer strong yet reversible adhesion, with great durability and self-cleanness. Biological adhesive systems in various creatures, such as insects, spiders, and lizards, have similar structures: hierarchical fibrillar structures. The microstructures used by beetles, flies, spiders, and geckos are shown in Figure 5.1.[3] As the size (mass) of the creature increases, the radius of the terminal attachment elements decreases.[4] This allows a greater number of setae to be packed into an area, hence increasing the linear dimension of contact and adhesion strength.

The gecko is a "king of climbers" among animals using their foot pads. The gecko's ability to "run up and down a tree in any way, even with the head downwards" was firstly described by Aristotle in his *Historia Animalium*, almost 25 centuries ago. The gecko has evolved one of the most effective adhesives known in nature. Among insects and animals, geckos have both the highest body mass and the greatest density of adhesive elements, and can generate the strongest adhesion to support its weight with a high factor of safety. While beetles and flies increase adhesion by secreting liquids at the contacting interface, geckos produce dry adhesion. In fact geckos can climb and maneuver on almost any surfaces at any angle, regardless of unpredictable surface conditions, whether wet or dry, smooth or rough, and contaminants. Their extraordinary climbing ability stems from their hierarchical toe pad structure, which consists of millions of microfibrils (i.e., setae) and nano-sized branches (i.e., spatulae).[5-7] Mimicking this hierarchical fibrillar structure could lead to the development of a new class of advanced adhesives useful in various applications, including but not limited to climbing robots, reusable tapes, super-grip tires, high-efficiency breaks, and rapid patch repairs on military vehicles.

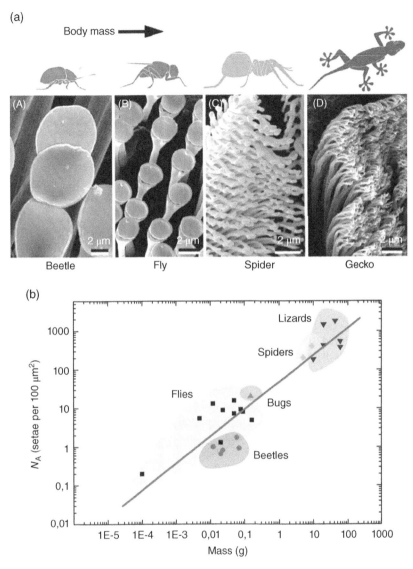

Figure 5.1   (a) Terminal elements of the hairy attachment pads of (A) beetle, (B) fly, (C) spider, and (D) gecko shown at different scales (left and right). (b) Dependence of the terminal element density (N_A) of the attachment pads on the body mass in hairy-pad systems of diverse animal groups. Source: Arzt *et al.* (2003).[3] Reproduced with permission of the National Academy of Sciences.

5.2.2   *Hierarchical Fibrillar Structure of Gecko Toe Pads*

Gecko attachment pads (toes) are composed of complex hierarchical structures from macro down to nano levels.[5–7] As illustrated in Figure 5.2, the first level of hierarchy is lamellae, soft, overlapping stripes (Figure 5.2b) that are 1–2 mm in length and easy to compress so that contact can be made with rough bumpy surfaces. Usually 15–20 lamellae occupy the

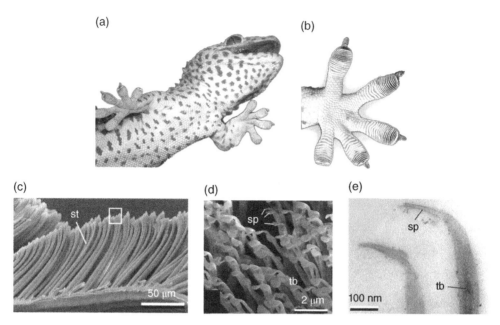

Figure 5.2   (a) Tokay gecko looking and the hierarchical structures of (b) a gecko toe. Source Autumn & Peattie (2002).[8] Reproduced with permission of Oxford University Press. Each toe contains hundreds of thousands of setae and each seta contains hundreds of spatula. SEM micrographs (at different magnifications) of (c) the setae, and (d) and (e) the spatula. st, seta; sp, spatula; tb, branch. Source: Persson & Gorb (2003).[9] Reproduced with permission of AIP Publishing LLC.

entire area of a single toe and can be easily spotted by the naked eye. The second level of hierarchy is seta, a tiny hairy structure that is usually considered to be the basic adhesive unit (Figure 5.2c,d). Setae are distributed in an array on the outer portion of lamellae, which are distally narrowed, 0.3–0.6 mm across at the mesoscale (Figure 5.2b). Setae extend from the lamellae with an angle of approximately 30°. They are typically 30–130 µm in length and 5–10 µm in diameter, and are primarily composed of dry, hard β-keratin with some α-keratin components. A Tokay gecko seta averages ~100 µm in length and ~5 µm in diameter with a density of approximately 14,000 mm^2 (Figure 5.2d). Each seta further branches out at the tip and splits into 100–1000 spatulae (Figure 5.2e). The spatula is the finest (nano) level of hierarchy of the gecko toe pads and functions as the ultimate contacting element. The tips of the spatula are approximately 200–300 nm in width, 500 nm in length and 10 nm in thickness. It should be noted that setae are asymmetric. This asymmetry is evident in Figure 5.2c, which shows the curved shape and directional tilt of the setal stalks and their flat hoof-like ends.

5.2.3   Adhesive Properties of Gecko Toe Pads

Geckos can generate strong adhesion through their hierarchical toe pads. On the macroscale level, the attachment pads on two feet of the Tokay gecko have approximately 3×10^6 setae on their toes with an area of approximately 220 mm^2, and can produce a clinging ability of approximately 20 N (the vertical force required to pull a lizard down a nearly vertical (85°) surface). This allows them to climb vertical surfaces at speeds of over 1 m s^{-1}, with the ability to attach or detach their toes in milliseconds. The clinging force is roughly equal to 20 times that required to support their body weight. When properly engaged, the maximum adhesive forces of a single seta were ~200 μN in shear and ~20 μN in normal against a smooth clean glass substrate.[7] The normal pull-off force was also obtained for individual spatulae as ~10 nN by applying atomic force microscopy (AFM).[10] Thus, if all the setae were to engage at a time, roughly half million on one foot, a single foot of a gecko could generate 100 N of adhesive force.

The adhesion of single spatula and seta depends on substrate and environments. The adhesion force of single spatulae as well as setae measured in air is much higher than that under water for most substrates except for Teflon.[11] Interestingly, while the adhesion force of both spatulae and setae on Teflon in air is nearly zero, it is substantial under water over a wide range of pull-off velocities.[12] Since Teflon and plant leaves are known to be hydrophobic, this may explain why geckos living in rainforests can generate sufficient adhesion on wet leaves.

Van der Waals interactions have been proven to be the primary molecular origin for this surface phenomenon,[7] while humidity also has a strong effect on altering the measuring forces. Intermolecular capillary forces are the principal mechanism of adhesion in many insects, frogs, and even mammals. Unlike many insects, geckos lack glands on the surface of their feet, but this does not preclude the role of capillary adhesion since layers of water molecules are commonly present on hydrophilic surfaces.

The adhesion and friction of gecko adhesive pads are directional or anisotropic.[13] As discussed above, the gecko's toe structures are only adhesive when loaded in a particular direction. Moreover, the amount of adhesion sustained is a direct function of the applied tangential load. In other words, the gecko can control adhesion by controlling tangential forces. The anisotropic adhesion results from the gecko's lamellae, setae, and spatulae all being angled instead of aligned vertically. Only by pulling in the proper direction does the gecko align its microstructures to make intimate contact with the surface. During attachment, a gecko presses and pulls its toe pads along the direction toward its body to generate large friction and adhesion; the toe pads move in the opposite direction during detachment. Consequently, both forces fall to almost zero during detachment with little expenditure of energy by the gecko.

Recent studies have revealed that geckos move by involving both adhesion in the normal direction and friction in the lateral direction, and that the two are strongly coupled: the friction enhances the adhesion when geckos grip onto substrate surfaces, so-called frictional adhesion (Figure 5.3a).[2] Geckos control their adhesion with anisotropic microstructures consisting of arrays of setal stalks with spatula tips. Instead of applying high normal preloads, geckos increase their maximum adhesion by increasing tangential force, pulling from the distal toward the proximal ends of their toes. In conjunction with their hierarchical structures, this provides geckos with a coefficient of adhesion $\mu_0 = F_a/F_p$ between 8 and 16, depending on conditions, where F_a is the maximum normal pull-off force and F_p is the maximum normal preload force.

The adhesion of gecko toe pads is also rate dependent.[14] From classical frictional mechanics, friction between dry, hard, macroscopic materials typically decreases at the onset of sliding, and as velocity increases friction continues to decrease because of a reduction in the number

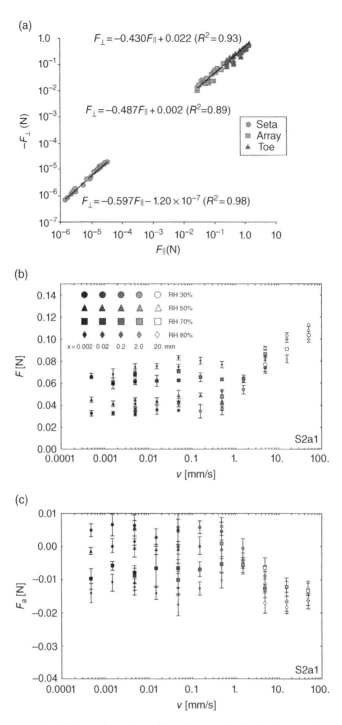

Figure 5.3 (a) Adhesion ($-F_\perp$) vs. shear force ($F_\parallel$) at three scales: isolated setae (circles), isolated setal arrays (squares), and live gecko toes (triangles). For all scales, $F_\perp = -F_\parallel \tan \alpha^*$, where $\alpha^* \approx 30°$. Source: Autumn *et al.* (2006).[2] Reproduced with permission of the Company of Biologists Ltd. (b) Frictional and (c) adhesive forces of a gecko's adhesive foot pads increase with increasing sliding velocity. Negative force in (c) indicates sticking to the surface. Source: Puthoff *et al.* (2013).[15] Reproduced with permission of the Royal Society of Chemistry.

of interfacial contacts, due in part to wear. However, gecko adhesives actually become stickier, increasing both their friction (shear) and adhesion (normal) forces as the sliding velocity increases (Figure 5.3b,c). There is no drop in frictional forces on the onset of sliding and the adhesive's performance can be maintained for over 30,000 repeated cycles of contact, sliding, and detachment. These properties can be captured in gecko synthetic adhesive (GSA) arrays composed of a hard silicone polymer. This behavior of both the gecko and the synthetic arrays occurs from the random stick-slip events of the many individual hairs of adhesive fibrils. This fibrillar structure enables a greater energy dissipation as the number of stick-slip events increases to a critical level. These results suggest that geckos do not face imminent failure of adhesion if sliding does begin. Rather they can simply maintain foot contact, and adhesion and friction will passively increase.

Besides strong adhesion, gecko feet possess the following two functions/properties that current manmade adhesives cannot replicate: (1) easy detachment (geckos can quickly attach and detach their feet on smooth and rough surfaces, enabling them walk and even run on a vertical wall or ceiling) and (2) highly re-useable and reliable adhesion (geckos use their toe pads over thousands of cycles on natural substrates but are able to keep them clean from everyday contaminants).

5.2.4   Mechanics of Fibrillar Adhesion

In addition to geckos, many insects, including beetles, flies, and spiders, show similar attachment pad design: fibrillar structures. Although the adhesion mechanisms involved are not the same (some are dry and some involve secretions, as shown in Figure 5.1), a general trend is that as the body size increases the radius of the terminals decreases among different wall-climbing species.[4] This morphological change, from large to small (micro to nano level), simple to complex (single level to hierarchy), has been theoretically explained by applying contact mechanics principles. Two important concepts, namely "contact splitting" and "equal load sharing", have been derived accounting for the increased adhesion from a divided no-continuum surface and relatively weak intermolecular origin. In fact, the size and weight challenges overcome by geckos utilizing highly branched hairs is merely one facet. Other traits such as angled setal stalks, the fiber aspect ratio, tip shapes, patterns, hierarchy, the materials properties as well as the attaching and releasing motions the animal triggers are equally crucial for the systems to be functional in their ecological settings with a broad range of surface conditions (e.g., roughness, contaminants, humidity, and temperature).[16] Several important design principles are discussed in the following sections.

5.2.4.1   Contact Splitting

A fibrillar surface shows stronger adhesion than a smooth surface when contacting another surface. In addition, the smaller the dimension of the fibers, the higher adhesion it can exhibit. These phenomena can be explained by contact splitting effects.[16] When a fibrillar element interacts a substrate via short-range molecular effects (such as van der Waals forces), there are several mechanisms that lead to the enhanced adhesion, as illustrated in Figure 5.4.[16]

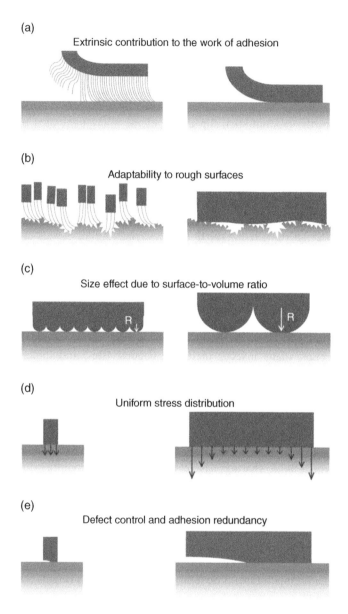

Figure 5.4 Schematic representation of adhesion mechanisms in fibrillar surfaces identified to date (contact-splitting effects): (a) extrinsic contribution to the work of adhesion; (b) adaptability to rough surfaces; (c) size effect due to surface-to-volume ratio; (d) uniform stress distribution; (e) defect control and adhesion redundancy. The overall effect on adhesion may be a superposition of some or all of these mechanisms. Source: Kamperman *et al.* (2010).[16] Reproduced with permission of John Wiley and Sons.

1. *Extrinsic contribution to the work of adhesion*: Fibrillar surfaces are more resistant to peeling than smooth surfaces since the strain energy stored in a fibril just before pull-off is not available to drive the detachment of the next fibril. From a crack propagation point of view, unlike the cracks in continuum media, which can continuously propagate due to the stress concentration, the detachment "crack" has to be re-initiated fibril by fibril, where re-nucleation of peeling is more difficult than its continuation.
2. *Adaptability to rough surfaces*: Long fibrillar elements can conform to the roughness of the substrate with more contact area. This effect is more significant in hierarchical fibril systems.
3. *Size effect due to surface-to-volume ratio*: For small contacts, the penalty associated with the distortion required for accommodation, being controlled by volume, vanishes more rapidly than the surface energy gain. This favors the adhesion of smaller contact elements.
4. *Defect control and adhesion redundancy*: It is well known that the smaller a structure, the smaller defect it will contain. When the surface is split up into finer fibrillar contacts, it will limit defect sizes and adhesion will become stronger.[16]

The contact splitting principle, or size effect, can be derived from the classic Johnson–Kendall–Roberts (JKR) concept, and the pull-off force of a pillar with radius R_0 is $P_0 = 3\pi R_0 w$, where w is the surface energy. Imagine the replacement of one large spherical contact of radius R_0 by N smaller ones of radius $R = R_0 / \sqrt{N}$ with the same total (Hertzian) contact area (i.e., similar to the idea above in which a single large contact is replaced by $N = (R_0/R)^2$ smaller ones). Considering the peeling effect, the pull-off force for simultaneous detachment of these N spheres is[16]

$$P_d = \frac{4}{3}\pi R_0 w \sqrt{N} = \frac{8}{9}\sqrt{N} P_0 \tag{5.1}$$

Thus, the force increases by a factor of $\sim \sqrt{N}$.

5.2.4.2 Equal Load Sharing

For a fibrillar contact, adhesive contacts below a critical size have been shown to develop a uniform stress distribution at maximum adhesion strength before pull-off occurs. Bundles of very small adhering fibrils can thus collectively reach theoretical adhesion strength; in addition, they are more defect tolerant.[17] For a fiber of radius R contacting with substrate, the pull-off force is no longer sensitive to variations in tip geometry when the fiber diameter is reduced below the critical value:

$$R_{cr} = \frac{8}{\pi} \frac{E^* \Delta\gamma}{\sigma_{th}^2} \tag{5.2}$$

where E^* is the modulus, $\Delta\gamma$ is the work of adhesion, and σ_{th} is the theoretical adhesion strength. The critical contact size for gecko spatula is estimated to be 225 nm, which is close to the radius of the gecko's spatula (typically around 100–250 nm). This suggests that the nanometer size of the spatula structure of geckos may have evolved to achieve optimization of adhesive strength in the tolerance of possible contact flaws.

5.2.4.3   Asymmetric Structure of Setal Array

A unique feature of biological attachment systems is that the adhesion must be easily overcome to allow rapid switches between attachment and detachment during the animal's motion. It seems that geckos achieve such reversibility via a unique design of their seta structure, which is one hierarchy above the spatula structure. Gao *et al.* calculated the pulling force at an angle with respect to the contact surface using a finite element model of a single seta, with cohesive elements between the contacting surface at the tip of the seta and the rigid substrate.[6] The computational results show that maximum attachment force is achieved when the force is exerted at around a 30° angle. The pull-off force at 90° (the peeling mode) is an order of magnitude smaller than the maximum attachment force at 30°, suggesting that the gecko would use peeling for detachment. It therefore appears that the asymmetrically designed seta structure (Figure 5.5) is particularly suitable for rapid switching between attachment (30° pulling) and detachment (90° peeling). Mimicking such hierarchic nanostructures would realize rapid switches between attachment and detachment, which is very important to many applications.

5.2.4.4   The Effects of Hierarchy

A number of theories have been applied to explain the gecko's hierarchic adhesive microstructure. From a mechanics point of view, the hierarchic structure may be critical to the realization of (1) adhesion on rough surface and (2) antimatting of the hair structures, etc.

First, a hierarchic fibrillar structure is softer and easier to attach to a rough surface. The hierarchic nature of the pad surface morphology reflects the fact that all natural surfaces (and

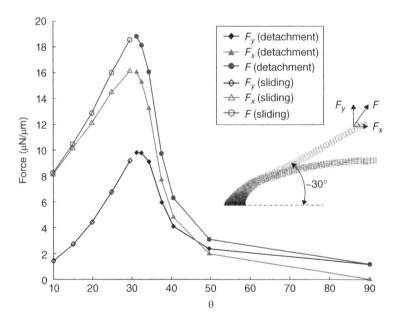

Figure 5.5   Finite-element analysis of the adhesive force of a single seta as a function of pull direction. Source: Gao *et al.* (2005).[6] Reproduced with permission of Elsevier.

most engineering surfaces) have surface roughness on many different length scales. Adhesion between two bodies is only possible if the surfaces are able to deform (elastically or plastically) to make direct (atomic) contact at a non-negligible fraction of the nominal contact area. For "hard" solids this is nearly impossible and as a result adhesion is usually negligible between hard rough surfaces. However, a fiber array made from the same material may be elastically extremely "soft", and hence can deform and make contact everywhere at the interface even when the substrate is very rough. The effective elastic modulus of a fibrous material can be expressed as

$$E_{\text{eff}} = CEnR^2 \left(\frac{R}{L} \right)^2 \tag{5.3}$$

where E is the Young's modulus, n is the number of fibers per area, R and L are the fiber radius and length, and C is a number which depends on the exact shape of the fiber but which is of order $C \approx 10$.[18] For the setal array, the effective elastic modulus reduces from $E \approx 4\,\text{GPa}$ to $E^* = 0.4\,\text{MPa}$, which is similar to that of relative soft (sticky) rubber. Clearly, at each level of hierarchy E_{eff} can be different, making the system more flexible. This may be one of the great innovations made by nature in the context of biological adhesive systems.

Second, hierarchic structures can adopt different surface roughnesses. On a large scale, the skin of the gecko toe pad is able to deform and follow the substrate roughness profile on length scales much longer than the thickness $d \approx 100\,\mu\text{m}$ of the elastic keratin film, say, beyond $\sim 1000\,\mu\text{m}$. At shorter length scales, the setae fiber array system is able to deform (without storing a lot of elastic energy) to follow the surface roughness in the wavelength range $10 < \lambda < 1000\,\mu\text{m}$. However, if the setae fiber tips are blunt and compact they will not be able to penetrate into surface cavities with diameters of less than a few micrometers. Thus, negligible atomic contact would occur between the surfaces and the adhesion would be negligible. For this reason, at the tip of each long (thick) fiber there is an array of ~ 1000 thinner fibers (diameter of order $\sim 0.1\,\mu\text{m}$). These fibers are able to penetrate the surface roughness, with cavities down to length scales of a few tenths of a micrometer. However, if the thin fibers have blunt and compact tips made from the same "hard" keratin as the rest of the fiber, there would still be a very small amount of adhesion, since a lot of elastic energy would be necessary to deform the surfaces of the thin fibers to make atomic contact with the substrate. The top of the thin fibers are therefore covered by a soft compliant layer, usually a liquid-like (high mobility) layer of polymer chains grafted to the tip of the thin fibers. This liquid-like layer, if thick enough, is able to adjust to the substrate roughness profile over lateral distances below $\sim 0.1\,\mu\text{m}$.

Third, the hierarchic structure is one of the most efficient ways to avoid bundling. In a dense fiber array, van der Waals interactions may cause clustering or bundling of adjacent spatulae due to their relatively large aspect ratios. The stability of spatulae against bundling is a necessary condition for their viability as an adhesion structure. Bunching leads to reduction in adhesive strength in the microfabricated artificial gecko structure made of polyimide microhairs. If the problem is treated as a crack problem with a detachment condition,[19] the bundling criteria can be written as

$$\frac{w}{R} = \left(\frac{L}{2R} \right)^2 \left(\frac{2\gamma}{3ER} \right)^{1/2} \tag{5.4}$$

where γ is the surface energy per unit area of each surface, and L, w, R are length, spacing and radius of the fiber, respectively.

According to the criteria, a dense array of very thin and long fibers would be unstable against bundling. Since long fibers are needed to make more contact with the rough surface by the deformation of the hairs, a hierarchic structure may be the best solution to antimatting.[6]

5.2.5 Bioinspired Strategies for Reversible Dry Adhesion

Gecko attachment pads exhibit a unique combination of properties: strength, reversibility, reusability, directionality, durability, and a self-cleaning mechanism. Obviously, a fully functional gecko-inspired adhesive could find many potential applications, as temporary fastenings in the construction industry, temporary labeling, optimization of surfaces for sports equipment, biomedical materials and devices, and fixtures for household items. For artificial surfaces to function as designed, the mechanical, structural, and chemical aspects all have to be optimized.

In the gecko adhesive system the nanoscale contact tip, which consists of dense spatulae with a unique platelet shape, is the key to creating strong reversible adhesives. The size, shape, and density (10^8–10^9 spatulae per animal) of the spatulas are essential for the gecko to generate enough adhesion force to defy gravity. Thus, the first biomimetic strategy is to mimic the spatula, including its shape, size, and density. According to contact splitting principles, smaller size and denser contacts will generate a stronger adhesion force. The nanoscale thin platelets at the contact ends are therefore important to the enhancement of the adhesion.

Directional adhesion and easy release are the major features of gecko attachment pads. Gecko setal arrays on their own already manifest the desirable behavior of strong adhesion and friction in the gripping direction and almost zero adhesion and low friction in the releasing direction. This suggests that the setae, spatulae, and other structures of the gecko adhesive mechanism need to be, and are, asymmetric to both attach strongly and detach easily and rapidly.

Gecko toe pads consist of hierarchical structures from microscale setae to nanoscale spatula. The hierarchy is also essential to achieve a number of functions such as adhesion on rough surfaces, rapid switches between attachment and detachment, antimatting of the hair structures, and self-cleaning. The hierarchical construction makes the gecko adhesive system elastically very flexible on all relevant length scales, from millimeter to nanometer, while minimizing the self-sticking effect. Thus, hierarchy is essential to the success of the biomimetic design of fibrillar dry adhesives.

The essence of learning design strategies from nature (i.e., biomimicry) is to determine how certain simple but intriguing structures generate functionality in a compatible and sustainable manner to cope with the constraining resources and environment. However, it is also intuitive to combine those strategies with other physical, chemical, and biological principles to tailor the functions for various ends. To obtain smart and responsive adhesives, one biomimetic design approach is to combine other physical, chemical, and biological principles with what nature already has.

5.3 Gecko-mimicking Design of Fibrillar Dry Adhesives and Processes

Over the past few decades extensive efforts have been made to create artificial gecko foot hairs via micro/nano fabrication techniques and chemical synthesis. Two main types of materials, polymers and carbon nanotubes, have been developed separately and offer different advantages.

More recently, hybrid adhesives offering both dry and wet adhesions as well as great durability have been proposed. External stimuli-regulated smart adhesives have also gained recognition for achieving reversible adhesion in a strictly controlled manner. Although at a very preliminary stage, self-cleaning similar to that found in gecko toe pads has been reported and demonstrated in some artificial systems.

A variety of materials, ranging from soft to hard polymers, and carbon nanotubes, have been used for the fabrication of biomimetic dry adhesives. Fabrication techniques involve template-assisted micro/nano cast molding and different lithographic methods for defining polymers and stochastic growth for both carbon nanotubes and polymers (i.e., chemical vapor deposition (CVD) and oxygen plasma induced growth, respectively). Relevant evaluation methods can be divided into three categories: (1) nanoscale normal pull-off tests implemented by tipless AFM cantilevers (e.g., silicon or Si_3N_4), (2) mesoscale normal pull-off, shear, and load-drag-pull tests utilizing spherical/hemispherical or flat-punch indenters (e.g., glass, sapphire, steel, and aluminum), and (3) macroscale normal pull-off, shear, and peeling measurements at various angles for adhesive patches that are a few square millimeters to a few square centimeters in size, where the opposing surfaces are usually flat and smooth glass, steel, silicon, or mica sheet.

Figure 5.6 shows the maximum adhesive strength vs fiber diameter evaluated at the nano and macro scales, respectively, for both polymer- and carbon nanotube (CNT)-based materials.[20] Mesoscale measurements are left out of this comparison because the apparent or real

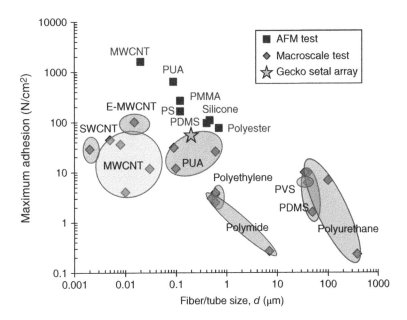

Figure 5.6 Maximum adhesion vs. fiber/tube diameter for various material candidates in biomimetic dry adhesive design. SWCNT, single-walled carbon nanotube; MWCNT, multi-walled carbon nanotube; E-MWCNT, entangled multi-walled carbon nanotube; PUA, polyurethane acrylate; PVS, polyvinyl siloxane; PDMS, polydimethylsiloxane; PMMA, poly(methyl methacrylate); PS, polystyrene. Source: Hu & Xia (2012).[20] Reproduced with permission of John Wiley and Sons.

contact area between the samples and spherical indenters varies significantly under different preloading. AFM pull-off tests indicate the adhesion forces obtained from individual fibrils, while macroscale normal pulling and shearing measurements give the information about the adhesive performance of an entire patch. Roughly speaking the smaller the fiber size, the larger the adhesion, especially for AFM tests where individual fibrils are in contact with the tipless cantilever. The material constituents seem to have a small effect on the adhesion force, presumably because the Hamaker constants are within a narrow range for different polymers, CNTs, and the opposing surfaces. In macroscale measurements, a collective effect allows other factors such as packing density, fiber compliancy, backing materials, and preloading come into play. A relatively scattered force distribution is observed, while the general trend is still consistent with nanoscale measurements. Apparently, CNTs render the highest adhesion within both nano and macro measurements.

Figure 5.7 summarizes the methodologies for creating fibrillar dry adhesive structures with respect to their smallest achievable terminal sizes.[20] Micro/nano cast molding and direct lithographic methods belong to top-down approaches. In cast molding, the smallest features are pre-determined by the specific molding templates chosen. Photolithography patterned silicon wafers are most commonly used, where the negative porosity of the templates could be either fairly large (micrometers) or relatively small (nanometers). Aluminum anodized oxide (AAO) templates and polycarbonate filters are used to mold nanofibrils, usually hundreds of nanometers in diameter comparable to gecko spatulae. AAO offers highly ordered hexagonally

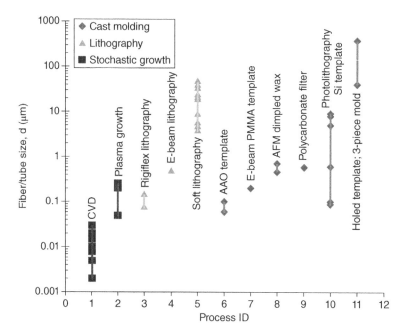

Figure 5.7 Fabrication/synthesization methods in biomimetic design of micro-/nanofibrillar structures for dry adhesion applications as a function of the minimum fiber/tube size. Cast molding and lithography belong to the top-down approaches, while stochastic growth is subject to bottom-up approaches. CVD, chemical vapor deposition; AAO, anodic aluminum oxide. Source: Hu & Xia (2012).[20] Reproduced with permission of John Wiley and Sons.

distributed straight pores, whereas polycarbonate filters possess degrees of randomness in the pore direction. Other nano-featured templates include electron-beam-defined poly(methyl methacrylate) (PMMA) and AFM dimpled wax. Softlithography is more frequently adopted than other direct lithographic methods. It offers great cost-effectiveness over conventional photolithography, which is crucial for massive production. The lithographic resolution, however, is sacrificed because of the loss of the mechanical integrities of the polymeric masters (e.g., SU-8 photoresist or polydimethylsiloxane (PDMS)) when the size is reduced. The smallest achievable fibril size is limited to the micro level. Finer fibrils could be obtained by either using more expensive and time-consuming lithographic methods such as photolithography and electron beam lithography, or improving the performance of imprinting masters. Rigiflexlithography is a modified version of softlithography, where a strong yet flexible polymer polysulfone (PUA) is adopted as master for consecutive processing. An additional drawing process at elevated temperatures could further reduce the fibril diameters down to ~80 nm. In contrast, stochastic growth method can be grouped with bottom-up approaches, which intrinsically generates nano-sized fibril structures ~100 times smaller than those of gecko spatulae. Single-wall carbon nanotube (SWCNT) and multi-wall carbon nanotube (MWCNT) arrays are synthesized by conventional, plasma-enhanced, or low-pressure CVD processes. On the other hand, dielectric polymers have been grown on single-pillar-supported silicon dioxide platforms as well as nickel-based micro cantilevers. The characteristic size of CNTs lies within the range of 2–30 nm in diameter, while the diameters of plasma growth polymers rangee from 50 to 250 nm.

5.3.1 Biomimetic Design Based on Geometric Replications of the Gecko Adhesive System

The first attempt to make gecko-inspired fibrillar dry adhesives was demonstrated via AFM-aided nano molding.[21] More complex fibrillar structures were then fabricated with various processes, as summarized in Figure 5.8.[20] These polymeric fibrillar structures include nano pillars from AAO templates and microfibers curved to one side, resembling the angled setal stalks of geckos. These finer pillars render higher pull-off forces for all tip geometries, possibly indicating contact splitting efficiency.

The moderate increase in adhesion strength in the early versions of polymeric fibrillar arrays is attributed to oversimplification in tip shape. Theoretical analysis also points out the importance of copying the tip geometries in the animal system. For this reason, micropillar arrays with various tip shapes were formed, including a control sample with flat punch tips and mushroom tips (Figure 5.8d).[22] Mesoscale pull-off tests show that pillar arrays with spatular tips have a higher adhesion than all other geometries, and finer pillars render higher pull-off forces for all tip geometries, possibly indicating contact splitting efficiency. The mushroom-shaped fibrillar geometries were also proven to have better performance. The capped micro pillars show high adhesion strength, good contamination tolerance, and are easy to clean so are suitable for repeated use. However, the frictional behavior found in mushroom-like fibrillar structures (symmetric caps) exhibits the opposite response to gecko setal arrays (asymmetric setal stalks and spatular plate). Shearing displacement reduces the normal pull-off resistance and makes it vanish beyond a critical point as oppose to gecko frictional adhesion, where high adhesion is shear induced and keeps increasing even after lateral slippage.

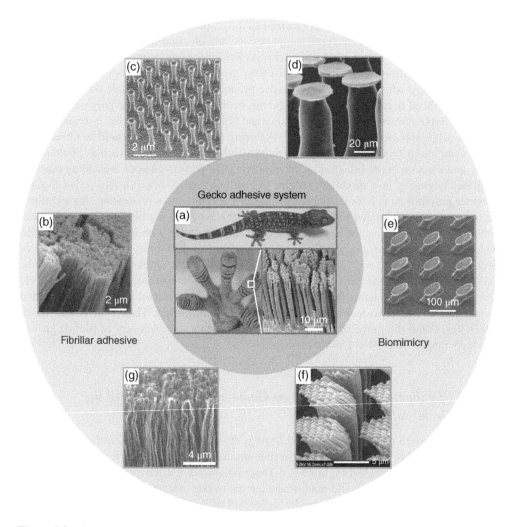

Figure 5.8 Comparison of (a) gecko hierarchical fibrillar adhesive system and (b–g) various biomimetic fibrillar dry adhesives.[20] Lower right picture in (a) shows the setal array with spatular braches. (b) Vertically aligned PMMA nanotube arrays. (c) Free-standing vertical polyimide nanohairs. (d) Mushroom-shaped PVS microfibrillar structure. (e) Tilted polyurethane microfiber arrays with slanted mushroom tips. (f) Hierarchical PUA hairs with angled bulge-tip nanofibers formed on straight micropillars. (g) Hierarchically structured vertically aligned MWCNTs with entangled top layer resembling spatulae. Source: Hu & Xia (2012).[20] Reproduced with permission of John Wiley and Sons.

Although adhesion enhancement was achieved by dividing bulk contacting surfaces into micro/nano fibrils, in certain cases with characteristic tip shapes, the reversibility and durability of the gecko's attaching system stimulated researchers to include anisotropic geometry in the artificial version, which could possibly allow switchable adhesion with directional dependency as well as additional conformability. With this in mind, various angled fibrillar adhesives with tilted stalks and angled wedge endings were fabricated (Figure 5.8e).[23] These angled structures show a significant improvement in adhesion over isotropic control when applied in wall-climbing robots. For example, wedge-shaped PDMS fibrillar arrays with tilted angles show force anisotropy between normal and shear directions, reported as 2.1 N and 13 N, respectively, for an 8.2 cm^2 patch. Around 67% of initial adhesion and 76% of initial friction forces were retained after 30,000 cycles of attachment and detachment.

According to gecko toe pad structure, two or more levels of hierarchy are necessary to prevent self-matting while maintaining maximal contact fraction by decreasing the terminal size. Several possible solutions for the building hierarchy were proposed involving modified photolithography AAO templates for polymeric cast molding, laser patterning, heated rolling, laminating and selective etching processes, and angled etching to make slanted nano holes out of poly Si substrate with an SiO$_2$ stopping layer. Different angles of the polymer nanorods have been made along with asymmetric tip shapes. These slanted nanorods were further transferred onto straight micro-sized pillars (Figure 5.8f).[24] This offers not only better performance on rough surfaces, but also greater anisotropy between a strong shear attachment along the fiber curvature ($\sim$26 N/cm^2) and an easy detachment against fiber curvature ($\sim$2.2 N/cm^2). Hierarchical structures with angled stalks but symmetric disc-like tips were fabricated to create two- and three-level structures with inclined/vertical polyurethane ($E = \sim$3 MPa) pillars at macro and meso scales, and vertical pillars at the micro scale with mushroom caps at each level. Greater adhesion strength was obtained for structured surfaces: two levels > single level > unstructured, especially in situations of high preloading. These hierarchical designs exhibit easy detachment when shear loading is removed, returning the structure to a non-adhesive default state just like that of gecko toe pad.

Mimicking the gecko adhesive system, vertically aligned MWCNT patterns on flexible polymer tapes were fabricated. Photolithography was used to define different square patterns, 50–500 μm in width, of the catalyst layers on silicon substrate, and then $\sim$8 nm in diameter vertically aligned MWCNTs were grown to 200–500 μm long under CVD. The MWCNT patterns were then transferred onto self-adhesive tape. A shearing force of 36 N for a 1 cm^2 patch was achieved under a preloading of 25–50 N/cm^2 via cylindrical roller pressing. Unlike conventional viscoelastic adhesive tapes, the "gecko tape" offers good time independency such that a strong and stable shear force could be maintained for 8–12 hours.

Hierarchically structured vertically aligned MWCNT arrays were synthesized through a low-pressure CVD process.[25] Analogous to gecko setal arrays (Figure 5.9a), this two-level MWCNT array features a straight aligning body mimicking setal stalks and a curly entangled end segment at the top mimicking spatulae (Figure 5.9b–d). Under shear loading, the entangled segment enables sidewall contact formation with the target surface, rendering a high macroscopic force of $\sim$100 N/cm^2, whereas normal pulling requires peeling of each entangled tip at the interface, leading to a drastic reduction of adhesion force down to $\sim$10 N/cm^2. Time-dependent adhesion measurements demonstrate 24-hour durability under a shear loading of 40 N/cm^2 and a normal pull-off force of 12 N/cm^2. This system offers strong shear bonding and easy normal lifting off capabilities that are promising for mimicking live gecko walking. The friction force was determined to be entangled length dependent, while the longer vertically aligned segment may also make contributions by inducing more CNT sidewall contacts to the substrates when dragged laterally.[26] When the entangled segment is removed, the structure does not show anisotropic adhesion any more, and the adhesion force also reduces, suggesting the importance of hierarchy in biomimetic dry adhesives.

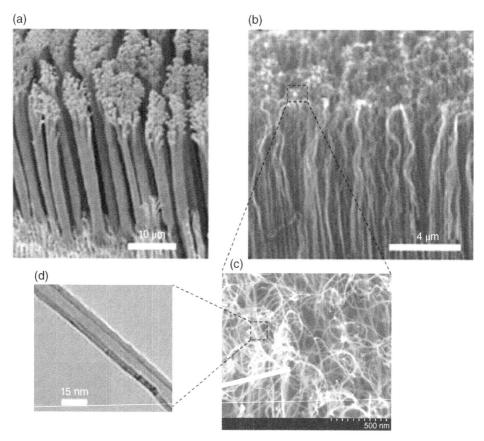

Figure 5.9   SEM micrographs of the structure similarity between the cross-section views of (a) gecko aligned elastic hairs and (b) hierarchically structured vertically aligned MWCNTs with (c) entangled top layer resembling spatulae. (d) TEM image of a single MWCNT at the top layer. Source: Qu *et al.* (2008).[25] Reproduced with permission of the American Association for the Advancement of Science.

5.3.2   Biomimetic Design of Hybrid/Smart Fibrillar Adhesives

The essence of learning design strategies from nature (i.e., biomimicry) is to look at how certain simple but intriguing structures generate functionality in a compatible and sustainable manner to cope with the constraining resources and environment. However, it is also intuitive to combine those strategies with other physical, chemical, and biological principles to tailor the functions for various ends. More and more gecko features are being successfully included in artificial fibrillar dry adhesives to improve their functionality.

An outstanding example of this is the integration of gecko setal structures with adhesive proteins found in mussels.[27] After a thin layer of mussel-mimicking adhesive coating, p(DMA-co-MEA), was applied to PDMS nanofiber arrays (Figure 5.10a), AFM pull-off forces were increased threefold in air and ~15-fold underwater. A similar idea was demonstrated in polyurethane microfiber arrays, where a continuous p(DMA-co-MEA) terminal film was added on top.[28] When fully submerged in water, the film-terminated microfiber arrays showed

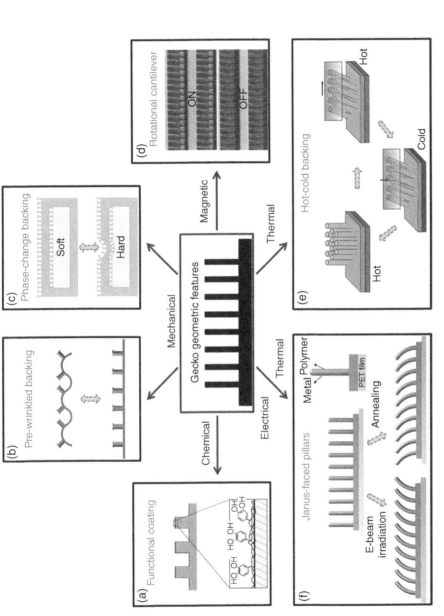

Figure 5.10 Extension from gecko geometric features in designing stimuli-responsive smart adhesive systems.[20] (a) Gecko-mussel inspired hybrid fibrillar adhesive for both dry and wet applications. (b) Stretchable wrinkled adhesive patch with caped micropillars. (c) Mushroom-shaped micropillar array with embedded phase-change backing substrate. (d) Nickel paddles and polymeric nanorod integrated hierarchical adhesive system, showing the ON and OFF states regulated by applying a magnetic field. (e) Microfibrillar structure fabricated from shape memory thermoplastic elastomer, deformed by heating of the original vertical pillars, followed by pulling of a glass slide over the top of the surface. Cooling the system in the deformed position stabilizes the non-adhesive, tilted pattern. Recovery occurs by heating the pattern above its transition temperature. (f) Double-faced fibrillar structure obtained by obliquely depositing a thin platinum layer on one side of free-standing PUA nanopillars, where the nanopillars could be bent against the metal side by annealing or the polymer side by E-beam treatment. Source: Hu & Xia (2012).[20] Reproduced with permission of John Wiley and Sons.

superior adhesive ability ~23-fold greater at best than the mushroom-shaped controls. Another step forward in this hybridization approach is the development of a biodegradable adhesive for the wet-tissue-like environments commonly encountered in biomedical procedures.[29] An adjustable biodegradable elastomer, known as poly(glycerol sebacate acrylate), was chosen as the structural material for creating nanofibrillar features, while the ingredient of functional coatings was changed to oxidized dextran. The performance of this hybrid adhesive was characterized not only *in vitro* but also *in vivo*, with supreme adhesion strength.

In addition, multiphysical stimuli-regulated smart adhesives have been designed and fabricated for achieving reversible/responsive adhesion in a strictly controlled manner (e.g., mechanical, magnetic, thermal, and electrical). A stretchable wrinkled adhesive patch with PDMS micropillars has been fabricated that shows great adhesion tunability under different straining conditions (Figure 5.10b).[30] 100 cycles of attachment and detachment were reported without detectable degradation. The mechanical properties of the backing substrate could also be actively controlled by inserting phase-change materials (Figure 5.10c).[31] To maximize adhesion, a soft backing was used to increase the conformability to irregular surfaces by initiating a large real contact area. The following phase transformation to a rigid state was locked in the deformed backing profile for equal load sharing. On release, the elastic properties (e.g., stiffness) could be switched backwards. One of the advantages of this actively switching mechanism is that it is free from building fibrillar hierarchy because even unstructured PDMS surfaces with phase-change backings were shown to have enhanced adhesion. Also in this vein, we speculate that integrating carbon nanotube arrays with stimuli-regulated backings may improve reversibility by eliminating the buckling issues resulting from large preloading.

A field-controlled mechanism can also be demonstrated in a nickel-polymer integrated system.[32] This hierarchical structure consists of vertically aligned photoresist nanorods, analogous to spatulae, coated on top of micro-sized nickel cantilever paddles, analogous to setae. The key structure, nickel paddles, offers not only additional conformability to the polymeric nanorods but also a magnetically controlled switching mechanism. Specifically, this adhesive system could be actuated through an externally applied magnetic field. The rotational motions of the cantilever paddles under magnetic signals could either expose or conceal the active polymeric nanorod-coated surfaces (Figure 5.10d). This produced a 40-fold drop in normal pull-off force from the ON to the OFF state.

Thermally stimuli-responsive adhesives have been made by taking advantage of shape-memory materials. Patterned vertical micropillar arrays, based on shape-memory thermoplastic elastomer, are able to bend into temporary angles (i.e., non-adhesive state) by hot pressing at the temperature above transition but below permanent deformation (Figure 5.10e).[33] Reheating to the same temperature range could reverse the temporary angled pillars to their original and permanent vertical position (i.e., adhesive state). More recently a double-faced fibrillar structure has been fabricated by obliquely depositing a thin layer of platinum on one side of free-standing polymeric nanopillars (Figure 5.10f).[34] This unique structure renders directional dependent frictional responses even when vertical. In addition, the distinctive physical properties of the metal and polymer make it possible to bend the pillars towards the metal side on thermal annealing. The thermally bent structure could also be switched back to its original upright position on electron beam irradiation and even be bent toward the polymer side within just a few seconds. Hence, for multiple and quick reactions, this type of controlling mechanism is more desirable than that of the pure shape memory thermoplastic fibrils.

5.4   Wet Adhesion: Biological Reversible Adhesive Systems Based on Soft Film

In addition to dry adhesion, nature also offers interesting examples of reversible attachment under wet conditions. Some amphibians, such as tree and torrent frogs and arboreal salamanders, are able to attach to and move over wet or even flooded environments without falling. Several families, including Hylidae, Rhacophoridae, Microhylidae, and Dendrobatidae have independently evolved adhesive toe pads, that is, disc-like, flattened enlargements of the tips of the digits, which all show a similar, specialized epidermal architecture and structure.[4] These architectures and associated adhesive mechanisms provide an excellent biological model to design adhesive devices that can function under wet conditions.

5.4.1   Tree Frog Adhesive System

Tree frogs have developed a distinct surface structure in their toe pads for safe attachment to wet surfaces. They can climb most surfaces, from sheer leaves to glass, with ease, although they do not fare so well on dry, rough materials. The toe pads can firmly grip on smooth leaf and branch surfaces. *Osteopilus septentrionalis* is one of the larger tree frog species (weight 28 g) with well-developed toe pads located on the expanded tips of each digit,[35] four on each forelimb and five on each hindlimb. Each toe pad is ovoid in the surface view (Figure 5.11b) and is raised above the digit's ventral surface with disc-like, flattened enlargements of the tips of the digits, which all show a similar, specialized epidermal architecture and structure. Each toe pad is completely surrounded by a circumferential groove, and lateral grooves continue down the margin of the ventral surface of each digit. Proximal to each toe pad are located subarticular tubercles (Figure 5.11d), whose epithelial surface appears identical to that of the toe pads. The toe attachment pads consist of a hexagonal array of flat-topped epidermal cells of approximately 10 μm in size separated by approximately 1 μm wide mucus-filled channels (Figure 5.11c). The flattened surface of each cell consists of a submicrometre array of nanopillars or pegs which have nanoscale diameter (diameter ≈ 100–400 nm) surrounded by small channels (width ≈ 40 nm, height ≈ 200 nm) (Figure 5.11e).[35,36] This subdivision of the contact zone into microscopic subunits is considered to be a general design principle in animal adhesive pads.

The toe pads are amongst the softest of biological structures (effective elastic modulus 4–25 kPa) and exhibit a gradient of stiffness, being stiffest on the outside.[37] This stiffness gradient results from the presence of a dense network of capillaries lying beneath the pad epidermis, which probably has a shock-absorbing function. The thickness of the pad, in combination with its flexibility, can generate small movements of the toes, for example when a frog changes its weight distribution, without resulting in movement of the pad epithelium with respect to the substratum. There are liquid-secreting glands that open into the channels between the blocks. The liquid viscosity $\eta \approx 0.0014 \, Pa\,s$ (i.e., about 40% larger than for water)[4] and the pads are permanently wetted by mucus secreted from the glands. Thus, they attach to mating surfaces by wet adhesion.[35] Like the dry adhesives of gecko toe pads, frog toe pads also exhibit velocity-dependent resistance to shear forces. This phenomenon is expected for any system employing a fluid as an adhesive mechanism. The largest adhesive forces that toe pads can generate is approximately 1.2 mN/mm^2.[38] Tree frogs are capable of climbing on wet rocks even when water is flowing over the surface.

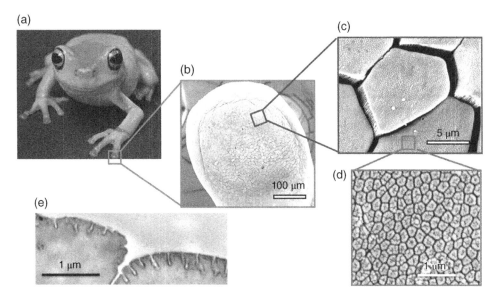

Figure 5.11 Morphology of tree frog toe pads. (a) White's tree frog (*Litoria caerulea*), SEMs of (b) toe pad, (c) epidermis with hexagonal epithelial cells, and (d) high-power view of the surface of a single hexagonal cell showing peg-like projections. (e) TEM of cross-section through cell surface, which is separated by narrow channels. Source: Scholz *et al.* (2009).[36] Reproduced with permission of the Company of Biologists Ltd.

5.4.2   Adhesive Mechanism of Tree Frog Toe Pads

The toe pads of tree frogs are hierarchical structures that include the nanoscale pillar on the microscale hexagonal array of epithelial cells separated by channels that are approximately 1 mm wide and full of fluid secretion. These blocks/pillars and channels are believed to produce high adhesion and friction by conforming to the mating rough surface at different length scales and by maintaining a very thin fluid film at the interface, responsible for animal locomotion and manoeuvrability. When the frog toe pad contacts a substrate surface, liquid is pulled out from the channels by capillary suction, as shown in Figure 5.12a. If the separation h between the toe pad and substrate is smaller than the width W of the channels, the pressure in the space between the toe pad and the substrate will be lower than in the grooves, resulting in the flow of liquid into that space. Because all the grooves are connected laterally, fluid will flow laterally within the network of grooves in such a way as to conserve the volume of fluid. For substrates with large enough surface roughness, there will be regions between the surfaces where $h > W$, and when the fluid reaches such a region the flow will stop (Figure 5.12a).[39]

Frogs remove their toe pads from surfaces by peeling with undetectable detachment forces. During this detachment process, the pads are removed from the rear forwards when the frog walks forward up to a vertical slope; it peels beginning from the front of the pad rearwards when the frog moves backwards down a vertical slope. During pad pull-off, the liquid (or part of it) may be pulled back into the grooves by capillary forces, that is, when $h > W$, the capillary force drives the liquid to flow back to the grooves, as shown in Figure 5.12b, where the fluid flow direction is indicated by the vertical arrows. This inflow of fluid back into the grooves

(a)   (b)

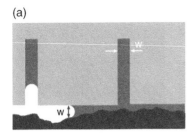
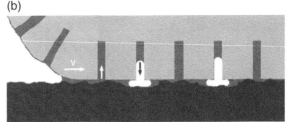

Figure 5.12 (a) When the frog toe pad comes into contact with a substrate surface, liquid is pulled out from the channels because of capillary suction. If the separation h between the solid walls at the toe-pad–substrate interface is smaller than the width W of the channels, the pressure in the film between the toe pad and the substrate will be lower than in the grooves, resulting in the flow of liquid into the space between the toe pad and the substrate. Since all the grooves are connected laterally, fluid will flow laterally within the network of grooves in such a way as to conserve the volume of fluid. (b) During pull-off an opening crack propagates (velocity v) at the pad substrate interface. Source: Persson (2007).[39] Reproduced with permission of IOP Publishing.

may be important during fast movement because the frog toe pads could dry out if the secreted liquid remained trapped on the substrate surface.[39]

The use of a fluid to produce adhesion between pads and substrates provides the frog with distinct advantages. The liquid joint between the pad and substrate is instantly fully functional and requires less force to break it since the sealant does not harden, as a glue would. The channel network is critical as it can act as a liquid reservoir to provide additional liquid if needed. This is particularly important to allow pads to quickly adhere to rough substrate surfaces. In addition, under the wet conditions, for example when it is raining, the channels will facilitate the squeeze-out of fluid between the toe pad and the substrate. This will lead to a suction-cup type of effective "adhesion" on flooded surfaces because the channels between the cells at the outer boundary of the toe pad may close during fast pull-off.[39]

The wetting adhesion is believed to occur primarily by a meniscus contribution because of the formation of menisci around the edges of the pads.[35] In this liquid adhesion, there is measurable static friction generated by the toe pads, therefore there must be some dry contact between the tips of the nanopillars and the mating surface. The dry contacts between the pad and the mating surfaces are produced by squeezing out the fluid film from the interface.

5.5 Artificial Adhesive Systems Inspired by Tree Frogs

The toe pads of tree frogs can be used as a biological prototype in the development of materials with reversible adhesion under wet or flooded conditions. Mimicking the surface structure of tree-frog toe pads, elastic, microstructured surfaces (hydrophobic and hydrophilic) were fabricated with different soft-moulding technologies.[40] The adhesion and friction behavior of the biomimetic wet adhesives in the presence of a liquid layer was evaluated and compared to flat controls. Figure 5.13 show the tree-frog-like patterns with hexagonal micropillars (15–19 μm diameter) terminated with flat tips, T-shaped tips (10 μm height, 3 and 5 μm channel width), and concave tips. In the presence of a liquid, the adhesion of tree-frog-mimicking microstructured surfaces involves capillary interactions and direct contact forces. The

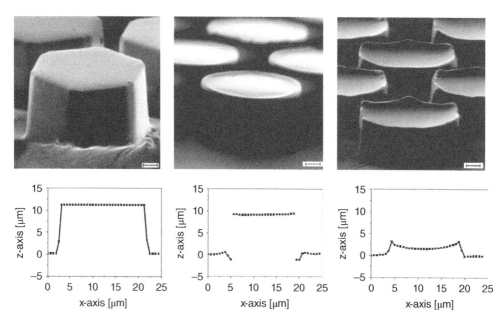

Figure 5.13 SEM images of PDMS arrays of hexagonal micropillars (15–19 μm diameter) terminated with flat tips (left), T-shaped tips (10 μm height, 3 and 5 μm channel width) (middle), and concave tips (3 μm height, 5 μm channel width) (right). Scale bars correspond to 2 μm. The corresponding profiles of the pillars were obtained with a confocal microscope. Note that this method cannot image the side walls of the pillars, especially if there are overhangs (T-shaped tips). Source: Drotlef et al. (2013).[40] Reproduced with permission of John Wiley and Sons.

magnitude of the contribution of these two forces to the net adhesion force depends on the wettability of the liquid to the surface. If the liquid wets the pillars, capillary interactions dominate, whereas viscous and suction forces do not make any significant contribution.[40] The resulting adhesion force does not depend on the microstructure or its geometrical design, and it decreases with increasing fluid volume. If the liquid does not wet the pillars, the surface patterns determine the adhesion, and the adhesive force induced from the direct contact force can surpass capillary interactions, depending on the pillar design. In this case, the design principles for gecko-like dry adhesives also apply to tree-frog-like wet adhesives.[40] In particular, T-shaped microstructures, being those most favored for dry adhesion, are also highly benefi-cial for wet adhesion. In addition to adhesion, the pillar patterns were observed to affect friction. Pillared surfaces show significantly higher friction than flat ones in the presence of a wetting liquid. The shear forces effectively drain the wetting liquid out of the contact area only if surface structures are present. If the pillars do not bend or collapse during shear, the surfaces containing low-aspect-ratio pillars can establish direct contact and generate much larger friction than the flat surface.[40]

Another example is the treads of tires used in transport vehicles. Inspired by the patterns on the toe pads of tree frogs, the channels on the contact surface are designed to run from the center to the edge of the tires. On wet roads, water/snow flows out through channels between the treads. This provides an intimate contact between the tire treads and the road, leading to high adhesion, which gives good grip while driving on a wet road.[39]

5.6   Slippery Surfaces and Friction/Drag Reduction

The previous sections give several examples that demonstrate the strategies used in nature for enhancing adhesion and friction under dry or wet conditions. There are various surfaces and structures designed by nature for the opposite purpose: reducing adhesion on a surface. The lotus leaf is one example where there is a significant reduction in adhesion and drag for water droplets, thus providing self-cleaning capability. These superhydrophobic surfaces (i.e., water-repellent surfaces), developed by well-designed chemistry and roughness, will be discussed in Chapter 7. Other examples include the *Nepenthes* pitcher plant, which uses its structure to lock in an intermediary liquid to form a slippery surface, shark skin, which relies on an ingenious antidrag design to reduce drag, and water striders, which use hydrophobic waxy microhairs on their legs to walk on water. Two typical mechanisms of slippery surfaces, the pitcher plant and shark skin, are discussed in the following sections.

5.6.1   Pitcher Plant: A Biological Model of a Slippery Surface

Nature offers a remarkably simple way to reduce friction and adhesion that is different from the method used by the lotus leaf. *Nepenthes* pitcher plants efficiently trap and retain insect prey in highly specialized jug-like leaves (Figure 5.14a). The rim of the pitcher (peristome) is completely soaked by nectar or rainwater. Well-matched solid and liquid surface energies, combined with the microtextural roughness, create a highly stable state in which the liquid fills the spaces within the texture and forms a continuous overlying film.[41] This homogeneous liquid films then acts as the repellent surface that impedes close contact between the plant and tarsal pad surfaces.[42] The film is effective enough to cause insects that step on it to slide from the rim into the digestive juices at the bottom by repelling the oils on their feet.[43] Besides a slippery peristome that inhibits adhesion of insects, the pitchers employ epicuticular wax crystals on the inner walls of the conductive zone of the pitcher to hamper insect attachment by adhesive devices.

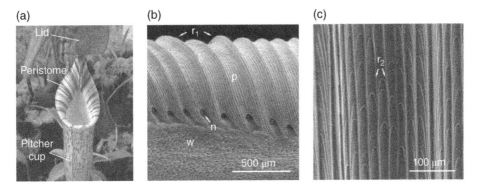

Figure 5.14   (a) Morphology of a pitcher of the carnivorous plant *Nepenthes alata*, macroscopic view. (b) Peristome surface (*p*) of *Nepenthes alata*, structured by first- (r_1) and second-order (r_2) radial ridges. In between tooth-like projections at inner edge of the peristome, pores of large extrafloral nectaries (n) can be seen. Below the peristome is wax-covered inner wall surface (w). (c) Second-order ridges (r_2) are formed by straight rows of overlapping epidermal cells. Source: Samaha & Gad-el-Hak (2014).[45]

The peristome of most *Nepenthes* species has a highly organized microstructure composed of first- and second-order radial ridges (Figure 5.14b,c).[42,44] The second-order ridges are smaller and formed by straight rows of overlapping epidermal cells, which form steps inside the pitcher (anisotropic topography). The zone adjoining the peristome toward the pitcher's inside is covered with waxy crystals, but the surface of the epidermal cell is smooth and free of wax. The microscopic topology facilitates the complete wetting of the peristome surface, and thus forms a superhydrophilic surface.[44] In addition, the absence of wax crystals and the presence of hygroscopic nectar increase the capillary forces. Water droplets can rapidly spread on the surface even against gravitational force. Thus, a continuous thin water film completely covers the peristome surface to form a slippery surface under humid conditions. This film prevents insects from standing on the surface by disabling the adhesion of their pads. On falling, the ridge structure makes it easier for victims to fall toward the pitcher cup. Additionally, the anisotropic surface topography prevents interlocking of claws when an insect slips into the pitcher cup.[45]

5.6.2   Shark Skin: A Biological Model for Drag Reduction

Many aquatic animals can move in water at high speeds, with a low energy input. Among them, sharks demonstrate surprisingly high drag reduction on their skin surface, which results in water moving very efficiently over their surface. Through its ingenious design, their skin turns out to be an essential aid in this behavior by reducing drag by 5–10% and self-cleaning ectoparasites from its surface.[46,47]

Shark skin is covered by small scales that are tiny compared with those of teleosts (bony fishes). These very small individual tooth-like scales, called dermal denticles (little skin teeth), are ribbed with longitudinal grooves (aligned parallel to the local flow direction of the water). The detailed structure varies from one location to another for a given shark. Figure 5.15 shows the scale structure on the right front of a Galapagos shark (*Carcharhinus galapagensis*). The scales are V-shaped, approximately 200–500 μm in height, and regularly spaced (100–300 μm).[48] Dermal denticles are composites that embed a hard, crystalline mineral called apatite inside a soft collagen matrix, providing the rigidity of the apatite without brittleness and the plasticity of the soft matrix without distortion. Due to their microstructure, dermal denticles are about as hard as granite and as strong as steel. These dermal denticles act like a built-in suit of chainmail armor, providing sharks with physical protection without sacrificing mobility. The dermal denticles of the white shark, for example, have crowns shaped like miniature horseshoe crabs, so tiny as to be barely visible to the naked eye. These crowns overlap tightly, providing protection from both large potential predators, including other great white sharks, and tiny skin parasites.

Grooves of tooth-like scales play an important role in reducing drag force. Because a shark swims relatively quickly (i.e., has a relatively high Reynolds number), turbulent flow occurs around the shark's body and the surface roughness of the body does not affect the skin drag (wall shear stress). Longitudinal scales on the surface lead to lower wall shear stresses and more efficient movement of water over the surface than on a smooth surface.[49] In principle, fast-moving water over smooth surfaces begins to break up into turbulent vortices, or eddies, in part because the water flowing at the surface of an object moves slower than water flowing further away from the object, with so-called low boundary slip. This difference in water speed

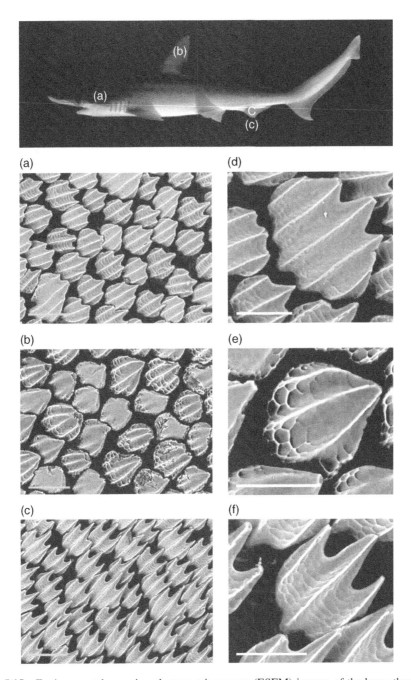

Figure 5.15 Environmental scanning electron microscope (ESEM) images of the bonnethead shark (*Sphyrna tiburo*) skin surface at different body locations. Wideview ESEM images were taken from skin pieces extracted at the positions of the head (a), the leading edge dorsal fin (b), and the anal fin (c), as indicated in the top panel. (d–f) Closer top-view ESEM images of the skin surface from regions a–c showing details of the three-dimensional structure at each position. Typical denticles along the trunk usually have an odd number of top-surface ridges. In (d) and (f), denticles that have either three or five top-ridges can be observed. In particular, denticles at the anal fin position (f) have sharp top-ridges. Non-typical denticle structures, such as denticles at the leading edge dorsal fin position (e), are teardrop shaped with a long mid-ridge and minimal side ridges. When the shark is swimming, the natural flow direction across the denticle surface is from lower left to upper right, from denticle base to tip. Scale bars 200 μm in (a), (b) and (c); Scale bars 100 μm in (d), (e) and (f). Source: Wen *et al.* (2014).[48]

causes the faster water to get "tripped up" by the adjacent layer of slower water flowing around an object, just as upstream swirls form along riverbanks.[49] The grooves in a shark's scales simultaneously reduce eddy formation in several ways: (1) the grooves reinforce the direction of flow by channeling it, (2) slower water is accelerated by the grooves at the shark's surface (as the same volume of water going through a narrower channel increases in speed), reducing the difference in speed of this surface flow and the water just beyond the shark's surface, (3) conversely, faster water is pulled by the grooves towards the shark's surface so that it mixes with the slower water, lessening this speed differential, and, finally, (4) the sheet of water flowing is divided up over the shark's surface, resulting in smaller, rather than larger, vortices where turbulent flow is created.[49]

In addition to the reduction in eddy formation, longitudinal scales also influence fluid flow in the transverse direction by limiting the degree of momentum transfer. The difference in the protrusion height in the longitudinal and transverse directions governs how much the scales impede the transverse flow. Thin, vertical scales result in low transverse flow and low drag; when the ratio of scale height to tip-to-tip spacing reaches 0.5, the drag reaches its lowest value.[50]

5.7   Biomimetic Designs and Processes of Slippery Surfaces

5.7.1   Pitcher-inspired Design of a Slippery Surface

Slippery surfaces are needed in many applications from refrigeration and architecture, to biomedical devices and consumer products. Inspired by the slippery surfaces and function of the *Nepenthes* pitcher plant, a biomimetic coating has been designed for slippery liquid-infused porous surfaces (SLIPS).[51] SLIPS repel the various liquid and solid materials they comes into contact with: water and ice, crude oil, blood, saltwater, wax, and more. It is expected that these slippery surfaces can be used in fluid handling and transportation, optical sensing, medicine, and as self-cleaning and antifouling materials operating in extreme environments.

SLIPS were designed based on three criteria:[51] (1) the lubricating liquid must wick into and stably adhere to the substrate, (2) the solid must be preferentially wetted by the lubricating liquid rather than by the liquid one wants to repel, and (3) the lubricating and impinging test liquids must be immiscible. The first requirement is satisfied by using micro/nanotextured, rough substrates whose large surface area, combined with chemical affinity for the liquid, facilitates complete wetting by, and adhesion of, the lubricating fluid.

SLIPS were fabricated by infiltrating an intermediary infused liquid lubricant (e.g., 3 M Fluorinert FC-70 or DuPont Krytox oils) into a fluorinated porous structure, as shown in Figure 5.16a. Because the oil covers the surface of the porous structure, it repels different kinds of liquids, such as water, blood, crude oil, alcohols, etc. This, in turn, could cause a slip effect. To achieve the best repelling and sliding effects, the key design principle is that the surface energies of the lubricant need to match that of the solid rough surface, although microfabrications could produce engineered omniphobic surfaces with ordered or disordered roughness. After the porous structure has been fabricated, the SLIPS were formed through liquid imbibition into the porous materials, resulting in a homogeneous and nearly molecularly smooth surface with a roughness of about 1 nm.[23] As the properties of such coatings are not sensitive to the precise micro/nanostructures of the surface, the process is relatively simple,

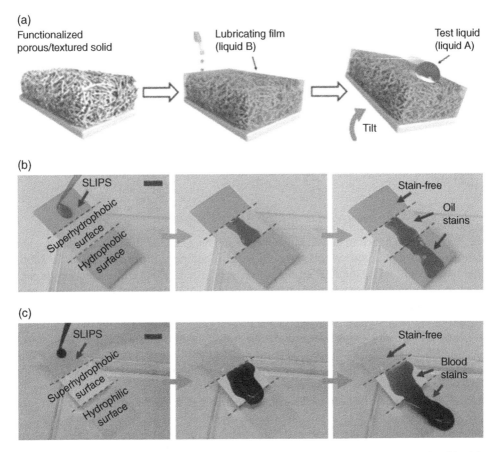

Figure 5.16 (a) Fabrication of a SLIPS by infiltrating a functionalized porous/textured solid with a low-surface-energy, chemically inert liquid to form a physically smooth and chemically homogeneous lubricating film on the surface of the substrate. (b) Movement of light crude oil on a substrate composed of a SLIPS, a superhydrophobic Teflon porous membrane, and a flat hydrophobic surface. Note the slow movement on and staining of the latter two regions. (c) Comparison of the ability to repel blood by a SLIPS, a superhydrophobic Teflon porous membrane, and a flat hydrophilic glass surface. Note the slow movement on and staining of the latter two regions. Source: Wong *et al.* (2011).[51] Reproduced with permission of Nature Publishing Group.

and cost-effective materials such as Teflon membrane and porous polyelectrolyte multilayers could be used in the fabrication. Other porous materials with functionalized ordered micro-structures, such as microposts, could also be used as substrates to prepare SLIPS. Fabrics such as cotton and polyester could be treated to fabricate SLIPS and used as stain-free fabrics since they can repel even low-surface-tension liquids.[23]

SLIPS are capable of repelling water and liquid hydrocarbons and function at low and high pressures. In addition to repelling liquids in their pure forms, SLIPS effectively repel complex fluids, such as crude oil and blood (Figure 5.16b,c), that rapidly wet and stain most surfaces.[51] These smart porous surfaces outperform their natural counterparts, with low contact angle hysteresis and resist ice adhesion, and can serve as antisticking, slippery surfaces for insects, a direct mimicry of pitcher plants. The omniphobic nature of SLIPS also helps to protect the surface from a wide range of particulate contaminants by allowing self-cleaning by a broad assortment of fluids that collect and remove particles from the surface. This lubricating film is also a self-healing coating that can rapidly restore the liquid-repellent function after it is damaged by abrasion or impact. After damage, the liquid simply quickly flows towards the damaged area by surface-energy-driven capillary action[52] and spontaneously refills the physical voids. As observed by high-speed camera imaging, it took only ~150 ms to self-recover the damage for a ~50-μm fluid displacement of the FC-70 lubricating layer on epoxy-resin-based SLIPS.[51]

5.7.2   Shark Skin-inspired Design for Drag Reduction

Drag is a major hindrance to movement in a fluid and on a surface. To reduce resistance and volume loss in solid–liquid contact, it is desirable to minimize the drag force at the solid–liquid interface. From a biomimetic point of view, drag-reduction can be achieved through two approaches drawn from nature: (1) shark skin–inspired surfaces and (2) superhydrophobic surfaces. Shark skin demonstrates surprisingly high drag reduction due to the unique structure of the skin surface. It is one of the perfect biological models for the biomimetic design and engineering of materials surfaces for drag reduction and antifouling. As discussed in section 5.6.2, shark skin is embedded with small individual tooth-like scales (dermal denticles), which are ribbed with longitudinal grooves (aligned parallel to the local flow direction of the water). The surface architecture of the scales is critical to reduce water drag by reducing the formation of vortices and allowing water to move more efficiently over the shark skin surface.

Inspired by natural shark skin, a flexible biomimetic shark skin was fabricated for detailed study of hydrodynamic function.[48] In the first step of the fabrication, a micro-CT imaging technology was used to construct a three-dimensional model of shark skin denticles of the shortfin mako (*Isurus oxyrinchus*). Three-dimensional printing was then applied to produce thousands of rigid synthetic shark denticles on flexible membranes in a controlled, linear-arrayed pattern (Figure 5.17). This flexible three-dimensional printed shark skin model was tested in water using a robotic flapping device that can either hold the model in a stationary position or move it dynamically at a self-propelled swimming speed. Compared with a smooth control model without denticles, the biomimetic shark skin reduced the static drag by 8.7% at slower flow speeds, but increased drag at higher speeds. Under swimming conditions, the biomimetic shark skin accelerated swimming speed under specific kinematic conditions up to 6.6% and helped to reduce swimming energy for most kinematic conditions up to a maximum of 5.9%. In addition, a leading-edge vortex with greater vorticity than the smooth control was generated by the three-dimensional printed shark skin, which may explain the increased swimming speeds. A significantly enhanced leading edge vortex was generated by the three-dimensional printed shark skin foil.[48]

Boat, ship, and aircraft manufacturers are trying to mimic shark skin to reduce friction drag and minimize the attachment of organisms on their surfaces. In mimicking the surface architecture, riblets can be created on the surface by painting or attaching a film. Skin friction contributes to about half of the total drag in an aircraft. Transparent sheets with a ribbed

(a)

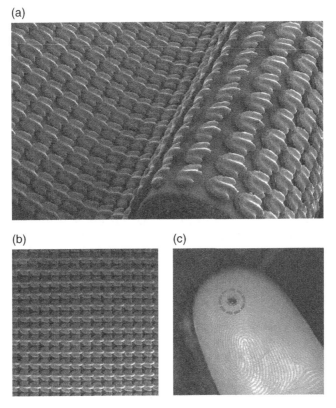

(b)   (c)

Figure 5.17 SEM images of the fabricated synthetic shark skin membranes. Rigid denticles were fabricated on a flexible substrate membrane using 3D printing technology. Membranes in curved and flattened states are shown in (a) and (b), respectively. Note the changes in spacing among the denticles in the convex and concave portions of the curved membrane (a), and how denticles overlap each other in the concave region and when the membrane is flat (b). A single synthetic denticle (enclosed by the dashed circle) on a human finger is shown in (c). Each denticle measures ca. 1.5 mm in length. Source: Wen *et al.* (2014).[48] Reproduced with permission of the Company of Botanists Ltd.

structure in the longitudinal direction have been used on the commercial Airbus 340 aircraft. It is thought that riblet film on the body of the aircraft can reduce drag of the order of 10% with a fuel saving of approximately 1.5%.[53]

Inspired by the riblet effect of shark skin, a variety of commercial swimsuits have been produced employing new fibers and weaving techniques. These have been used in swimming competitions, especially in the Olympic Games. In 2006, Speedo created a whole-body swimsuit, the Fastskin bodysuit (TYR Trace Rise), for elite swimming. This suit is made of polyurethane woven fabric with a texture based on shark scales. The outstanding performance of the swimsuit quickly received intensive public attention. In the 2008 Summer Olympics, two-thirds of the swimmers wore Speedo swimsuits, and a large number of world records were broken. The key feature is the fabric, which was designed to mimic the structure of shark skin by superimposing vertical resin stripes. The stripes are designed to produce vertical vortices or spirals of water, which keep the passing water closer to the swimmer's body and reduce the formation of separation bubbles and consequently reduce drag force. Based on passive drag tests, the drag reduction could reach up to 4.0% when wearing the Fastskin.[54] However, in order to be effective the drag riblets have to be aligned with fluid flow, otherwise resistance can increase. Thus, the Fastskin could actually hinder performance when swimmers roll from side to side, as in crawl and backstroke, or move vertically in butterfly and breaststroke.

References

1. Peattie, A.M. Functional demands of dynamic biological adhesion: an integrative approach. *Journal of Comparative Physiology B* **179**, 231–239 (2009).
2. Autumn, K., Dittmore, A., Santos, D., Spenko, M. & Cutkosky, M. Frictional adhesion: A new angle on gecko attachment. *Journal of Experimental Biology* **209**, 3569–3579 (2006).
3. Arzt, E., Gorb, S. & Spolenak, R. From micro to nano contacts in biological attachment devices. *Proceedings of the National Academy of Sciences of the United States of America* **100**, 10603–10606 (2003).
4. Federle, W. Scaling of animal attachment devices: implications on adhesive mechanisms, pad design and performance. *Comparative Biochemistry and Physiology A* **143**, S86–S87 (2006).
5. Autumn, K. How gecko toes stick – The powerful, fantastic adhesive used by geckos is made of nanoscale hairs that engage tiny forces, inspiring envy among human imitators. *American Scientist* **94**, 124–132 (2006).
6. Gao, H.J., Wang, X., Yao, H.M., Gorb, S. & Arzt, E. Mechanics of hierarchical adhesion structures of geckos. *Mechanics of Materials* **37**, 275–285 (2005).
7. Autumn, K. *et al.* Adhesive force of a single gecko foot-hair. *Nature* **405**, 681–685 (2000).
8. Autumn, K. & Peattie, A.M. Mechanisms of adhesion in geckos. *Integrative and Comparative Biology* **42**, 1081–1090 (2002).
9. Persson, B.N.J. & Gorb, S. The effect of surface roughness on the adhesion of elastic plates with application to biological systems. *Journal of Chemical Physics* **119**, 11437–11444 (2003).
10. Huber, G., Gorb, S.N., Spolenak, R. & Arzt, E. Resolving the nanoscale adhesion of individual gecko spatulae by atomic force microscopy. *Biological Letters – UK* **1**, 2–4 (2005).
11. Xu, Q., Wan, Y., Hu, T.S., Liu, T.X., Tao, D. *et al.* Robust self-cleaning and micromanipulation capabilities of gecko spatulae and their bio-mimics. *Nature Communications* **6**, 8949 (2015).
12. Stark, A.Y. *et al.* Surface wettability plays a significant role in gecko adhesion underwater. *Proceedings of the National Academy of Sciences of the United States of America* **110**, 6340–6345 (2013).
13. Gravish, N., Wilkinson, M. & Autumn, K. Frictional and elastic energy in gecko adhesive detachment. *Journal of the Royal Society Interface* **5**, 339–348 (2008).
14. Autumn, K. *et al.* Frictional adhesion of natural and synthetic gecko setal arrays. *Integrative and Comparative Biology* **46**, E5–E5 (2006).
15. Puthoff, J.B. *et al.* Dynamic friction in natural and synthetic gecko setal arrays. *Soft Matter* **9**, 4855 (2013).
16. Kamperman, M., Kroner, E., del Campo, A., McMeeking, R.M. & Arzt, E. Functional adhesive surfaces with 'gecko' effect: The concept of contact splitting. *Advanced Engineering Materials* **12**, 335–348 (2010).
17. Gao, H.J. & Yao, H.M. Shape insensitive optimal adhesion of nanoscale fibrillar structures. *Proceedings of the National Academy of Sciences of the United States of America* **101**, 7851–7856 (2004).
18. Persson, B.N.J. On the mechanism of adhesion in biological systems. *Journal of Chemical Physics* **118**, 7614–7621 (2003).
19. Hui, C.Y., Jagota, A., Lin, Y.Y. & Kramer, E.J. Constraints on microcontact printing imposed by stamp deformation. *Langmuir* **18**, 1394–1407 (2002).
20. Hu, S.H. & Xia, Z.H. Rational design and nanofabrication of gecko-inspired fibrillar adhesives. *Small* **8**, 2464–2468 (2012).
21. Geim, A.K. *et al.* Microfabricated adhesive mimicking gecko foot-hair. *Nature Materials* **2**, 461–463 (2003).
22. Gorb, S., Varenberg, M., Peressadko, A. & Tuma, J. Biomimetic mushroom-shaped fibrillar adhesive microstructure. *Journal of the Royal Society Interface* **4**, 271–275 (2007).
23. Murphy, M.P., Aksak, B. & Sitti, M. Gecko-inspired directional and controllable adhesion. *Small* **5**, 170–175 (2009).
24. Jeong, H.E., Lee, J.K., Kim, H.N., Moon, S.H. & Suh, K.Y. A nontransferring dry adhesive with hierarchical polymer nanohairs. *Proceedings of the National Academy of Sciences of the United States of America* **106**, 5639–5644 (2009).

25. Qu, L.T., Dai, L.M., Stone, M., Xia, Z.H. & Wang, Z.L. Carbon nanotube arrays with strong shear binding-on and easy normal lifting-off. *Science* **322**, 238–242 (2008).
26. Hu, S., Xia, Z. & Gao, X. Strong adhesion and friction coupling in hierarchical carbon nanotube arrays for dry adhesive applications. *ACS Applied Materials and Interfaces* **4**, 1972–1980 (2012).
27. Lee, H., Lee, B.P. & Messersmith, P.B. A reversible wet/dry adhesive inspired by mussels and geckos. *Nature* **448**, 338–U334 (2007).
28. Glass, P., Chung, H., Washburn, N.R. & Sitti, M. enhanced reversible adhesion of dopamine methacrylamide-coated elastomer microfibrillar structures under wet conditions. *Langmuir* **25**, 6607–6612 (2009).
29. Mahdavi, A. *et al.* A biodegradable and biocompatible gecko-inspired tissue adhesive. *Proceedings of the National Academy of Sciences of the United States of America* **105**, 2307–2312 (2008).
30. Jeong, H.E., Kwak, M.K. & Suh, K.Y. Stretchable, adhesion-tunable dry adhesive by surface wrinkling. *Langmuir* **26**, 2223–2226 (2010).
31. Krahn, J., Liu, Y., Sadeghi, A. & Menon, C. A tailless timing belt climbing platform utilizing dry adhesives with mushroom caps. *Smart Materials and Structures* **20** (2011).
32. Northen, M.T., Greiner, C., Arzt, E. & Turner, K.L. A gecko-inspired reversible adhesive. *Advanced Materials* **20**, 3905–3909 (2008).
33. Reddy, S., Arzt, E. & del Campo, A. Bioinspired surfaces with switchable adhesion. *Advanced Materials* **19**, 3833–3837 (2007).
34. Yoon, H. *et al.* Adhesion hysteresis of Janus nanopillars fabricated by nanomolding and oblique metal deposition. *Nano Today* **4**, 385–392 (2009).
35. Federle, W., Barnes, W.J., Baumgartner, W., Drechsler, P. & Smith, J.M. Wet but not slippery: Boundary friction in tree frog adhesive toe pads. *Journal of the Royal Society Interface* **3**, 689–697 (2006).
36. Scholz, I., Barnes, W.J.P., Smith, J.M. & Baumgartner, W. Ultrastructure and physical properties of an adhesive surface, the toe pad epithelium of the tree frog, *Litoria caerulea* White. *Journal of Experimental Biology* **212**, 155–162 (2009).
37. Barnes, W.J., Goodwyn, P.J., Nokhbatolfoghahai, M. & Gorb, S.N. Elastic modulus of tree frog adhesive toe pads. *Journal of Comparative Physiology A: Neuroethology, Sensory, Neural, and Behavioral Physiology* **197**, 969–978 (2011).
38. Hanna, G. & Barnes, W.J.W. Adhesion and detchment of toe pads of tree frogs. *Journal of Experimental Biology* **155**, 103–125 (1991).
39. Persson, B.N.J. Wet adhesion with application to tree frog adhesive toe pads and tires. *Journal of Physics: Condensed Matter* **19**, 376110 (2007).
40. Drotlef, D.-M. *et al.* Insights into the adhesive mechanisms of tree frogs using artificial mimics. *Advanced Functional Materials* **23**, 1137–1146 (2013).
41. Bauer, U. & Federle, W. The insect-trapping rim of Nepenthes pitchers: surface structure and function. *Plant Signaling & Behavior* **4**, 1019–1023 (2009).
42. Bohn, H.F. & Federle, W. Insect aquaplaning: Nepenthes pitcher plants capture prey with the peristome, a fully wettable water-lubricated anisotropic surface. *Proceedings of the National Academy of Sciences of the United States of America* **101**, 14138–14143 (2004).
43. Federle, W., Riehle, M., Curtis, A.S.G. & Full, R.J. An integrative study of insect adhesion: Mechanics and wet adhesion of pretarsal pads in ants. *Integrative and Comparative Biology* **42**, 1100–1106 (2002).
44. Bauer, U., Willmes, C. & Federle, W. Effect of pitcher age on trapping efficiency and natural prey capture in carnivorous *Nepenthes rafflesiana* plants. *Annals of Botany* **103**, 1219–1226 (2009).
45. Samaha, M. & Gad-el-Hak, M. Polymeric slippery coatings: Nature and applications. *Polymers* **6**, 1266–1311 (2014).
46. Bechert, D.W., Bruse, M., Hage, W., VanderHoeven, J.G.T. & Hoppe, G. Experiments on drag-reducing surfaces and their optimization with an adjustable geometry. *Journal of Fluid Mechanics* **338**, 59–87 (1997).
47. Bechert, D.W., Bruse, M., Hage, W. & Meyer, R. Fluid mechanics of biological surfaces and their technological application. *Naturwissenschaften* **87**, 157–171 (2000).
48. Wen, L., Weaver, J.C. & Lauder, G.V. Biomimetic shark skin: design, fabrication and hydrodynamic function. *Journal of Experimental Biology* **217**, 1656–1666 (2014).
49. Bhushan, B. Biomimetics: lessons from nature – an overview. *Philosophical Transactions of the Royal Society A* **367**, 1445–1486 (2009).
50. Barthlott, W. & Neinhuis, C. Purity of the sacred lotus, or escape from contamination in biological surfaces. *Planta* **202**, 1–8 (1997).

51. Wong, T.S. *et al.* Bioinspired self-repairing slippery surfaces with pressure-stable omniphobicity. *Nature* **477**, 443–447 (2011).
52. Ishino, C., Reyssat, M., Reyssat, E., Okumura, K. & Quere, D. Wicking within forests of micropillars. *Europhysics Letters* **79** (2007).
53. Fish, F.E. Limits of nature and advances of technology: What does biomimetics have to offer to aquatic robots? *Applied Bionics and Biomechanics* **3**, 49–60 (2006).
54. Vizard, F. & Lipsyte, R. *Why a Curveball Curves: The Incredible Science of Sports* (Popular Mechanics, Hearst; 2009).

6

Self-healing Materials

6.1 Introduction

Self-healing is defined as the capability of a material to heal (recover/repair) damages automatically and autonomously, that is, without any external intervention.[1] From a materials science point of view, self-healing materials are polymers, metals, ceramics, and their composites that, when damaged through thermal, mechanical, ballistic, or other means, have the ability to heal and restore the material to its original set of properties.[2] In fact, there are several different ways to describe the self-healing properties of materials, such as self-repairing, autonomous healing, and automatic repairing, depending on the mechanism of healing functionality, such as properties and shapes. Here, we focus on materials with the ability to heal the damage and restore their mechanical properties automatically.

Materials under static or cyclic loadings can degrade through microcracks and other types of micro-damage such as delamination in fiber-reinforced composites. In fact, virtually all materials are susceptible to natural or artificial degradation and deteriorate with time. The damage inevitably reduces mechanical, thermal, electrical, and acoustical properties, and eventually cause catastrophic failure in structural materials. Under cyclic loadings, the failure may occur spontaneously even below the materials' maximum critical load. Since the damage is usually difficult to detect, especially for inaccessible components, and often nearly impossible to repair by conventional methods, it remains one of the most significant factors limiting reliability and leads to the conservative design of material structures. Clearly, stopping or even healing the damage may inhibit or at least decelerate crack expansion. This approach may significantly increase the lifetime and reliability of critical components where catastrophic failure is unacceptable.

In nature, living organisms autonomously heal themselves using various repair strategies, for example mammals have a highly developed vascular network that is able to heal any injury anywhere at anytime. When a body is wounded, blood in the cut quickly clots and protects wound, and there follows a series of reactions that restore the tissue to its original physical and mechanical properties.[3] Mussel byssal thread is capable of regaining strength after yielding,[4] whilst lection in abalone possesses microstructures that are broken during loading and reformed during relaxation.[5] Mimicking these microstructures and self-healing mechanisms could create new self-healing materials and provide biomimetic design principles for continual improvement of materials performance.

Inspired by natural self-healing systems, engineering materials with self-healing abilities have been developed based on biological self-healing mechanisms applied using broadly traditional engineering approaches. Various biomimetic and non-biomimetic self-healing methods have been developed for polymers, metals, ceramics, and their composites. Biomimetic self-healing methods can be classified into three types: vascular, compartmentalization, and regrowth/remodeling systems. In vascular systems, fluid agents are transported through microchannels to the damaged area to heal the cracks. This approach has the ability to heal large amounts of damage for multiple healing with high efficiency. Compartmentalized systems contain microcapsules or hollow fibers filled with a healing agent, which are embedded into a matrix. When the material is damaged, these microcapsules or hollow fibers/ microtubes break, allowing the healing agent to enter the lesion and heal the damage. In the regrowth approach, active microparticles distributed in the material react with air or gases and form deposits to heal the cracks through oxidation or bacterial activity. This approach mimics bone wound healing through mineralization and turgescent cellular plant structures that seal and repair fissures caused by internal growth processes or external lesions. These bioinspired approaches do not typically include mimicry of the biological processes involved because in many cases they are too complex.

Introducing natural materials repair strategies could promote engineering materials design from a conventional conservative damage tolerance philosophy to a more modern design concept. Current material design in engineering follows the concept of damage prevention. With self-healing materials the design principle can be transformed to the concept of damage management, which would lead to reduced weight and increased service life.[6] Although damage is sometimes inevitable, it does not necessarily cause problems if it is subsequently healed. In many cases, self-healing materials could provide new solution to the systems that are difficult to repair during their function (e.g., spaceships), once materials possess the ability to heal defects of any size, multiple times, completely and autonomously.

6.2   Wound Healing in Biological Systems

6.2.1   Self-healing via Microvascular Networks

Vascular networks, such as blood vessels in animals and veins in plants, are a vital system that transports fluid or other substances to promote growth and healing. In particular, blood vessels have evolved to form major system that is able to heal the body when it is injured. This remarkable and complex process starts as soon as an injury occurs and automatically stops when the healing is complete. In addition to injuries, this healing process also takes care of normal everyday wear and tear. Through the system, damaged or dead cells are cleaned in great numbers daily from our skin, mouth, intestines, and blood. Minor damage to these areas can be healed repeatedly because of the vascular nature of this supply system.

The vascular network in animals is a complex tree structure consisting of heart, arteries, arterioles, capillaries, venules, and veins. A key feature of the system is that fluid is pumped from a point reservoir (heart) through networks to an area where the body needs. To supply fluid to whole body, arteries from the heart branch out into smaller arterioles while arterioles further branch out into the capillaries. The branching and size of these vessels have evolved to minimize the power required to distribute and maintain the supporting fluid within many other constraints.[7–9] Capillaries are tiny blood vessels of approximately 5–20 μm diameter that distribute blood and connect to form a network of capillaries in most of the organs and tissues of the body. Blood containing nutrients is supplied by arterioles and drained by venules through these capillaries. Capillary walls are only one cell thick, which permits the exchange of materials between the contents of the capillary and the surrounding tissue. The extensive network of capillaries in the human body is estimated to be between 50,000 and 60,000 miles long.[10] Thoroughfare channels allow blood to bypass a capillary bed. These channels can open and close by the action of muscles that control blood flow through the channels. The system is also reconfigurable in response to circumstances by adjusting the radius of individual vessels by vasoconstriction and dilation in mature tissue, or by growth in embryonic blood vessels.[10]

Plants have similar vascular networks (veins) (Figure 6.1) that distribute water from the roots up to all the cells within the leaf, and also bring resources from the leaf back to the rest of the plant after photosynthesis. These networks are also multifunctional. Like a nervous system, the networks transmit chemical signals to the leaves from other parts of the plant through the liquid in the veins. Similar to the vascular system in animals, they also provide wound-healing capabilities.

Human skin is composed of multiple sublayers that can continually rebuild the surface of the skin. The outer epidermal layer is protecting layer, whereas the underlying dermal layer supplies the epidermis with nutrient-laden blood and regulates temperature. Because skin serves as a protective barrier, it must rapidly and efficiently self-heal any damage to it.[3] Once the skin is cut deep into the underlying dermal layer containing blood vessels, and the protective barrier of skin is broken, the normal process of wound healing is immediately set in action. The process can be divided into four sequential, yet overlapping, phases: hemostasis, inflammation, proliferation, and remodeling, as shown in Figure 6.2. When an injury takes place anywhere in our bodies, under blood pressure blood flows from the capillary network in the dermal layer to the wound site, and at the same time the blood vessels at the site contract

Figure 6.1 Distribution pattern of vein network in leaves.

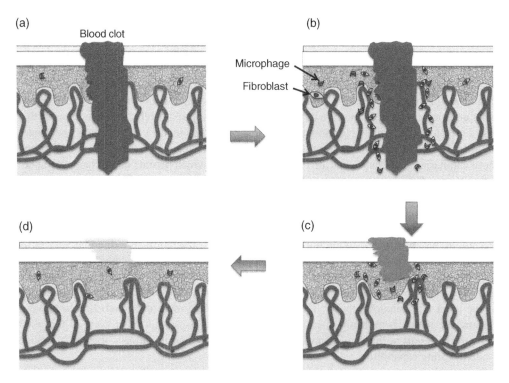

Figure 6.2 Illustration of wound healing marked by a spectrum of four overlapping and closely coupled stages: (a) hemostasis; (b) inflammation; (c) proliferation; and (d) tissue remodeling. Source: Toohey et al. (2007).[3]

to slow the bleeding. Blood platelets rapidly form a blood clot at the cut to completely stop bleeding and serve as a matrix for further healing. Specialized cells (e.g., white blood cells) then accumulate at the cut, and destroy and digest dead cells by secreting special enzymes stored in small packets in the cells called lysosomes. The area around the wound is made clean by removing the dead-cell debris. Almost simultaneously with the cleaning, the process of new cell formation begins in the cleaned spot. While older cells are pushed to the site of the injury, these new cells gradually fill the cut.[3]

6.2.2 Self-healing with Microencapsulation/Micropipe Systems in Plants

The healing process in plants occurs in different ways. Ficus, for example, is a typical plant that can quickly repair and heal injured bark and stems. When ficus tree stems are injured, they secrete pre-made latex at the site of the wound. On being exposed to air, this complex emulsion coagulates into an elastic polymer that serves several defensive functions, including halting any further tearing and sealing the wound from infection until cell growth can permanently mend the injury (Figure 6.3a).[11] In addition to the prevention of further tearing and the entry of pathogens (fungi, bacteria, and viruses) into underlying tissues, it is also crucial to restore the mechanical properties. Since stresses in plant stems are largest in the wound area

(a)																				(b)

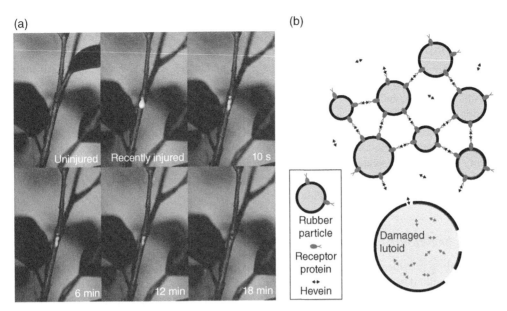

Figure 6.3 (a) Macroscopic observation showing the process of latex coagulation after injuring the bark of a weeping fig (*Ficus benjamina*). Directly after injury the fresh latex droplet is white due to total reflection of the fresh latex emulsion.[15] (b) The coagulation and healing mechanism of plant lattices. During coagulation progress the latex droplet becomes increasingly translucent, mirroring the chemical processes taking place in the latex during coagulation. The protein hevein is stored in vacuolar structures called lutoids. Rubber particles and lutoids are present in the latex. After injury the lutoids burst due to the pressure drop caused by the pressure difference between latexbearing micro-pipes and air pressure. Thus they release the hevein, which cross-links the rubber particles via receptor proteins enveloping the rubber particles.[15,16] Source: Binder *et al.* (2013).[15] Reproduced with permission of John Wiley and Sons.

of the stem and lesions are at risk of expanding due to such stresses, it is particularly important for the tree to recover the mechanical properties of the peripheral region (e.g., the bark) of a plant stem. This ability to repair quickly enables trees to quickly recover partial properties and better protect the wound for complete recovery of these properties by cell and tissue growth in later repair phases. Experiments show that 30 minutes after damage is sustained by the ficus, enough latex has already coagulated at the damage site for about 55% of the original, unwounded tensile strength to be restored.[11] The latex that covers the wound acts as a "crack stopper", hindering crack propagation in the lesion and thus increasing the tensile strength. This protection and strengthening mechanism is immediately established and maintained until cellular growth can restore the complete strength.

Coagulation is a smart, autonomic self-healing system that functions without any external stimulus.[12] Several plant species, such as *Hevea brasiliensis*, seal fissures by coagulation of lattices, which occur as healing agents in branched micro-pipe systems. In this plant species latex is stored in the laticifer under high pressure of 7–15 bar. The latex of *H. brasiliensis* contains, among other substances, rubber particles and vacuolar structures, called "lutoids", comprising the protein hevein. On injury, due to the pressure drop from 7 bar or more in the intact laticifers to ambient pressure (ca. 1 bar), the lutoids burst and hevein is released. Hevein dimers are formed under the influence of Ca^{2+}. They crosslink rubber particles which have binding sites for this protein on their surface, causing an autonomous latex coagulation (Figure 6.3b).[13,14]

The coagulation and healing mechanisms of plant lattices provide a new approach and biological model for the biomimetic design of self-healing materials. In particular, it may be possible to use the healing concept of plant lattices for the design and development of self-healing elastomeric materials for technical purposes. According to the healing mechanism of plant lattices, the bioinspired approaches may include (1) embedding of microcapsules filled with healing agents in technical self-healing elastomers that burst on injury, releasing a healing agent (inspired by vesicular lutoids found in *H. brasiliensis* and *Ficus benjamina* latex) and (2) the development of technical ionomeric elastomers (inspired by the function of Ca^{2+} ions in *H. brasiliensis* latex during coagulation).[11]

6.2.3   Skeleton/Bone Healing Mechanism

Bone healing provides an alternative self-healing model in hard tissue that could be used for the biomimetic design of self-healing hard materials. Bone fracture healing involves several phases of recovery for the proliferation and protection of the areas surrounding fractures and dislocations. The recovery time varies depending on the extent of the injury. It takes 2–3 weeks for our bodies to repair most upper body fractures, and more than 4 weeks to heal lower body injuries.

There are three distinct but overlapping phases of fracture healing: (1) the early inflammatory stage, (2) the repair stage, and (3) the late remodeling stage.[17] In the first stage, after bone fracture, blood cells from the blood vessel network quickly flow to the tissues adjacent to the injury site, followed by the constriction of blood vessels, stopping any further bleeding. Within a few hours of fracture, the extravascular blood cells form a blood clot (Figure 6.4b). In this same area the fibroblasts replicate, resulting in the formation of granulation tissue, ingrowth of vascular tissue, and migration of mesenchymal cells.

The repair phase can be divided into two stages: the formation of soft and hard callus. During the first stage, several days after the fracture, special cells replicate, transform, and

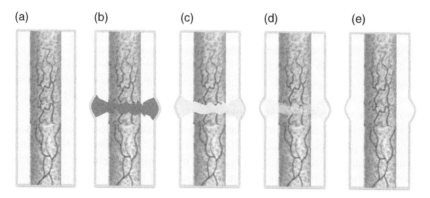

Figure 6.4   Major phases of fracture healing of bone: (a) normal bone, (b) inflammation, (c) soft callus, (d) hard callus, and (e) remodeling.

form woven bone. The fibroblasts within the granulation tissue develop into chondroblasts, which also form hyaline cartilage. These two new tissues grow in size until they unite with their counterparts from other parts of the fracture. These processes culminate in a new mass of heterogeneous tissue, known as the soft callus. Eventually, the fracture gap is bridged by the hyaline cartilage and woven bone, restoring some of its original strength.[17] In the second stage, lamellar bone replaces the hyaline cartilage and woven bone formed in the first stage. Once collagen matrix becomes mineralized, the lamellar bone begins growing by forming channels in the mineralized matrix; each channel contains a microvessel and numerous osteoblasts. This new lamellar bone is in the form of trabecular bone. Eventually, all of the woven bone and cartilage of the original soft callus is replaced by trabecular bone, restoring most of the original strength of the bone.[17]

In the remodeling stage the trabecular bone is substituted with compact bone. In this process, osteoblasts first resorb the trabecular bone, creating a shallow resorption pit, and then deposit compact bone within the resorption pit. As the fracture site is exposed to an axial loading force, bone is generally laid down where it is needed and resorbed from where it is not needed. Eventually, the fracture callus is remodeled into a new shape that closely duplicates the original shape and strength. It may take 3–5 years to complete the remodeling, depending on factors such as age or general condition.

6.2.4 Tree Bark Healing Mechanism

Another interesting self-healing mechanism is the tree bark cracking-induced injure and healing process. Trees with smooth bark (e.g., American beech, blue beech, striped maple) have an ability to generate continuous phellogen that persists, barring injury, for the whole life of the tree. For species with rough bark, however, the phellogen does not persist indefinitely as a continuous tissue; segments of it periodically cease to function and eventually die. As trees grow in a radial direction, tensile stress is added to the dead bark, resulting in cracks in the bark and exposure of live cells within. Repair of these natural gaps is accomplished through a remarkable sequence of events known as phellogen restoration (Figure 6.5).[18]

There are several overlapping steps in phellogen restoration. The first visible step is that phelloderm and/or phloem cells near a crack swell, and their walls become darker and thicker than the surrounding cells. In the second step, fungitoxic chemicals are deposited in these cells due to their physiological changes and at least one layer of cells on the inner margin becomes impervious to dyes and, presumably, other liquids. Since this first impervious layer does not contain suberin, it is different from normal cork and is referred to as non-suberized, impervious tissue (NIT). After NIT is formed, phelloderm and/or phloem parenchyma subjacent to the NIT are transformed to phellogen through a process of dedifferentiation. In the final step, these new cells form a sheet beneath the senescent region and beyond its edges, and join with previously existing phellogen to restore continuity.[18]

Sometimes, the dried bark can cause many deep cracks in tree trunks. The trees react to deeper wounds in similar way to that already described near the phellogen, but additional changes may also take place. If there is not enough room for phellogen restoration to occur between the innermost point of injury and the vascular cambium, the vascular cambium may also become inactive, and adjacent xylem vessels may become plugged with gums, tyloses, and other materials.[18] In this way, xylem dysfunction can occur even though xylem is not

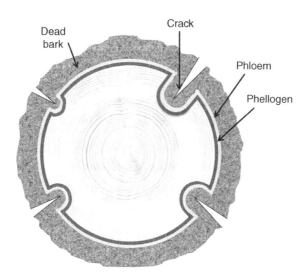

Figure 6.5   When bark is wounded, the repair response is similar to that occurring during normal growth. However, if a wound is close to the vascular cambium, it and subjacent xylem may be killed or plugged. Source: Hudler (1985)[18] (http://agricola.nal.usda.gov/cgi-bin/Pwebrecon.cgi?Search_Arg=GU A84131578&DB=local&CNT=25&Search_Code=GKEY&STARTDB=AGRIDB).

directly wounded or invaded by a pathogen (Figure 6.5). These plant systems have been used successfully as biomimetic models for the development of self-repairing foams for pneumatic technical structures. The concept of phellogen restoration is also appealing when developing biomimetic protective coatings that could heal themselves on cracking.

6.2.5   Bioinspired Self-healing Strategies

The ability of biological materials to heal has inspired new ideas in developing self-healing engineering materials. Mimicking the complex integrated microstructures and micromechanisms found in biological organisms offers considerable scope for the improvement in the design of future multifunctional materials. Different healing concepts have been proposed to offer the ability to restore the mechanical performance of the material. Various self-healing methods have been considered and assessed from an engineering perspective. These self-healing mechanisms and biomimetic strategies are summarized in Table 6.1.[19] This table is by no means exhaustive but gives a general overview of the characteristic similarities of biology and engineering systems.

Complex microvascular networks are widely observed in biological systems, such as leaf venation and blood vascularization. The vessels in the network function together in a branched system to supply blood to all points in the body simultaneously. These highly developed, multifunctional vascular networks serve as an excellent model for self-healing, distribution of fuel, and control of internal temperature, etc. Although exact replication of these microvascular systems remains a significant challenge for those pursuing synthetic analogs due to their complex architecture, these diverse functions in living organisms can be mimicked by embedding a network of interconnected microchannels in synthetic materials. Several

Table 6.1 Some biomimetic self-healing strategies for engineering materials (reprinted with permission from IOP Publishing).[19]

Biological attribute	Self-healing mechanisms	Biomimetic self-healing strategy
Vascular network	Channel transport fluid Fluid fills the cuts Chemical reactions catalyzed by enzymes to form clotting and tissue	Built-in two- or three-dimensional network Healing agents to be replenished and renewed during the life of the structure
Encapsulation	Microtube/channel store agents Bleeding when injured Chemical reactions catalyzed by enzymes to coagulation	Built-in encapsulated spheres, tubes or hollow fibers Two-phase polymeric cure process Action of bleeding from a storage medium housed within the structure.
Skeleton/bone healing	Blood clotting formed at fracture Clotting transform via mineralization Mineralized structure remodeling	Deposit particles that are stable but react with air when damaged. Promote reactions to form minerals similar to matrix when crack forms
Tree bark healing	Formation of internal impervious boundary walls to protect the damaged structure from environmental attack	Build internal impervious boundary walls to protect the damaged structure from environmental attack
Elastic/plastic behavior in reinforcing fibers	Repeated breaking and reforming of sacrificial bonds	Introduce reversible bounding to promote bond break and formation

emerging applications have recently been demonstrated, including self-healing materials, active cooling networks, and tissue cultures, for example self-healing materials with micro-vascular networks can have repetitive healing and use new healing chemistries. One of the main advantages of vascular systems is their ability to heal larger damage volume with multiple healing events. Numerous healing cycles can be achieved by providing a material with a quasicontinuous flow of healing agent.

The development of biomimetic self-healing materials using vascular networks relies on two key factors: (1) the development of a continuous healing network embedded within a material that delivers healing agent from a reservoir to regions of damage and (2) the replenishment and renewal of the healing agent during the life of the structure to permit the repair of all types of failure modes. For composites, self-healing must restore the matrix material properties and the structural efficiency of fractured fibers.

Coagulation is another efficient autonomic self-healing system, in which two or more healing agents are separately stored in encapsulated microtubes or microspheres. The tree seals lesions after injury by crosslinking rubber particles with hevein.[18] To mimic this healing mechanism, encapsulated microtubes or microspheres containing agents first need to be distributed into the matrix. Catalysts are then distributed in the matrix and finally encapsulated microtubes are broken, producing self-healing agents when the material is damaged.

Ceramic materials are generally characterized by strong and directional chemical bonds, and, even more so than metals, have a very limited atomic mobility, making self-healing behavior very difficult to achieve.[20] The concept of bone healing with mineralization could be applied to ceramic materials. For systems in which oxidative reactions at high operating temperatures lead to reaction products, healing can be achieved by filling cracks with the reaction products. The cracks can be healed and the mechanical properties of the ceramics restored if the fine-grained reaction product has sufficient adhesion to the parent ceramic and has decent mechanical properties itself. Even if the reaction products are poor in mechanical properties compared to the ceramic, they are still useful in hindering crack propagation if they can block the original crack. When the ceramic is used as an antioxidant coating on top of an underlying metallic substrate, the reaction products are even more desirable as they can "heal" the protective character of this ceramic material.

6.3 Bioinspired Self-healing Materials

6.3.1 Self-healing Materials with Vascular Networks

Mimicking natural vascular systems, a synthetic vascular network for self-healing has been successfully demonstrated in fiber-reinforced composites (FRCs). FRCs are widely used in aerospace, automotive, naval, civil, and even sporting goods owning to their high strength-to-weight ratio. However, because woven laminates are stacked in layers, the structure is easy to delaminate between the layers, significantly reducing its mechanical properties and reliability. Introducing a self-healing system to FRCs is a promising solution to this long-standing problem and could greatly extend their lifetime and reliability. The market for this kind of self-healing material is huge. Approximately 20 million tons of composite material are used every year in engineering, defense projects, offshore oil exploration, electronics, and biomedicine. A self-healing material could be used in many everyday items, including polymer composite circuit boards, artificial joints, bridge supports, and tennis rackets.

The early versions of self-healing composites were created by constructing an interconnected microvascular microchannel network in the coating to allow the healing agent to flow through an epoxy polymer block.[21] Following this work, a simple vascular network within a composite sandwich structure was built, consisting of channels approximately 1.5 mm in diameter within a polymethacrylimide (Rohacell) core capped with glass-fiber-reinforced epoxy skins.[22] Recently, three-dimensional microchannel structures have been developed in woven fiber-composite laminates.[23] The biomimetic fiber-reinforced composites can repeatedly heal damage with high efficiency. The fabrication of FRCs involves several steps. In the first step, sacrificial fibers (poly(lactic acid) monofilaments) are woven in a precise pattern into a fiberglass fabric. The preforms of the glass fibers and sacrificial fibers are then infiltrated with an epoxy resin using vacuum-assisted resin transfer molding. During curing, the temperature is raised high enough to vaporize the sacrificial fibers, leaving behind evacuated, undulating microchannels in the composite (Figure 6.6a). Finally, two healing agents (an epoxy resin and a hardener) are separately added to the vascular network using pressurized fluid pumping. These two healing agents reside in their own microchannel networks, which are interpenetrating, but do not interconnect.

The healing agent is a key factor in controlling self-healing. The selected components should possess low viscosity to ensure low pressurized delivery, adequate coverage of fracture surface(s), the ability to polymerize under non-stoichiometric ratios at ambient temperature, and sufficient bonding/fracture toughness to restore structural integrity. Additionally, each

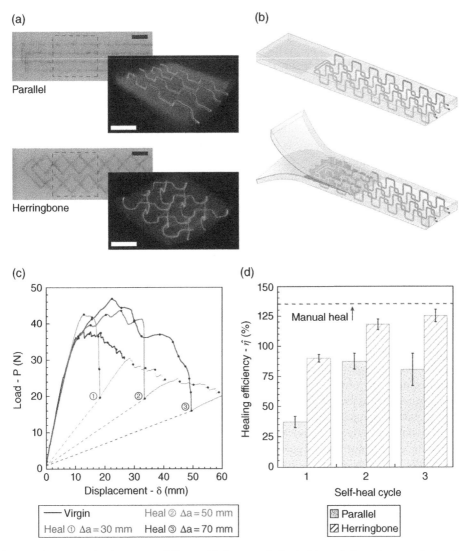

Figure 6.6 *In situ* healing agent delivery. (a) Pre-vascularized, fiber-reinforced composite laminate samples showing sacrificial PLA stitching patterns (scale bars = 10 mm) and post-vascularized, X-ray computed microtomographic reconstructions of vascular networks filled with eutectic galliumindium alloy for radiocontrast (scale bars = 5 mm). (b) Schematic of microvascular double cantilever beam (DCB) fracture specimen with dual channel vascular network where fracture triggers release of liquid healing agents from ruptured microchannels. (c) Representative multiple heal cycle (30 °C, 48 h) load–displacement data for an *in situ* self-healing DCB specimen (herringbone vasculature). (d) Average healing efficiencies obtained using the area method for each *in situ* vascular pattern (parallel, herringbone) at a component delivery ratio of 2 parts epoxy resin (R) to 1 part amine based hardener (H) by volume. Source: Patrick *et al.* (2014).[23] Reproduced with permission of John Wiley and Sons.

healing agent should exhibit excellent compatibility and chemical stability when sequestered in the separate vascular networks of a cured epoxy matrix. A two-part healing chemistry is preferred on the basis of rheology, reaction kinetics, and post-polymerized mechanical properties.[23] The selected components (epoxy resin (EPON 8132) and hardener (EPIKURE 3046)) possess low viscosity (<10 P) and meet the requirements for viscosity.

The artificial self-healing system is accessed using standard double cantilever beam samples (Figure 6.6b). In *in situ* self-healing tests the dual-vascular networks are filled with their respective healing agents. The double cantilever beam samples are loaded and unloaded repeatedly, and the fracture energy is measured. Because of crack-tip blunting by releasing agents, both parallel and herringbone networks increase virgin fracture resistance (3% and 10%, respectively) compared to control composites. After healing at 30 °C for 48 hours, the next cycle begins and the sample is reloaded from the same initial pre-crack length until the delamination front again reaches virgin material and pressurized delivery of healing agents resumes through the newly ruptured vasculature. After each healing cycle, higher loads are required to propagate the crack (Figure 6.6c,d). For the herringbone network, the recovered fracture energy increases with each cycle, leading to healing efficiencies greater than 100%. The increased performance of the interpenetrating herringbone vasculature over the isolated parallel configuration, particularly in the first heal cycle, is attributed to improved fluid interspersion in the fracture plane. Importantly, the herringbone "mixing" geometry with 2 parts resin to 1 part hardener closely approximating stoichiometry, approaches the maximum values established in pre-mixed, manual (reference) tests.[23]

This system demonstrates that *in situ* self-healing can be achieved in structural FRCs via biomimetic microvascular delivery of sequestered, reactive healing chemistries. Vascular architectures not only provide efficient and repetitive delivery of healing agents, but also provide increased resistance to delamination initiation and propagation. The unique design of vascular systems ensures that delamination damage ruptures the internal vasculature, causing the microchannels to effectively release their healing agents. The epoxy resin and hardener bleed into the crack plane and polymerize when they meet, forming a kind of structural glue that reinforces the composite at the fracture site.

6.3.2 Biomimetic Self-healing with Microencapsulation Systems

Inspired by the coagulation mechanism in ficus trees, synthetic self-healing materials have been developed by introducing microencapsulation systems. Similar to ficus trees, the healing system comprises two major parts:

1. a microencapsulated healing agent, which is the liquid glue that fixes the microcracks formed in the material and is encapsulated in tiny bubbles that are distributed throughout the material
2. a catalyst, which is the hardening agent that polymerizes the healing agent; the healing agent must come into contact with a catalyst in order to heal the cracks.

Like hevein dimers and rubber particles in ficus trees, the catalyst and healing agent remain separated in the material until they are needed to seal a crack. When a microcrack forms in the material, it will propagate through the material. The propagating crack will encounter and

(a)

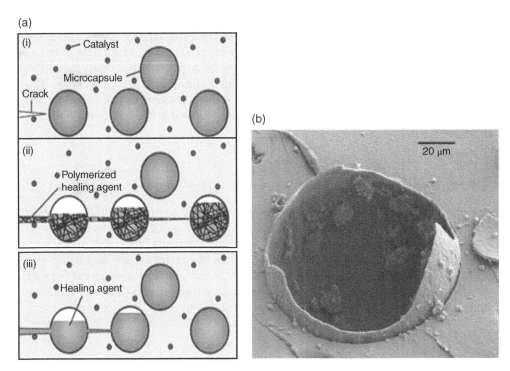

Figure 6.7   (a) Basic method of the microcapsule approach. Autonomic healing concept incorporating encapsulated healing agent and embedded catalyst particles in a polymer matrix: (i) damage event causes crack formation in the matrix, (ii) crack ruptures the microcapsules, releasing liquid healing agent into crack plane, and (iii) healing agent polymerizes upon contact with embedded catalyst, bonding crack closed. (b) ESEM image showing ruptured microcapsule. Source: White *et al.* (2001).[24] Reproduced with permission of Nature Publishing Group.

rupture the microcapsules, releasing the healing agent. The healing agent will flow down through the crack and come into contact with the catalyst, which initiates the polymerization process. This process will eventually glue the microcracks (Figure 6.7).

The first microencapsulation self-healing system used dicyclopentadiene (DCPD), a monomer stored in urea-formaldehyde microcapsules dispersed within a polymer matrix.[24] When ruptured by a progressing crack, the microcapsules release the monomer that flows along the crack. When the fluid comes into contact with a dispersed particulate catalyst (ruthenium-based Grubbs catalyst), it initiates polymerization and thus healing (Figure 6.7). After healing, the components are clearly shown to have restored some of the loss in mechanical properties due to microcracking within the polymer matrix. Although the microcapsules within a bulk polymer matrix material can affect mechanical properties their effect is not detrimental to stiffness. More recently, microcapsule self-healing techniques have been applied to improve the fatigue life of an epoxy bulk polymer. The performance of the materials is enhanced by either manual infiltration of pre-mixed DCPD monomer and catalyst or *in situ* healing using monomer-filled microcapsules and dispersed catalyst.[25,26] With *in situ* healing, crack arrest phenomena were observed in the self-healing composites and fatigue life was improved up to 213% of that of the control specimens in high-cycle fatigue (>10^4 cycles).

The microencapsulation self-healing can be easily incorporated within a bulk polymer material, but the microcapsules need to be fractured and the resin bursting out from the micro-capsule needs to encounter the catalyst prior to any repair occurring. In addition, problems arise in FRC materials because the size of microcapsules (typically 10–100 µm) disrupts the fiber architecture (i.e., fiber waviness and fiber volume fraction). A good dispersion of the catalyst is needed to provide uniform healing functionality. Furthermore, microcapsules have only limited resin volume, which results in the creation of a void in the wake of the crack after consumption of healing resin.

6.3.3   Biomimetic Self-healing with Hollow Fiber Systems

In addition to microencapsulation, hollow fibers are used to store agents. The liquid-filled hollow fibers are embedded within an engineering structure (Figure 6.8), similar to the arteries in a natural system. On mechanical stimulus (damage-inducing fracture of the fibers), this kind of polymer will "bleed" into the damage site to initiate repair, not unlike biological self-healing mechanisms.[27] The hollow fibers not only gain the desired structural improve-ments, but also introduce a reservoir suitable for holding a healing agent in different engineering materials, including concretes, polymers, and polymeric composites.

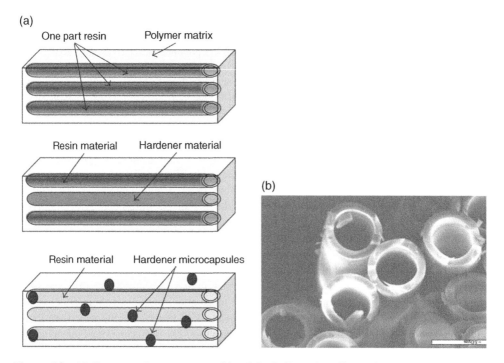

Figure 6.8   (a) Smart repair concepts considered for hollow glass fiber polymer matrix composites using single-part, two-part resin, and hardener, or resin with a catalyst/hardener. (b) Typical scanning electron microscopy (SEM) image of the hollow glass fiber.[27,28] Source: Aissa *et al.* (2012).[27]

Hollow glass fibers were first used to achieve a self-healing mechanism in a composite laminate.[28] Because of their high strength, hollow glass fibers are shown to improve the structural performance of materials without creating sites of weakness within the composite.[29] These hollow fibers offer increased flexural rigidity and allow for greater custom tailoring of performance, by adjusting, for example, both the thickness of the walls and the degree of hollowness.[30] In composite fabrication, commercially available hollow fibers were consolidated in lamina-like structure fibers and then embedded in composite laminates. The advantages of the hollow fiber self-healing concept are that the orientation, volume fraction, and distribution of the fibers can be controlled to match the surrounding reinforcing fibers, thereby minimizing Poisson ratio effects. The hollow fibers can be placed at any location within the stacking sequence to address specific failure threats.[27] The hollow fibers can be filled with different healing resins, depending on the operational requirements of the structure, various polymerization methods can be used to cure the resin, and, crucially, a significant volume of healing agent can be made be available. The disadvantages of the hollow fiber self-healing approaches are that the fiber diameter is relatively larger than the reinforcement, which could lead to unreinforced gaps between the fiber and the reinforcement. The fiber fracture is critical to release the liquid resin to heal the cracks; it is important to choose the correct hollow fibers to match the mechanical properties of reinforcement so that the healing agent is released at right time and place. In addition, low viscosity resin systems are need to facilitate fiber and damage infusion, and an extra processing stage is needed for fiber infusion.[19]

The hollow fiber self-healing approach has been reported to successfully recover the mechanical properties of various fiber-reinforced polymer composites, for example hollow glass fibers have been embedded in a carbon fiber-reinforced polymer (CFRP) to form an autonomic self-healing composite. In this composite, a resin-filled hollow glass-fiber system was distributed at specific interfaces within a laminate, minimizing the reduction in mechanical properties whilst maximizing the efficiency of the healing event.[31] The strength recovery of the self-healing composite was significant and reached over 90%.

6.3.4 Self-healing Brittle Materials Mimicking Bone and Tree Bark Healing

The healing of fractured bone is influenced by a variety of biochemical, biomechanical, cellular, hormonal, and pathological mechanisms.[17] The healing process is a continuous state of bone deposition, resorption, and remodeling. One of key difference between soft and hard tissue is that bone involves subsequent mineralization. Minerals are transferred or formed at the site of the injury through complex chemical reactions to bridge the fractured bone. This process takes a long time as the collagen matrix has to be transformed into hard tissue and the healing bone remodeled to restore its original shape, structure, and mechanical strength. In biomimetic self-healing, mineralization can be introduced to heal the microcracks in ceramic materials.

The biomineralization process can be mimicked through oxidative reactions at high operating temperatures if reaction products are used to fill cracks of modest dimensions and the fine-grained reaction product has sufficient adhesion to the parent ceramic. Although this concept resembles the self-healing concept of microencapsulation systems,[20] the healing agents are directly added to the materials and react with air/water or other agents to heal the cracks when exposed to air, water, or special environments. This approach could be applied to protective coatings under extreme conditions such as high temperature, strong corrosion, and aqueous environments. One example is self-healing thermal barrier coatings made of yttria-stabilized zirconia (ZrO_2 with 6–8 wt% Y_2O_3).[32] These coatings are applied to, for example, combustion chambers and the blades and vanes of gas turbine engines to increase the operating temperature and thereby enhance turbine efficiency. During operation, cracks can develop in these coatings. If these cracks run parallel to the interface, delamination of the coating takes place, which leads to deterioration of the coated component. As a mineralizing ingredient, a high-temperature-resistant material (e.g., intermetallic) can be added to the coating, which, upon oxidation, forms oxide products that heal the crack gap. Intermetallic compounds based on Mo–Si are selected as a healing agent for this purpose because the reaction product, SiO_2, can strongly adhere to the parent materials and its volume increase is in part compensated by volatile MoO_3 species leaving the system.[32] Similar oxidation approaches can be applied to other ceramics. Compared to oxide-based ceramics, carbide- or nitride-based ceramics have the advantage of autonomous self-healing by oxidation. For example, microcracks in SiC, Si_3N_4, and their composites can be healed effectively with SiO_2 formed on high-temperature oxidation.[33] This crack-healing ability makes these ceramics attractive as high-temperature structural components.

Compared to binary systems, ternary ceramics shows even more promise of autonomous self-healing by oxidation. These ceramics are composed of layered compounds denoted by $M_{n+1}AX_n$, with $n = 1–3$, where M is an early transition metal (e.g., Ti, V, Zr, Nb, Hf, Ta), A is an element such as Al or Si, and X is C or N. Recently, oxidation-induced crack healing in ternary ceramics such as Ti_3AlC_2 has been demonstrated.[34,35] As shown in Figure 6.9, a 5-µm wide crack in this ceramic is fully healed with the formation of mainly Al_2O_3 and some TiO_2 through the oxidation of Ti and Al in the ternary ceramic at high temperature. In this healing process, the outward diffusion of the weakly bonded Al atoms is much faster than that of the strongly bonded Ti atoms in the Ti_3AlC_2 structure. The preferential nucleation of Al_2O_3 occurs mainly at the edges of the fractured lamellar Ti_3AlC_2 grains as well as on the hexagonal basal surfaces, thereby ensuring a good adhesion with the parent matrix. The reaction products have an ultrafine grain size.[20] Furthermore, the hardness, Young's modulus, and coefficient of

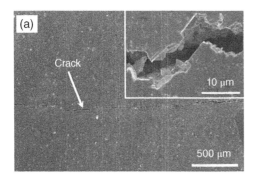
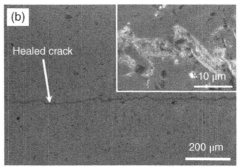

Figure 6.9   Oxidation-induced crack healing in Ti_3AlC_2 ceramic: (a) open crack just after fracture and (b) the same crack after the healing process.[20,34] Source: Ando *et al.* (2005).[33] Reproduced with permission of Elsevier.

thermal expansion of the healing product all are comparable with those of the Ti_3AlC_2 matrix. This makes it likely that the mechanical strength is largely restored, but this is the subject of further study.[20]

6.3.5 Bacteria-mediated Self-healing Concretes

While the self-healing of ceramic materials requires high temperatures, bacteria-based self-healing for concretes occurs at room temperature. Cracks in concrete structures are commonly seen and may lead to reduced performance. To heal the cracks in concrete it is desirable to precipitate minerals into the cracks that can firmly combine the concrete and seal the cracks. One possible healing mechanism is based on mineral-producing bacteria. Spores of specific alkali-resistant bacteria related to the genus Bacillus, for example, are able to seal cracks when they are added to the concrete mixture.[36] Interestingly, these bacteria are able to form spores, which are specialized spherical thick-walled cells somewhat homologous to plant seeds. These spores are viable but dormant cells that can withstand mechanical and chemical stresses, and remain in a dry state viable for periods of over 50 years. The spores hibernate in the dry state but germinated after activation by crack ingress water and produced copious amounts of crack-filling calcium carbonate-based minerals through the conversion of precursor organic compounds that were purposely added to the concrete mixture. It was observed that surface cracks were sealed efficiently by mineral precipitation when bacteria-based solutions were externally applied by spraying onto damaged surfaces or by direct injection into cracks.[20] This process resembles the mineralization of wound healing in bone, although the bone healing involves much more complex reactions.

To keep the bacteria ready for healing after incorporation in the concrete, the bacterial spores must be protected so that they can convert various natural organic substances to copious amounts of calcium carbonate. To this end, organic mineral precursor compounds are packed with the spores in porous expanded clay particles prior to addition to the concrete mixture. The function of the clay particles is very similar to that of the microcapsule in self-healing polymer matrix composites. The two-component biochemical self-healing agent embedded in porous expanded clay particles acts as a reservoir particle and replaces part of the regular concrete aggregates, significantly extending the viability period. On crack formation, this agent, which consists of bacterial spores and calcium lactate, is released from the particle by crack-ingress water. The water acts as a catalyst to trigger the biomineralization. The micro cracks are gradually sealed by formation of bacterially mediated calcium carbonate within the cracks. Experimental results show that the 0.46-mm wide cracks in bacterial concrete can be healed in 100 days (Figure 6.10).[37]

This microbial-enhanced crack-healing ability stems from combined direct and indirect calcium carbonate formation (Figure 6.10a,b): (1) direct $CaCO_3$ precipitation through metabolic conversion of calcium lactate and (2) indirect formation due to reaction of metabolically produced CO_2 molecules with $Ca(OH)_2$ minerals present in the concrete matrix, leading to additional $CaCO_3$ precipitation.[37] In addition, as the metabolically active bacteria consume oxygen, the healing agent may act as an oxygen diffusion barrier, protecting the steel reinforcement against corrosion. The anticipated potential advantages of this bacteria-based self-healing concrete are primarily a reduction in maintenance and repair costs, and an extension of the service life of concrete constructions.[37]

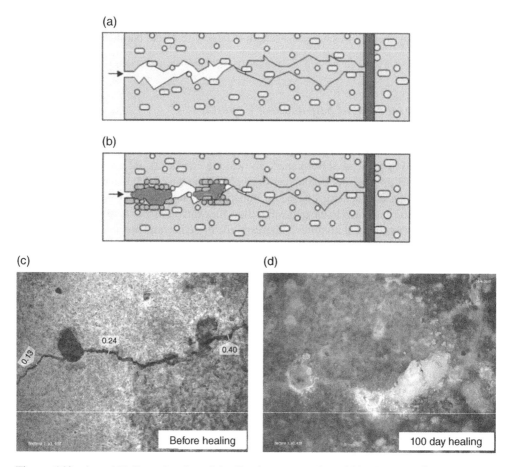

Figure 6.10   (a and b) Scenario of crack-healing by concrete-immobilized bacteria.[38] Ingress water activates bacteria on fresh crack surfaces, bacteria start to multiply and precipitate calcium carbonate, which eventually seals and plugs the crack, protecting the embedded steel reinforcement from further external attack. Stereomicroscopic images of the crack-healing process in biochemical agent-based specimen (c) before and (d) after 100 days of healing. Before and after pictures of the surface of a slab of self-healing concrete. The crack is visible in the left-hand image and on the right the white limestone has filled up the gap. Source: Wiktor & Jonkers (2011).[37] Reproduced with permission of Elsevier.

References

1. Ghosh, K.S. *Self-healing Materials Fundamentals, Design Strategies, and Applications*, 1st edn (Wiley-VCH Verlag GmbH & Co. KGaA.; 2009).
2. Wool, R.P. Self-healing materials: a review. *Soft Matter* **4**, 400 (2008).
3. Toohey, K.S., Sottos, N.R., Lewis, J.A., Moore, J.S. & White, S.R. Self-healing materials with microvascular networks. *Nature Materials* **6**, 581–585 (2007).
4. Vaccaro, E. & Waite, J.H. Yield and post-yield behavior of mussel byssal thread: A self-healing biomolecular material. *Biomacromolecules* **2**, 906–911 (2001).
5. Smith, B.L. *et al.* Molecular mechanistic origin of the toughness of natural adhesives, fibres and composites. *Nature* **399**, 761–763 (1999).

6. van der Zwaag, S. Self healing materials: an alternative approach to 20 centuries of materials science. (Springer, Rotterdam; 2007).

7. McCulloh, K.A., Sperry, J.S. & Adler, F.R. Water transport in plants obeys Murray's law. *Nature* **421**, 939–942 (2003).

8. Riva, C.E., Grunwald, J.E., Sinclair, S.H. & Petrig, B.L. Blood velocity and volumetric flow-rate in human retinal-vessels. *Investigative Ophthalmology & Visual Science* **26**, 1124–1132 (1985).

9. Sherman, T.F. On connecting large vessels to small – The meaning of Murray's law. *Journal of General Physiology* **78**, 431–453 (1981).

10. Taber, L.A., Ng, S., Quesnel, A.M., Whatman, J. & Carmen, C.J. Investigating Murray's law in the chick embryo. *Journal of Biomechanics* **34**, 121–124 (2001).

11. Bauer, G. & Speck, T. Restoration of tensile strength in bark samples of *Ficus benjamina* due to coagulation of latex during fast self-healing of fissures. *Annals of Botany* **109**, 807–811 (2012).

12. Wititsuwannakul, R., Pasitkul, P., Jewtragoon, P. & Wititsuwannakul, D. *Hevea latex* lectin binding protein in C-serum as an anti-latex coagulating factor and its role in a proposed new model for latex coagulation. *Phytochemistry* **69**, 656–662 (2008).

13. Wititsuwannakul, R., Rukseree, K., Kanokwiroon, K. & Wititsuwannakul, D. A rubber particle protein specific for *Hevea latex* lectin binding involved in latex coagulation. *Phytochemistry* **69**, 1111–1118 (2008).

14. Dauzac, J., Prevot, J.C. & Jacob, J.L. Whats new about lutoids – a vacuolar system model from *Hevea latex*. *Plant Physiology and Biochemistry* **33**, 765–777 (1995).

15. Binder, W.H., Speck, T., Mulhaupt, R. & Speck, O. In: *Self-Healing Polymers: From Principles to Applications* (ed. W.H. Binder), 61–89 (Wiley-VCH Verlag GmbH & Co. KGaA, Weinheim; 2013).

16. Bauer, G.N., Nellesen, A., Sengespeick, A. & Speck, T. Fast self-repair mechanisms in plants: biological latices as role models for the development of biomimetic self-healing, mechanically loaded polymers, in: *6th Plant Biomechanics Conference 2009*, S.367–373 (UMR EcoFoG, Kourou; 2009).

17. Kalfas, I.H. Principles of bone healing. *Neurosurgical Focus* **10**, 1–10 (2001).

18. Hudler, G.W. Wound healing in bark of woody plants. *Journal of Arboriculture* **10**, 241–245 (1985).

19. Trask, R.S., Williams, H.R. & Bond, I.P. Self-healing polymer composites: mimicking nature to enhance performance. *Bioinspiration & Biomimetics* **2**, P1–P9 (2007).

20. van der Zwaag, S., van Dijk, N.H., Jonkers, H.M., Mookhoek, S.D. & Sloof, W.G. Self-healing behaviour in man-made engineering materials: bioinspired but taking into account their intrinsic character. *Philosophical Transactions Series A: Mathematical, Physical, and Engineering Sciences* **367**, 1689–1704 (2009).

21. Toohey, K.S., Sottos, N.R., Lewis, J.A., Moore, J.S. & White, S.R. Self-healing materials with microvascular networks. *Nature Materials* **6**, 581–585 (2007).

22. Williams, H.R., Trask, R.S. & Bond, I.P. In: *Fifteenth United States National Congress of Theoretical and Applied Mechanics* (Boulder, CO; 2006).

23. Patrick, J.F. *et al.* Continuous self-healing life cycle in vascularized structural composites. *Advanced Materials* **26**, 4302–4308 (2014).

24. White, S.R. *et al.* Autonomic healing of polymer composites. *Nature* **409**, 794–797 (2001).

25. Brown, E.N., White, S.R. & Sottos, N.R. Retardation and repair of fatigue cracks in a microcapsule toughened epoxy composite – Part 1: Manual infiltration. *Composites Science and Technology* **65**, 2466–2473 (2005).

26. Brown, E.N., White, S.R. & Sottos, N.R. Retardation and repair of fatigue cracks in a microcapsule toughened epoxy composite – Part II: In situ self-healing. *Composites Science and Technology* **65**, 2474–2480 (2005).

27. Aïssa, B., Therriault, D., Haddad, E. & Jamroz, W. Self-healing materials systems: Overview of major approaches and recent developed technologies. *Advances in Materials Science and Engineering* **2012**, 1–17 (2012).

28. Bleay, S.M., Loader, C.B., Hawyes, V.J., Humberstone, L. & Curtis, P.T. A smart repair system for polymer matrix composites. *Composites Part A: Applied Science and Manufacturing* **32**, 1767–1776 (2001).

29. Trask, R.S., Williams, G.J. & Bond, I.P. Bioinspired self-healing of advanced composite structures using hollow glass fibres. *Journal of the Royal Society Interface* **4**, 363–371 (2007).

30. Hucker, M., Bond, I., Bleay, S. & Haq, S. Experimental evaluation of unidirectional hollow glass fibre/epoxy composites under compressive loading. *Composites Part A: Applied Science and Manufacturing* **34**, 927–932 (2003).

31. Williams, G.J., Bond, I.P. & Trask, R.S. Compression after impact assessment of self-healing CFRP. *Composites Part A: Applied Science and Manufacturing* **40**, 1399–1406 (2009).

32. Kochubey, V. & Sloof, W.G. In: *Proceedings of the International Thermal Spray Conference* (DVS-Verlag, Düsseldorf; 2008).

33. Ando, K., Furusawa, K., Takahashi, K. & Sato, S. Crack-healing ability of structural ceramics and a new methodology to guarantee the structural integrity using the ability and proof-test. *Journal of the European Ceramic Society* **25**, 549–558 (2005).

34. Song, G.M. *et al.* Oxidation-induced crack healing in Ti3AlC2 ceramics. *Scripta Materialia* **58**, 13–16 (2008).

35. Song, G.M. *et al.* Early stages of oxidation of Ti3AlC2 ceramics. *Materials Chemistry and Physics* **112**, 762–768 (2008).

36. Jonkers, H.M., Thijssen, A., Muyzer, G., Copuroglu, O. & Schlangen, E. Application of bacteria as self-healing agent for the development of sustainable concrete. *Ecological Engineering* **36**, 230–235 (2010).

37. Wiktor, V. & Jonkers, H.M. Quantification of crack-healing in novel bacteria-based self-healing concrete. *Cement and Concrete Composites* **33**, 763–770 (2011).

38. Jonkers, H. Self-healing concerte *Ingenia* **46**, 39–43 (2011).

Part II

Biomimetic Functional Materials and Processing

7

Self-cleaning Materials and Surfaces

7.1 Introduction

Nature exhibits many remarkable self-cleaning surfaces, such as plant leaves, insect wings, the feathers of water birds, rose petals, and gecko feet. Although encountering daily contaminants, even growing in muddy habitats or living in dirty environments, the surfaces of these organisms stay very clean. Plants, insects, and animals capitalize on basic physical characteristics in the way surfaces of materials interact, achieving cleanliness effortlessly and without detergents.[1] In fact, self-cleaning is of great importance for plants as protection against pathogens like fungi or algae growth; it is also important for animals (e.g., geckos, flies) to protect their adhesive pads from contamination. Nature also provides many excellent solutions to control fouling in aquatic environments. In aquatic environments, biofouling is a widespread and serious problem for both aquatic organisms and artificial underwater structures. Underwater organs such as some shark skin and whale skin possess natural antifouling defenses.

Self-cleaning is a desirable property for humans and makes the dream of a contamination-free surface come true in our daily life. The ability to effortlessly self-clean is very appealing in broad engineering applications, from solar panels to boat hulls and medical devices. The story of natural self-cleaning surfaces begins with the sacred lotus (*Nelumbo nucifera*), which has been a symbol of purity in Asia for more than 2000 years, but only recently has its self-cleaning ability been studied and mimicked based on materials science principles at the micro/nanoscale.[2] Mimicking natural self-cleaning surfaces can generate new types of materials and surfaces for a wide range of practical applications in daily life, industry, agriculture, and the military. Biofouling, for example, results in high functional and monetary costs for both military and commercial vessels because of increased fuel costs due to drag, the cost of dry docking for cleaning, and loss of hull strength due to biocorrosion.[3] It is desirable to design and manufacture universal, environmentally friendly coatings with both antifouling and

Biomimetic Principles and Design of Advanced Engineering Materials, First Edition. Zhenhai Xia.
© 2016 John Wiley & Sons, Ltd. Published 2016 by John Wiley & Sons, Ltd.

fouling-release properties. The extraordinary properties and functions of biological systems provide a new paradigm for the biomimetic design and fabrication of advanced engineering materials.

Bioinspired self-cleaning can be classified into four mechanisms: lotus effect (superhydrophobicity), superhydrophilicity, unbalance contact, and underwater antifouling. The current self-cleaning technology, whether hydrophilic or hydrophobic, cleans surfaces by the action of water. In hydrophilic coatings, the water is made to spread (sheeting of water) over surfaces to carry away dirt and other impurities, whereas in the hydrophobic technique, the water droplets slide and roll over surfaces, thereby cleaning them. Geckos, frogs, and insects, on the other hand, self-clean their toe pads in a way that involves direct contact. Contaminated attachment pads are restored after walking only a few steps. Shark skin, pilot whale skin, carp scale, and other biomaterials also demonstrate self-cleaning properties in aqueous environments. These self-cleaning mechanisms, together with bioinspired strategies and synthetic self-cleaning materials, are discussed in detail in the following sections.

7.2   Fundamentals of Wettability and Self-cleaning

The wettability of solid surfaces by a liquid is important in understanding wetting phenomena and the design of self-cleaning surfaces. When sitting on a solid substrate, a liquid droplet will form a cap-like shape, and under extreme conditions it either retains its spherical shape or spreads out on the surface to form a thin liquid film, depending on surface tensions. Whatever shape the water takes on the solid surface, there is a contact angle θ_Y at the liquid, gas, and solid three-phase boundary. For a liquid droplet on an ideal flat film (Figure 7.1a), the wettability can be described by Young's equation:[4]

$$\gamma_{sg} = \gamma_{sl} + \gamma_{lg} \cos\theta_Y \qquad (7.1)$$

where θ_Y is Young's contact angle and γ_{sg}, γ_{sl}, and γ_{lg} refer to the interfacial surface tensions of solid, liquid, and gas, respectively. The contact angle is a quantitative measure of the wettability of a solid surface by a liquid. The hydrophobicity of a surface can be measured by its contact angle. The larger the contact angle, the higher the hydrophobicity of a surface. A surface with a contact angle $<90°$ is referred to as hydrophilic and that with an angle $>90°$ as hydrophobic. The Young's equation is only valid for ideal flat films. Surface roughness significantly affects the wettability of solid surfaces. In the case of rough surfaces, Wenzel and later Cassie and Baxter provided different expressions for the relationship between the contact angle and the roughness of a solid surface. In Wenzel's theory[5] the relationship between surface roughness and contact angle is given by:

$$\cos\theta_W = r\cos\theta_Y \qquad (7.2)$$

where θ_W is the apparent contact angle in the Wenzel model, θ_Y is Young's contact angle, and r is the surface roughness factor given by the ratio of rough to planar surface areas. It is predicted from the Wenzel model that surface roughness enhances wettability, depending on the nature of the corresponding flat surface (Figure 7.1b). For $r>1$, a hydrophobic surface ($\theta_Y>90°$) becomes more hydrophobic ($\theta_W>\theta_Y$) when rough, whereas a hydrophilic surface ($\theta_Y<90°$) shows

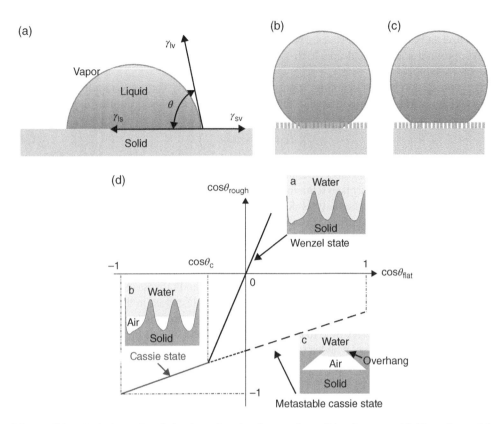

Figure 7.1 Typical wetting behavior of a droplet on the solid substrates. (a) Young's model, (b) Wenzel's model, (c) Cassie's model, and (d) for a moderate hydrophobicity ($90° < \theta < \theta_c$). The apparent contact angle $\theta*$ should be given by the Wenzel mode. If $\theta > \theta*$, air remains trapped below the drop which sits on a composite surface made of solid and air. Source: Cao et al. (2007).[9] Reproduced with permission of the American Chemical Society.

increased hydrophilicity ($\theta_w < \theta_Y$). Although these tendencies are generally (but not always) observed, the Wenzel model is not sufficient when one is dealing with a heterogeneous surface. Cassie and Baxter derived an equation to describe the effect of chemical heterogeneities on the contact angle of solid surfaces.[6] It is assumed that liquid only contacts the solid through the top of the asperities, underneath which air pockets are trapped. This gives a composite surface (Figure 7.1c). In this composite state, air parts of the surface can be considered perfectly non-wetting. If only air is present between the solid and the liquid, the cosine of the contact angle is −1; in this case, the Cassie–Baxter equation can be written as:

$$\cos\theta_{CB} = f_s \left(\cos\theta_Y + 1 \right) - 1 \tag{7.3}$$

where θ_{CB} is the apparent contact angle in the Cassie–Baxter model and f_s is the surface fraction of the solid. The Cassie–Baxter equation can be used to estimate the contact angle of heterogeneous surfaces. Although both the Wenzel and Cassie–Baxter models describe the relationship between the contact angle and surface roughness, neither can be shown to be superior.

Recently, some models have been proposed to describe the coexistence and transition between the Wenzel and Cassie–Baxter states. These new models, together with the basic theory discussed above, provide a theoretical approach for the design and predict functional surfaces with special wettability.[7,8]

For a surface to be self-cleaning, it must shed water easily. The force required to move a drop across a surface is proportional to the contact angle hysteresis, according to[10,11]

$$\gamma_{LV}\left(\cos\theta_{rec} - \cos\theta_{adv}\right) \tag{7.4}$$

where θ_{adv} and θ_{rec} are the advanced contact angle and the receding contact angle, respectively. Hence, the key to repellency is in reducing the hysteresis and not necessarily the actual surface energy. Here, superhydrophobicity is defined as the ability of surfaces to exhibit near-zero hysteresis of water and, therefore, have drops that move readily.[12] In the transition between the Wenzel and Cassie equations (Figure 7.1d), the physical heterogeneity of the roughness acts to increase hysteresis, whereas the composite nature of the surface lessens this effect, as air has no hysteresis. This region with the composite nature is the superhydrophobic region, where a droplet has almost no energetic barrier to motion and moves easily, and self-cleaning originates from the superhydrophobic effect.[13] As will be discussed in the following sections, when the nearly spherical water droplets roll around, they encounter and "grabs" debris and other particulates that loosely bind to the surface because the structuring provides few contact points. Eventually, the droplet slides off the surface with ease, thus carrying the debris off the surface as well.

7.3 Self-cleaning in Nature

7.3.1 Lotus Effect: Superhydrophobicity-induced Self-cleaning

Some biological materials exhibit a superhydrophobicity-induced self-cleaning property, and of these the lotus (*Nelumbo nucifera*) leaf (Figure 7.2a) is one of the best self-cleaning models in nature. Because of its self-cleaning characteristics, the lotus has been a symbol of purity in religions and cultures in Asia for more than 2000 years. Although lotus roots are embedded in muck, its leaves are seemingly never dirty. When water droplets falling onto the leaves bead up and roll off, they wash dirt away from the leaves. This self-cleaning ability is the so-called lotus effect. This effect is also demonstrated in many other plants, for example nasturtium (*Tropaeolum*), prickly pear (*Opuntia*), and cane (*Alchemilla*).[14] Apart from the plants, many insects, such as butterflies, dragonflies and others, also have excellent self-cleaning properties and are able to clean all their bodies. Similar superhydrophobic self-cleaning properties can also be found in lady's mantle leaves, taro leaves, cicada wings, termite wings, and other biological materials.[14]

Lotus leaf surfaces possess randomly distributed micropapillae with diameters ranging from 5 to 9 µm (Figure 7.2c(i)). Each papilla is covered by hundreds of branch-like nanofibers with diameter of approximately 120 nm (Figure 7.2c(ii)). In addition, there are hydrophobic three-dimensional epicuticular waxes with a tubule structure on lotus leaf surfaces (Figure 7.2c(iii)).[15] This structure allows air-pocket formation, resulting in an extremely low contact area between lotus leaf surfaces and water droplets. Furthermore, epicuticular waxes

(a)   (b)

(c)

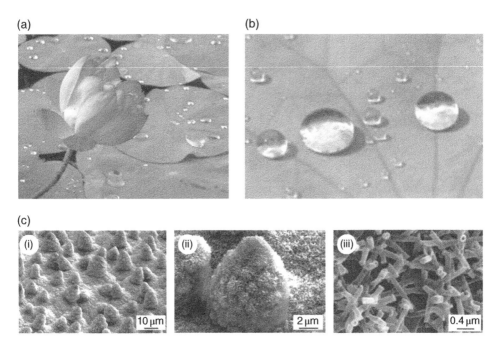

Figure 7.2 Optical image of (a) lotus leave and (b) water drop on the leave, and (c) SEM images (shown at three magnifications, (i)–(iii)) of lotus (*N. nucifera*) leaf surface, which consists of a microstructure formed by papillose epidermal cells covered with three-dimensional epicuticular wax tubules on the surface. Source: Koch *et al.* (2009).[15] Reproduced with permission of the Royal Society of Chemistry.

also are hydrophobic. This, together with micro- and nanostructure roughness, reduces the contact area between water droplets and the leaf surface even further.[16,17]

The hierarchical structures described above play an important role in increasing the contact angle of water on the leaves. As schematically shown in Figure 7.3a–c, the cooperation of surface micro/nanoscale hierarchical structures and hydrophobic epicuticular waxes confers a high water contact angle and a small sliding angle, leading to superhydrophobic and low-adhesion characteristics. When a drop of water falls onto lotus leaves, it tends to minimize its surface area by achieving a spherical shape due to high surface tension, but adhesion forces on water tend to be maximized on smooth surfaces because the liquid-to-solid contact area is large. However, as trapped air in the interstitial spaces of the roughened surface results in a reduced liquid-to-solid contact area, water self-attraction can occur more fully, leading it to form a sphere. The water contact angle and the sliding angle of lotus leaves are approximately 160° and 2°, respectively. The actual contact area of a droplet on the lotus leaf surface is only 2–3%.[14] Water droplets on the surface are almost spherical and can roll freely in all directions. Since a ball rolls more easily than a flattened bump, the role of gravity now becomes significant: the slightest angle in the surface of the leaf (e.g., caused by a passing breeze) causes balls of water to roll off the leaf surface, carrying away attached dirt particles (Figure 7.3d,e) without the leaf having to expend any energy or use any harmful chemicals.

To achieve self-cleaning, it is essential for a surface to have a very high static water contact angle and a very low sliding angle. Owning to the superhydrophobic and low-adhesion characteristics of

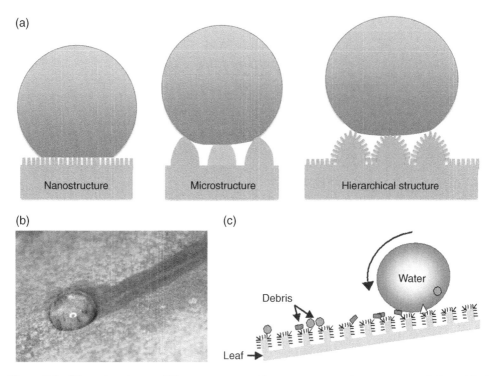

Figure 7.3 Water dropping on different structures: (a) nanostructure, microstructure, and hierarchical structure. (b) Image of water rolling off the leaf surface, carrying away the attached dirt particles. (c) The self-cleaning effect on a lotus leaf. Source: Nishimoto & Bhushan (2013).[14] Reproduced with permission of the Royal Society of Chemistry.

lotus leaves, water droplets on them can pick up dirt particles and thus easily clean them off the surface. In this case, the adhesion between the dirt particle and the droplet is higher than between the particle and the surface. However, the self-cleaning mechanism based on the high surface tension of water does not work with organic solvents as the surface tension of most organic solvents is much lower than that of water, therefore the lotus effect is no protection against graffiti.

7.3.2   Slippery Surfaces: Superhydrophilicity-induced Self-cleaning

Superhydrophilic surfaces, the opposite extreme of superhydrophobic surfaces in terms of wettability, have also received attention for the design of self-cleaning surfaces. A surface is called superhydrophilic when the contact angle between a water droplet and the surface is nearly zero. On such a surface, water droplets quickly spread into a thin film and run off the surface with considerable velocity. Due to the rapid spreading and quick evaporation of water on superhydrophilic surfaces, some plants in tropical rainforests have developed this ability. For example, *Ruellia devosiana* has an outstanding water-spreading property; water film can run up the leaves against gravity.[18] Such superhydrophilic properties enable a self-cleaning ability. When rain or a light spray of water flows onto such surfaces, it can wedge into the space between the substrate and any dust that is present, washing the dust away.

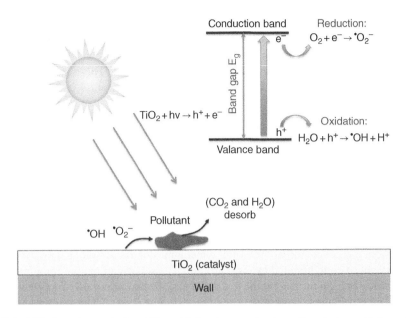

Figure 7.4 TiO_2-based superhydrophilic self-cleaning mechanism. On photo-excitation, electrons overcome the band gap barrier and move to the conduction band. Activated photocatalysts generate electron–hole pairs, where highly active electrons (e⁻) form in the conduction band and positive holes (h⁺) in the valence band. The electrons diffuse to the surface of the photocatalyst and react with adsorbed oxygen molecules, producing reactive oxygen radicals ($^{\bullet}O^-$, $^{\bullet}O_2^-$, and $^{\bullet}O_3^-$) (reduction), while the holes oxidize water molecules or adsorbed hydroxide ions, forming highly oxidizing hydroxyl radicals ($^{\bullet}OH$) (oxidation). The radicals promote photo-electrochemical reactions that decompose attached dirt, harmful microorganisms, and organic contaminants into carbon dioxide and water.

Superhydrophilic surfaces can be induced by a photocatalyst based on natural photosynthesis with sunlight and water. In particular, photocatalytic water splitting under light radiation can overcome energy barriers to convert photon energy into chemical energy, accompanied by a largely positive change in the Gibbs free energy.[19] These reactions give the surfaces superhydrophilicity, which can be utilized to clean surfaces. For example, when exposed to light, titanium oxide (TiO_2) reacts with water in the air, producing a hydrophilic group (–HO) layer that blends easily with water on the surface (Figure 7.4). In addition, under UV radiation the titanium oxide also produces electrons and holes. These react with oxygen and water in the air and produce activated oxygen and hydroxyl radicals (O_2^-, OH), which decompose various organic substances and bacteria. These dirt particles (the organic substances and bacteria) can then be cleaned by the action of water.

Some biological materials also show superhydrophilic properties that make surfaces slippery for prey capture. As described in Chapter 5, pitcher plants possess pitcher-shaped leaves to capture prey (Figure 5.14). Like insect-pollinated flowers, the pitchers attract insects by presenting visual and olfactory signals, and offering food rewards. Once the insects are trapped on the pitcher rim (peristome), they fall to the bottom part of the pitcher, which is filled with a digestive fluid. The captured prey drown and subsequently decompose. One fascinating weapon that the pitcher plant uses to capture the insects is its slippery surface.[20]

Because of its unique surface micro-topography and secretion of hygroscopic nectar, the surface of the peristome is superhydrophilic and is usually covered by stable water film under humid conditions. When insets step onto the rim, this water film repels the oils on their feet and they slide from the rim into the bottom of the pitcher plant.[16] This slippery surface helps to capture prey, as well as cleaning any unwanted debris from the surface.

7.3.3   Self-cleaning in Fibrillar Adhesive Systems

Gecko and insect toes are sticky on almost any surfaces yet stay remarkably clean while encountering day-to-day contaminant. Gecko setae are the first known self-cleaning dry adhesive in nature. Recent experiments have shown that when their sticky footpads are under extreme exposure to clogging particles, after only four steps gecko footpads recover ~80% of their original adhesion to support body weight by a single toe (Figure 7.5a).[22] However, when geckos walked wearing specially designed shoes (Figure 7.5b) to restrict hyperextension, it recovers only ~40% of its shear adhesion force in four steps on glass after their feet have been contaminated by dirt particles. The self-cleaning rate for a free-walking gecko was twice as high as that for the restricted case or that previously reported for arrays of setae isolated from gecko and intact gecko toes (Figure 7.5a).[22] The extraordinary ability of gecko feet to be both sticky and clean presumably stems from their unique fibrillar adhesive system. The structure of gecko attachment pads is described in detail in Chapter 5. Briefly, the gecko foot is made up of well-aligned fine microscopic hairs called setae (approximately 110 μm in length and 5 μm in diameter), which are split into hundreds of smaller nanoscale ends called spatulae. Contact between the gecko spatulae and an opposing solid surface generates van der Waals forces that are sufficient to allow geckos to climb vertical walls or across ceilings. The van der Waals mechanism implies that gecko adhesion depends

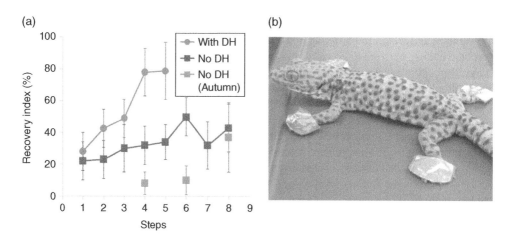

Figure 7.5   (a) Recovery indices for trials with and without digital hyperextension (DH) by steps. For comparison, the results for simulating gecko walking[21] are also plotted. (b) Photo of a gecko wearing specially-designed aluminum shoes to restrict hyperextension during walking. Source: Hu *et al.* (2012).[22] Reproduced with permission of The Royal Society.

more on surface geometry than on surface chemistry. Thus, the self-cleaning mechanism of gecko feet is completely different from that of the lotus effect discussed above. Understanding this non-fouling property is fundamental to the success of gecko-inspired alternatives over traditional pressure-sensitive adhesives.

7.3.3.1   Static Self-cleaning Mechanism of Gecko Feet

Self-cleaning is considered to be a property intrinsic to the hierarchical attachment pads of geckos. Tokay gecko feet contaminated with microspheres recovered their ability to cling to vertical surfaces after only a few steps on clean glass. Similarly, isolated setal arrays self-cleaned by repeated contact with a clean surface. It was suggested the contact self-cleaning mechanism occurs when small solid particles bind more strongly to the substrate than to the toe pads. Thus, when pressing a contaminated toe pad against a clean surface multiple times, the particles are removed from the toe pad via a force imbalance. It was observed that when simulating gecko walking, gecko setal arrays can show ~40% force recovery after four steps.[23] According to contact mechanics, when pulling a contaminated toe pad against a clean surface there is an energetic disequilibrium between the adhesive forces attracting a dirt particle to the substrate and those attracting the same particle to one or more spatulae (Figure 7.6).

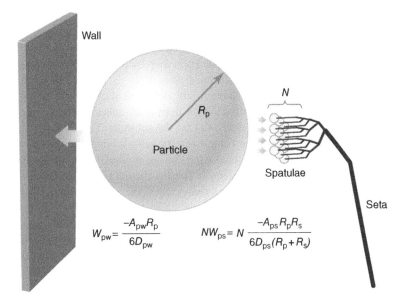

Figure 7.6   Model of interactions between N (number of particle–spatula interactions) gecko spatulae of radius R_s, a spherical dirt particle of radius R_p, and a planar wall. Van derWaals interaction energies for the particle–spatula (W_{sp}) and particle–wall (W_{wp}) systems are shown. When $N \times W_{sp} = W_{wp}$, equal energy is required to detach the particle from wall or N spatulae. N is sufficiently large that self-cleaning results from energetic disequilibrium between the wall and the relatively few spatulae that can attach to a single particle. Source: Hansen & Autumn (2005).[21] Reproduced with permission of the National Academy of Sciences.

Self-cleaning occurs when the adhesion force between the seta and the particles (F_{sp}) is smaller than that between the substrate and the particles (F_{wp}). Setal self-cleaning is dependent largely on particle and spatula size, and spatula material properties. From the static self-cleaning point of view, to rationally design dry self-cleaning adhesive nanostructures, an array of spatulae should have the following properties: (1) surface area smaller than that of dirt particles, (2) a composition of relatively hard, non-tacky materials, and (3) low surface energy.[23]

7.3.3.2   Dynamic Self-cleaning Mechanism of Gecko Feet

A dynamic self-cleaning mechanism for gecko spatulae has been observed in experiments.[24] It was shown that strongly normal-velocity-dependent adhesion (F_{wp}) occurs when a microsphere is pulled off a substrate in normal directions. By contrast, the setae show weakly normal-velocity-dependent adhesive forces (F_{sp}) with the substrate. The difference between the velocity-dependent particle-wall adhesion and the velocity-independent spatula-particle dynamic response leads to a self-cleaning capability, allowing geckoes to more efficiently dislodge dirt during their locomotion.

When pulling off a seta that is in contact with a particle on a substrate, there are two scenarios for the particle: either drop-off (case I, Figure 7.7b) or pick-up (case II, Figure 7.7c) by the seta. In case I, the particle is dropped off from the seta if it binds more strongly to the substrate than to the seta ($F_{wp} > F_{sp}$); conversely, in case II, the particle is picked up by the seta when it binds more strongly to the seta than to the substrate ($F_{wp} < F_{sp}$). In case I, the seta remains clean naturally, but in case II it is contaminated by the particles. Figure 7.7d shows a case (case II) in which F_{wp} is estimated to be twice as small as F_{sp} at a low speed of $V_n < 1\,\mu\mathrm{m\,s^{-1}}$. In this case, the microparticles on the substrates are successfully picked up by the seta. With increasing V_n, F_{wp} increases and eventually surpasses F_{sp} at a critical point V_c (e.g., $V_c \sim 2000\,\mu\mathrm{m\,s^{-1}}$ for a 10-μm SiO$_2$ particle on polystyrene substrate), leading to the detachment of the particle from the seta. Thus, self-cleaning occurs at high pull-off velocities even if the adhesive force of the seta to the particle is initially stronger than that of the particle to the substrate. This dynamic self-cleaning mechanism has an efficiency that is twice as high as static self-cleaning mechanism.

It is well known that before taking each step during animal locomotion, most geckos hyperextend their toes to disengage their feet by peeling the toe pads from distal to proximal directions.[22,25] During digital hyperextension, individual setae are sequentially pulled off the substrates at a relatively high speed, as schematically shown in Figure 7.7d. Each seta undergoes a normal pull-off displacement Δy, as well as a lateral displacement Δx at its root, before its tip completely detaches from a substrate. Using a scrolling model[22] it is estimated that normal velocity $V_n = 10,000\text{–}30,000\,\mu\mathrm{m\,s^{-1}}$, whereas the shear velocity is in the range of $170\text{–}450\,\mu\mathrm{m\,s^{-1}}$ but is minimal compared to normal velocity. The hyperextension-induced normal velocity is high enough to cause the particle-substrate adhesive force (F_{wp}) to overcome the seta-particle adhesive force (F_{sp}), resulting in effective self-cleaning. Thus the pull-off velocity induced by hyperextension (gecko hyperextension velocity) is higher than these critical values V_c, indicating that gecko setae are always in a "self-cleaning regime" during gecko locomotion, as illustrated in Figure 7.7e. During gecko locomotion, active digital hyperextension generates high normal pull-off speed as well as shearing speed in each step, effectively and efficiently dislodging dirt from the toes in a progressive manner, which keeps the gecko's feet sticky yet clean.

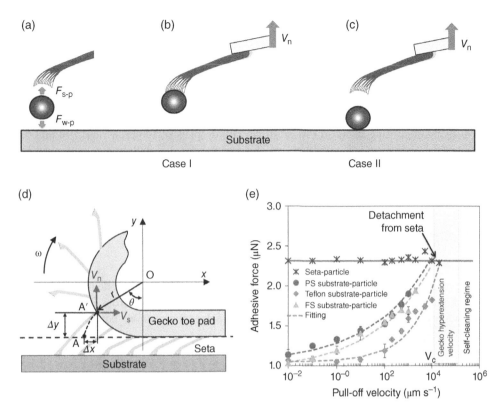

Figure 7.7 (a) Schematic of seta–particle–substrate system and adhesion forces. (b) Particle detaches from seta (case I). (c) Particle attaches from seta (case II). (d) Schematic of gecko toe pad scrolling motion under digital hyperextension, which is modeled as a rolling motion of a circle, with a radius r, along a horizontal plane. The trajectory of a setal root (e.g., from point A to A′) follows the corresponding cycloid curve (e.g., dashed line from A to A′) depending on the specific location of each seta. The hyperextension generates normal velocity (V_n) as well as shear velocity (V_s) on the setal roots. (e) Dynamic adhesion responses between single seta and glass microsphere (F_{sp}) and those between SiO_2 microparticles ($d = 10 \mu m$) and fused silica (FS), polystyrene (PS), and Teflon substrates (F_{pw}) with increasing pull-off velocity (V_n). Source: Xu et al. (2015).[24] Reproduced with permission of Nature Publishing Group.

7.3.3.3 Self-cleaning of Fluid-based Pads of Insects

Insects possess adhesive pads that allow attachment to diverse surfaces. Efficient adhesion must be retained throughout their lifetime. Since the insects live in a dusty world, when pads are exposed to contamination, there must be mechanisms for insects to clean their sticky pads. Many insects groom their adhesive structures, but they may also possess a self-cleaning ability to clean themselves during locomotion, like geckos do. It was found that both smooth pads (stick insects, *Carausius morosus*) and hairy pads (dock beetles, *Gastrophysa viridula*) exhibit self-cleaning.[26] Contaminated pads recovered high levels of adhesion after only eight simulated steps. Self-cleaning was strongly enhanced by shear movements, and only beetle pads showed the ability to self-clean during purely perpendicular pull-offs. The smooth and hairy pads of stick insects and beetles recovered 53.4% and 98.4%, respectively, of lost shear force over eight simulated steps after contamination, therefore hairy pads self-cleaned more efficiently than smooth pads.

   Compared to gecko feet, an important difference is that insects secrete small volumes of fluid into the contact zone. This fluid can increase adhesive forces on rough surfaces by filling in small gaps between asperities due to its hydrophobic nature, which allows good wetting on diverse substrates. The fluid secretion could also protect the pad from wear, "stick-slip" or damage due to high shear stress, and enhance self-cleaning.[27] Insects remove particles from their pads by depositing them on the substrate with droplets of footprint fluid.[26] While the amount of fluid present on the pad shows no effect on the pad susceptibility to contamination, the recovery of adhesive forces after contamination was faster when higher fluid levels were present. The fluid may aid the recovery of adhesive forces by filling in the gaps between contaminating particles. With the gaps filled, the pads may regain contact area faster, thereby increasing adhesion and friction forces, despite the presence of contaminating particles.[27]

7.3.4   Self-cleaning in Soft Film Adhesive Systems

The toe pads of tree frogs not only have strong adhesion but also excellent self-cleaning ability. Figure 7.8 shows the adhesion force change during walking of frogs.[28] Adhesive and friction forces were significantly reduced after initial contamination, but recovered with

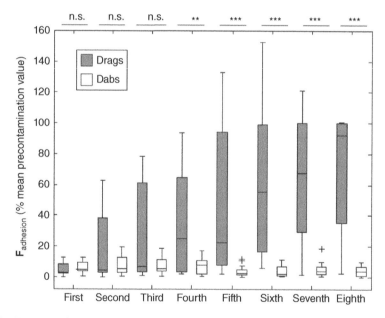

Figure 7.8   Recovery of forces following full contamination. Boxplot comparing adhesive force (normal component of pull-off force) as a percentage of the mean pre-contamination value for both the drag and dab forms of the experiment during recovery from full contamination (2 mN) in single toe pads. The boxes represent consecutive simulated "steps" taken by the toe pad on the force plate following pad contamination. *t* tests were conducted on each pair, with the results for each shown above the plots. n.s., not significant. Source: Crawford *et al.* (2012).[28] Reproduced with permission of the Company of Biologists Ltd.

subsequent measurements. By the fourth step, the frog was able to utilize 91.9% of all original adhesive forces and 98.5% of all original friction forces. In whole-animal for experiments, frog walking significantly increases the recovery of contaminated feet for both adhesion and friction forces. The result is similar to a live gecko that regains adhesion at a much fast rate after contamination when walking. Although exact comparisons cannot be made because of the different sizes and material properties of the contaminants used in different self-cleaning studies,[21,22,26] it does appear that tree frogs self-clean as efficiently as other organisms studied to date.

The significant reduction in adhesion after initial contamination is attributed to a reduction in effective contact area because contaminants reduce the area of close contact between pad and surface. The recovery of the adhesion forces (or self-cleaning) in tree frogs depends on several factors, including mucus and shear movement. Since toe pad secretion is essential for adhesion in both insects and tree frogs, it could play a role in self-cleaning. As discussed in Chapter 5, mucus, secreted at the edge of the toe pads, generates capillarity and viscous forces.[29] As mucus production can be sporadic, recovery of forces will occur faster or slower depending on mucus volumes. In addition, dirt particles move towards the distal end of the pad during pad sliding, and are finally dislodged in the mucus footprint left after each step. In some sense, the mucus acts as a "flushing" effect, aiding the movement of dirt particles to the tip of the pad and their subsequent removal from it with mucus footprint.[28]

Shear movements of the pad over the surface also play an important role in self-cleaning. When the simulated steps included a drag movement, the pre-contaminated adhesive force recovered after about eight steps, but if the pad was simply pressed against the surface (dabs) the force showed little if any recovery. In fact, some slipping of pads on smooth surfaces is a common feature of walking frogs, particularly on vertical surfaces, and slipping will be enhanced when contaminating particles reduce the area of close contact. Frogs will continually reposition their slipping pads on vertical or overhanging surfaces, therefore the drag movement is a behavior used by the frog that contributes to self-cleaning as well as maintaining adhesion to a surface.[28]

7.3.5 Underwater Organisms: Self-cleaning Surfaces

Many natural surfaces of underwater organisms resist biofouling in marine environments. These marine organisms include sharks, mussels, crabs, etc., among which shark skin is a typical model for self-cleaning and low adhesion/drag. Shark skin is composed of embedded placoid scales that have a vascular core of dentine surrounded by an acellular "enamel" layer similar to human teeth. There are ridges and grooves on the dentine surfaces, which are oriented transversely to the direction of flow; the size, number, and spacing of ribs vary slightly between species (Figure 7.9a,b). The natural wavelength of the ridges and grooves is 0.3–0.4 mm, with a trough to crest wave height of about 10 μm.[30] The chemical secretion makes the shark skin highly hydrophilic. Self-cleaning is thus achieved by these topographic features and a mucosal coating secreted by epidermal cells, which prevents sea plants and organisms from adhering to it. In addition, drag force is reduced by the unique topographic features, enabling the shark to swim faster and more efficiently.[30] Overall, the self-cleaning capability can be attributed to the following factors:[28] (1) the ridges and grooves structure accelerates water flow on the shark's surface, consequently reducing the contact time of fouling organisms,

(2) the rough nanotexture of shark skins make it hard for organisms to adhere to it due to the reduced surface area available for adhesion, and (3) as the shark moves through water, the perpetual realignment of the dermal scales creates a moving target, making it more difficult for fouling organisms to adhere to the target in response to changes in internal and external pressure.

The shells of blue mussels are also effective self-cleaning and antifouling surfaces. Like shark skin, the microtopographically structured periostraca on the shells are critical in preventing the shell from fouling. The periostraca surfaces have periodic distributions of grooves and ridges 1–2 μm wide with an average depth of 1.5 μm (Figure 7.9c). To verify the role of these unique topological structures, the blue mussel was exposed to barnacle larvae for a period of time. During a 14-week field exposure trial, it was observed that the shells significantly reduced settlement of barnacle larvae.[32] However, if the test used microtopography replicates cast in epoxy resin from the blue mussel, edible crabs, the egg-case of the lesser-spotted dogfish, and the brittle star, the antifouling ability of the replicas only lasted for 3–4 weeks,[31] therefore natural antifouling must be a combination of chemistry and microtopography.

Fish also have outstanding self-cleaning and antifouling ability in an aquatic environment. The fish body is covered by scales that protect it from plankton and keep it clean even though the sea can be polluted by oil due to shipwrecks. Grass carp, for example, can effectively resist

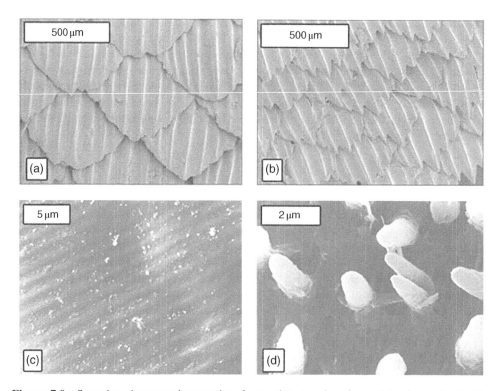

Figure 7.9 Scanning electron micrographs of natural textured surfaces: (a) spinner shark skin, (b) Galapagos shark skin, (c) mussel shell (*M. edulis*) and (d) crab shell (*C. pagurus*). Source: Bers & Wahl (2004).[31] Reproduced with permission of Taylor & Francis.

oil pollution or biofouling underwater. The self-cleaning ability of grass carp stems from the surface hydrophilic mucus and multiscale structures of fish scales Topologically, fish scales have a hierarchical structure consisting of sector-like scales with diameters of 4–5 mm covered by papillae 100–300 μm in length and 30–40 μm wide.[33] Chemically, the fish scales are composed of calcium phosphate, protein, and a thin layer of mucus, which leads to their hydrophilic behavior. Owning to their unique structure, fish scales possess superoleophilicity in air and superoleophobicity in water. They demonstrate self-cleaning capability as well as a low-adhesive superoleophobic interface due to the combination of their hierarchical structures and chemical cues.

7.3.6 Biomimetic Strategies for Self-cleaning

The ability to effortlessly self-clean is very appealing for many engineering applications, ranging from cleaning solar panels to boat hulls. There are many different self-cleaning systems in nature, which provide excellent biological models for the design of bioinspired self-cleaning materials and surfaces. As discussed above, there are a variety of approaches for self-cleaning in different environments. Sometimes, the solutions to the self-cleaning given by nature are completely different but the function is the same: automatic cleaning. The strategies used in nature for self-cleaning are summarized in Table 7.1.

Table 7.1 Overview of self-cleaning strategies used in nature.

Surface characteristics	Self-cleaning strategies	Biological prototypes	Multifunctionality
Superhydrophobic surface	Use hierarchical structures to increase hydrophobicity Action of water droplets	Lotus leaf Rice leaf Cicada wing Butterfly wing	Low adhesion, low drag Anisotropic wetting, low drag Antireflection Directional adhesion, low drag
Superhydrophilic surface	Use surface liquid to increase hydrophilicity Action of water films	Pitcher plant	Low drag (slippery)
	Use photocatalyst to increase hydrophilicity Action of water films	Photosynthesis	Photocatalysis, sterilizing, antifogging
Fibrillar surface	Split contact Static/dynamic mechanics Action of contact	Gecko feet	Dry adhesion High adhesion
Soft films	Secretion of liquid Shear force/surface contact	Frog feet	Dry adhesion High adhesion
Hydrophilic and oleopholic surface	Use topologic surface to change surface fluid dynamics Make surface hydrophilic/oleophobic Secretion of chemical agents	Shark skin Whale skin Mussel Grass crap	Low drag, antifouling Low drag, antifouling Low adhesion, antifouling Low adhesion, antifouling

7.3.6.1 Superhydrophobicity vs. Superhydrophilicity

The most common strategy for self-cleaning is to minimize the surface energy of the materials. One approach to minimize the surface energy for self-cleaning is to form a superhydrophobic surface. The formation of special superhydrophobic surfaces can be achieved by combining surface chemistry and roughness on multiple scales. A self-cleaning ability can also be achieved with superhydrophilicity. Water droplets easily spread out and form a thin film on superhydrophilic surfaces. When rain or a light spray of water flows onto such surfaces, it can infiltrate the space between the substrate and any dust, and wash the dust away. The natural extensions of superhydrophobicity and superhydrophilicity are superoleophobicity and super-oleophilicity. These types of materials are appealing because of their important applications in self-cleaning, corrosion resistance, antifogging, antibacterial applications, smart coatings, and other fields. Moreover, smart surfaces whose wettability can be modulated reversibly between superhydrophobicity and superhydrophilicity or superoleophobicity and superoleophilicity could find more applications.

7.3.6.2 Smooth vs. Hairy Systems

Smooth and hairy adhesive pads occur frequently throughout the animal kingdom. Both smooth and hairy adhesives possess superior self-cleaning ability. For example, when live geckos or frogs walks on surfaces, the adhesive forces of both frog (smooth) and gecko (hairy) adhesive pads can recover from contamination after four steps. However, hairy pads seem to outperform smooth pads in their ability to self-clean without shear movement, and they also recover more rapidly from contamination. The excellent ability of hairy pads to recover from contamination may be an important factor explaining the widespread appearance of hairy pad morphology across different taxa.[26]

7.3.6.3 Land vs. Aquatic Environments

In dry or wetland environments the self-cleaning ability of most plants, insects, and animals seems to rely on the physical properties and topologies of their surfaces. For example, the lotus effect is based on hierarchical micro/nano-structures, and gecko dry self-cleaning uses the splitting principal and dynamics. Thus, the strategies for self-cleaning should be focused on the wetting ability and topological change of bioinspired surfaces. On the other hand, bio-fouling in marine environments is a dynamic process that spans numerous length scales and involves a complex variety of molecules and organisms. Self-cleaning/antifouling strategies, therefore, must include both chemical and physical concepts. Nature provides examples of antifouling and fouling-release surfaces that emphasize the importance of these factors. Physical cues, such as surface roughness and fluid hydrodynamics, can act singularly or in concert with surface chemistry to enhance or inhibit the attachment of organisms to a surface. Chemical cues, especially surface energy, influence not only the ability of an organism to initially attach to a surface, but also the degree of fouling-release from the surface once adhesion has been established. At this point, no single technology has been demonstrated to be universally effective at either antifouling or fouling-release.[30] In marine environments, the above strategies can be used to construct self-cleaning antifouling coatings through the following two approaches:[34] (1) mirror-imaging the lotus effect by designing a surface that repels biological entities and (2) designing a surface that minimizes the water-wetted area when submerged in water (by keeping an air film between the water and the surface).

7.3.6.4 Static vs. Dynamics

In gecko adhesive systems, self-cleaning occurs when small solid particles bind more strongly to the substrate than to the toe pads. Consequently, when pressing a contaminated toe pad against a clean surface multiple times, the particles are removed from the toe pads via a force imbalance. This static or passive self-cleaning shows ~40% force recovery for gecko setal arrays when simulating gecko walking.[21] In contrast, dynamic self-cleaning can still occur under dynamic conditions, that is, the toe pads are pulled off at high pull-off velocities. This dynamic self-cleaning mechanism increases the efficiency of self-cleaning by a factor of 2 compared to the static one.[24] For a live gecko, the force recovering rate reaches nearly 80% of its original stickiness in only four steps.[22] Thus, a dynamic self-cleaning mechanism can significantly enhance the self-cleaning efficiency. In aqueous environments, apart from physical and chemical cues, fluid hydrodynamics is also an important factor in anitfouling surfaces.

7.3.6.5 Single vs. Multifunctionality

Although artificial materials are designed mostly to fulfill a single task or function, most biological materials are multifunctional, for example gecko feet that exhibit strong adhesion and self-cleaning, shark skin that possesses antifouling and drag reduction, and moth eyes show self-cleaning, antireflection, water repellence, and antifogging. Nature has evolved different solutions to achieve efficient multifunctional integration. The fusion of two or more functions into a unique composite is an exciting direction for the fabrication of novel multifunctional self-cleaning surfaces. These multifunctional smart surfaces will be useful in fluid handling and transportation, optical sensing, and medicine, and as self-cleaning and antifouling materials operating in extreme environments. Optimized biological solutions provide biomimetic design principles for the rational design and construction of multifunctional smart surfaces. It is therefore necessary to have interdisciplinary cooperation in biomimetically designing multifunctional smart self-cleaning surfaces and advanced engineering materials.

Even when a single function is focused in the design of biomimetic micro/nanostructures, many other factors should be considered for the applications. For example, although lotus leaf-like superhydrophobic coatings have been successfully fabricated on the small scale in the laboratory, they are not widespread in markets. Apart from self-cleaning, the mechanical durability or scratch resistance of the coatings, optical transparency, and cost are all key points in applications such as windows and glass doors.

7.4 Engineering Self-cleaning Materials and Processes via Bioinspiration

A variety of bioinspired self-cleaning materials and surfaces have been fabricated by different synthesis methods. According to the mechanisms of self-cleaning, these synthesis approaches can be divided into four categories: (1) lotus effect self-cleaning (superhydrophobicity) in which dirt particles on low-adhesive superhydrophobic surfaces are removed by spherical water droplets, (2) superhydrophilic surfaces, including pitcher plant-inspired slippery surfaces, on which debris slips off the surface due to the presence of water thin film, and TiO_2-based self-cleaning arising from photocatalysis and photo-induced superhydrophilicity, (3) gecko setae-inspired, dry self-cleaning in which particulate contamination is cleaned solely by contact with a dry surface without using water droplets or other liquids, and (4) underwater organisms (shark skin, pilot whale skin, carp scale, etc.) inspired, antifouling self-cleaning in which underwater biofouling organisms cannot settle and grow on bioinspired antifouling surfaces with special chemical and physical structures.

7.4.1 Lotus Effect–inspired Self-cleaning Surfaces and Fabrication

The hierarchical surface of the lotus leaf provides a perfect model for the construction of biomimetic self-cleaning surfaces. These geometric scale effects contribute to the superior self-cleaning performance of nanostructured surfaces. Inspired by lotus leaves, many different process approaches have been developed to fabricate superhydrophobic self-cleaning surfaces. There are two main synthetic methods: morphology templating and biomimicking. The templating method can achieve near-perfect mimicry of the lotus leaf-like morphology, whereas other physical and/or chemical methods partially mimic the surface structure but with more freedom in using different synthetic materials and processing.

7.4.1.1 Morphology Templating of the Lotus Leaves

Templating is an easy and versatile method for biomimetics, especially for mimicking surface morphologies. In the typical templation method, a bioprototype is directly used as template to prepare a negative replica by molding, and this replica is then pressed onto a soft polymer surface to copy the desired features, particularly the morphological characteristics. This method is also known as soft lithographic imprinting and needs neither sophisticated instruments nor complex chemical reactions. The templating method can be used to transfer a desired morphology to a range of polymer surfaces at a large scale via roller-printing. This method has been used to demonstrate that the lotus effect can be duplicated by copying lotus leaf surface morphology to achieve similar superhydrophobic wetting properties.

A lotus leaf-like surface has been fabricated by using the lotus leaf itself as a biological template. The fabrication method include liquid poly(dimethylsiloxane) (PDMS) casting, polyether (PE) replication, etc.[35] After removing the stamp (replica from lotus leaf), micro-structured morphology was obtained on the substrate. This lotus leaf-like surface showed strong superhydrophobicity. Figure 7.10 shows an example for templating hierarchical lotus leaves that were recreated to characterize the influence of hierarchical roughness on superhy-drophobicity and adhesion.[15] In the fabrication of the hierarchical structures, negative replicas of lotus leaf surface structures were fabricated by a fast and precise molding of the lotus leaf microstructure. The natural lotus wax was deposited on the negative replica by thermal evaporation to create the wax tubules nanostructures. The tubules were formed by exposure of the specimens to a solvent vapor phase at a selected temperature. The artificial lotus leaf with hierarchical structure showed improved wetting properties, having a static contact angle of 171°, contact angle hysteresis (2°) and tilt angles of 1–2° as compared to natural lotus leaves.

(a)

(b)

(c)

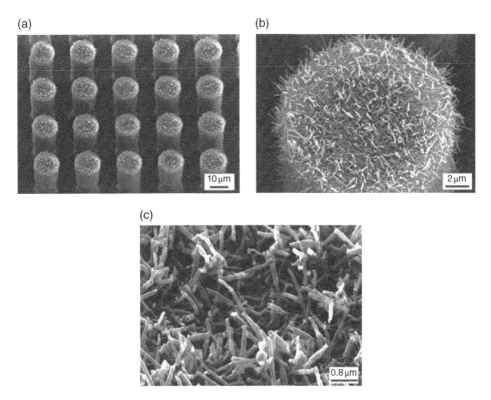

Figure 7.10 SEM micrographs taken at 45° tilt angle shown using three magnifications of hierarchical structure using lotus and micropatterned Si replicas: (a) low, (b) middle, and (c) high. Nano- and hierarchical structures were fabricated with lotus wax after storage for 7 days at 50°C with ethanol vapor. Source: Koch *et al.* (2009).[15] Reproduced with permission of the Royal Society of Chemistry.

This can be attributed to the wax tubule lengths, which are 0.5–1 µm longer in the artificial lotus leaf (Figure 7.10c). The hierarchical structure leads to more effective air trapping and consequently less surface contact area with water droplets and smaller contact angle hysteresis and adhesion forces.[15]

7.4.1.2 Morphology Mimicking of Lotus Leaves

Artificial biomimetic superhydrophobic surfaces that mimic the hierarchical structure of lotus leaves are fabricated by various processes, including atom transfer radical polymerization (ATRP), chemical etching, chemical vapor deposition, electrospinning, a hydrothermal approach, spin coating, etc. The materials used in the fabrication include polymers, metals, and ceramics. Superhydrophobic self-cleaning cellulose surfaces with hierarchical structures have been fabricated through the combination of ATRP of glycidyl methacrylate followed by pentadecafluorooctanoyl modification.[36] After contamination of the surface with carbon black powder, water droplets adsorbed the dirt and rolled off the surface. Superamphiphobic (both superhydrophobic and superoleophobic) silicon surfaces on a large area (4-inch wafer) were fabricated through a wet chemical etching method followed by a fluorinated self-assembled monolayer modification.[37] This superamphiphobic surface can remove sticky dirt by glycerin droplets, and more interestingly could be turned superamphiphilic by simply exposing it to UV light. A superamphiphobic Ca-Li-based bulk metallic glass surface was also constructed by etching and then modified with a fluoroalkylsilane coating. Pyramidal hierarchical structures were also prepared on wafer-scale silicon surfaces by employing potassium hydroxide etching and silver catalytic etching on a crystalline silicon wafer. The surfaces show both self-cleaning and antireflective characteristics due to the formation of the hierarchical structures. Furthermore, flexible superhydrophobic surfaces can be fabricated by imprint lithography using the obtained hierarchical structures as molds. In addition to self-cleaning properties, the surface is mechanically durable and maintains superhydrophobicity even after abrasion.[2]

In the fabrication of artificial hierarchical lotus leaves, nanofibers/particles are frequently used to construct the structures at the nanoscale while substrates are processed to form the micro-level structures. Multifunction could be easily introduced into the superhydrophobic film by using functionalized nanofiber/particles. Mimicking lotus leaves, superhydrophobic polystyrene film was fabricated by combining porous microspheres and nanofibers. The porous microspheres contributed to the superhydrophobicity by increasing surface roughness, while nanofibers interwove to form a three-dimensional network that reinforced the composite film. In another example of using nanofibers, a hierarchical biomimetic (lotus leaf-like) surface was fabricated from the combination of polystyrene (PS) and carbon nanotubes (CNTs), where the CNTs were decorated on a monolayer PS colloidal crystal by a wet chemical self-assembly method.[38] Dense single-walled carbon nanotubes (SWCNTs) were adsorbed like an interlaced "net" structure on the hexagonally close-packed PS microspheres and the sphere joints (Figure 7.11a). This hierarchical surface exhibited strong superhydrophobic behavior after surface chemical modification. This lotus leaf-mimicked hierarchical surface has unique electrical and electrochemical properties and a large specific area, and can be used as a gas sensor with good selectivity and great sensitivity.

Superhydrophobic self-cleaning has also been applied to solar cells.[39] Mimicking the hierarchical structure of the lotus leaf, a nanostructured, transparent, antireflective coating with superhydrophobic self-cleaning properties was formed on organic solar cells. The multifunctional surface resulted in improved photovoltaic power conversion efficiency because of the reflection suppression and transmittance enhancement. Similar microshell structures with ordered microshell arrays were fabricated on a transparent and flexible PDMS elastomer surface (Figure 7.11b). The microshell PDMS exhibited excellent superhydrophobic and water-repellent properties without chemical modification. The self-cleaning PDMS arrays can effectively reduce degradation of solar cell efficiency by airborne dust. In addition to the self-cleaning characteristic, nanodome solar cell devices also exhibit effective antireflection and light trapping over a broad spectral range. This film is transparent over the entire UV visible near-infrared range and is applicable to surfaces other than glass. The reflector coatings are expected to provide as much as a 90% reduction in mirror cleaning and maintenance costs, and about a 20% improvement in the average amount of reflected solar energy.[39]

Moths' eyes have multiscale ommatidial arrays offering multifunctional properties, including antireflectivity and self-cleaning, the microstructure of which will be described in detail in Chapter 9. Inspired by the structure of moth's eye, multiscale ommatidial arrays over large areas have been fabricated by a distinct approach called sacrificial layer mediated nanoimprinting, which involves nanoimprinting aided by a sacrificial layer.[40] The nanoimprinting technique is scalable and has a high-throughput in the fabrication

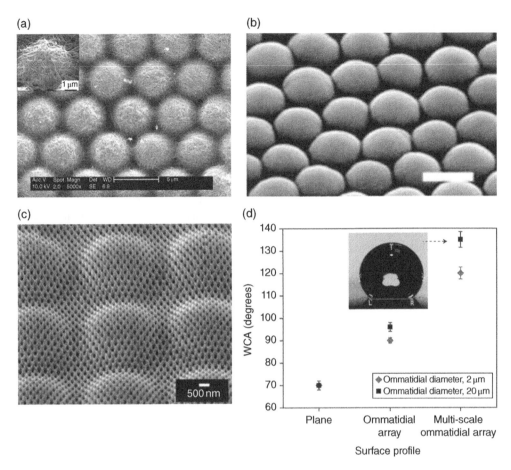

Figure 7.11 (a) Surface morphology of the bionic surface prepared with PS microsphere/SWCNTs composition arrays, inset shows the higher magnification image of SWCNTs decorated microsphere. Source: Li *et al.* (2007).[38] (b) a-Si:H nanodome solar cells after deposition of multilayers of materials on nanocones. Scale bar 500 nm. Source: Zhu *et al.* (2010).[39] (c) SEM images of the multiscale ommatidial arrays and (d) water contact angle (WCA) on surfaces with progressively multiscale texturing. Source: Raut *et al.* (2015).[40] Reproduced with permission of the American Chemical Society.

of multiscale ommatidial arrays. The process produced biomimetic ommatidial arrays with excellent pattern uniformity over the entire patterned area (Figure 7.11c). This biomimetic coating can achieve a broadband suppression of reflectance to a minimum of ~1.4% and omnidirectional antireflection for highly oblique angles of incidence up to 70°. In addition, the multiscale ommatidial arrays show the highest increase in water contact angle (Figure 7.11d). The apparent advancing and receding angles for the multiscale ommatidial arrays are 152.0° and 150.8°, respectively, indicating a low contact angle hysteresis. Because of their superhydrophobicity and antifogging characteristics, the devices could work in wet and humid outdoor environments. These properties could potentially enhance the performance of optoelectronic devices and minimize the influence of in-service conditions.

7.4.2   Superhydrophilically-based Self-cleaning Surfaces and Fabrication

7.4.2.1   Pitcher Plant-based Self-cleaning Surfaces

Superhydrophilicity can lead to the formation water film flowing on a surface, which can penetrate into the space between the substrate and any dust on the substrate, weakening the contact and washing the dust away. This type of synthetic surface is inspired by the pitcher plant slippery surface. As described in Chapter 5 (Figure 5.16), this slippery liquid-infused porous surface(s) (SLIPS) exhibits self-cleaning capability.[41] Briefly, SLIPS consists of a film of a lubricating liquid locked in place by a micro/nanoporous substrate. The materials were fabricated based on the following criteria:[41] (1) the lubricating liquid should wick into and stably adhere to the porous substrate, (2) the porous substrate should preferentially be wetted by the lubricating liquid rather than by the liquid one wants to repel, and (3) the lubricating and impinging test liquids must be immiscible. Since SLIPS were made using the concept of the slip surface, the surface is not necessarily superhydrophilic. Compared with superhydrophobic surfaces, the bioinspired pitcher plant surface exhibits excellent self-cleaning characteristics with a self-healing capability. Moreover, SLIPS also repel various liquids and ice, stabilize pressure, and enhance optical transparency, and the surface can serve as an antisticking, slippery surface, as seen in nature.

7.4.2.2   Photocatalyst-based Superhydrophilic Self-cleaning Surfaces

The superhydrophilicity of a surface can also be introduced by photocatalysis. Among the wide variety of superhydrophilic materials, titania (TiO_2) is one of the most promising for fabricating hydrophilic-induced self-cleaning surfaces. The self-cleaning capacity of TiO_2 stems from its excellent photocatalytic and photo-induced superhydrophilic properties. The band gap of bulk anatase TiO_2 is 3.2 eV, corresponding to wavelength 390 nm (near-UV light). When a TiO_2 film is irradiated under sunlight, excited charge carriers (i.e., electrons and holes) are generated on the surface of the film to catalyze water-splitting reactions. In addition to its photocatalytic property, TiO_2 induces superhydrophilicity upon light radiation. The combination of these features on TiO_2 surfaces results in the desired self-cleaning function, which endows TiO_2 with a wide variety of practical applications.

Different synthesis approaches have been used to fabricate TiO_2-based superhydrophilic self-cleaning surfaces on different substrates. A hierarchical amorphous TiO_2 nanocolumn array was fabricated with polystyrene colloidal monolayers as templates by pulsed laser deposition (Figure 7.12).[42] The hexagonal close-packed nanocolumn array showed superamphiphilicity with both water and oil contact angles of 0° without UV irradiation. Interestingly, amorphous TiO_2 arrays exhibited superior photocatalytic performance compared to anatase nanocolumn arrays due to large surface area and special microstructures. This photocatalytic activity, together with superamphiphilicity, led to the enhanced self-cleaning property. In addition to TiO_2, similar nanocolumn arrays of other materials, including SnO_2, Fe_2O_3, and carbon, can also be fabricated using this approach and then transferred to almost any substrate.

Self-cleaning technology can also be applied to textiles. Following the same idea as above, various self-cleaning fibers were made by introducing TiO_2 nanoparticles into fibers, for example cotton fabric surfaces were modified with TiO_2 nanosols via the sol-gel approach.

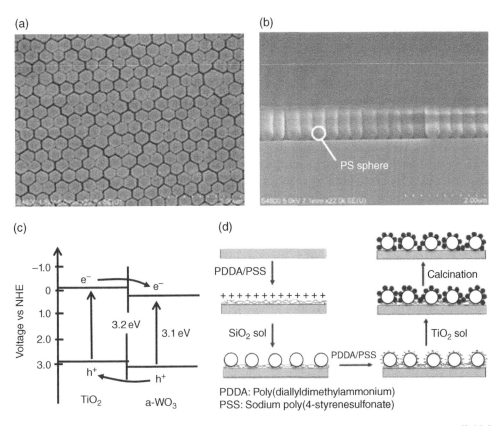

Figure 7.12   SEM images of a sample obtained by pulsed laser deposition using a polystyrene colloidal monolayer as the substrate. (a and b) Low-magnification images observed from the top and side. Source: Li *et al.* (2008).[42] Reproduced with permission of the American Chemical Society. (c) Energy diagram for the heterogeneous TiO$_2$/a-WO$_3$ film. Source: Miyauchi *et al.* (2000).[45] Reproduced with permission of John Wiley and Sons. (d) Flow chart for the preparation of TiO$_2$–SiO$_2$ self-cleaning coatings with antireflection properties. PDDA, poly(diallyldimethylammonium); PSS, sodium poly(4-styrenesulfonate). Source: Zhang *et al.* (2005).[46] Reproduced with permission of the American Chemical Society.

TiO$_2$-nanosol-treated cotton can effectively decompose coffee and red wine stains on it under UV radiation.[43] In addition to self-cleaning, TiO$_2$-nanosol treatment imparted excellent UV-radiation protection to the cotton fabric. Photocatalytic self-cleaning textile nanofibers were also coaxial electrospinning, with high surface-to-volume ratio. When cellulose acetate and TiO$_2$ were used as the core phase and sheath phase, respectively, the photocatalytic fibers obtained maintained their self-cleaning properties after repeated staining and washing.[44]

TiO$_2$ photocatalysts are wide-band semiconductors and mainly absorb UV photons. However, solar light contains only a small amount of UV photons (approximately 5%), therefore a variety of process approaches have been proposed to enhance the sensitivity of the catalyst or extend the spectral response of TiO$_2$ from UV light to visible light. One approach is to use heterogeneous systems to enhance the photocatalysis. The deposition of a thin film of WO$_3$ on the TiO$_2$ surface, for example, increases the sensitivity of the photo-induced superhydrophilicity.[45] Because of the heterogeneous TiO_2/WO_3 system, a photo-induced superhydrophilic conversion was achieved at a very weak UV light intensity of $1\,\mu W\,cm^{-2}$. Porous $Au/TiO_2/SiO_2$ nanocomposites were fabricated to further improve the visible-light self-cleaning performance.[47] When using $Au/TiO_2/SiO_2$ to coat cotton fabrics, their appearance does not change. These fabrics self-cleaned red wine stains and coffee stains after visible-light irradiation for 20 hours. This excellent self-cleaning performance can be attributed to the presence of Au species that improve the rate of electron transfer to oxygen and then decrease the rate of recombination between excited electron/hole pairs. Silica also increases the photocatalytic activity of TiO_2 because of its high dispersion and the structural effects of the amorphous silica. In addition to the superior photocatalytic and self-cleaning properties compared with pure TiO_2, TiO_2–SiO_2 self-cleaning coatings are also antireflective and have been applied to the cover glass of solar cells.[46] Although conventional TiO_2 coatings are highly reflective due to the large refractive index of TiO_2, SiO_2 combines with TiO_2 nanoparticles to produce coatings that exhibit transmission of 97% visible light.

TiO_2-based self-cleaning surfaces have been mainly applied to exterior construction materials, such as tiles, glass, plastic films, and tent materials, since these materials could be exposed to abundant sunlight and natural rainfall. TiO_2-based coatings with dual cleaning and wetting behavior are already commercially available. The photocatalytic, superhydrophilic Pilkington Activ™, for example, is now offered in North America and Europe as a standard option for windows.[48] Saint-Gobain has also launched a self-cleaning glass with similar dual-action functionality (SGG Bioclean). The transparent coating on Pilkington Activ is just 40 nm thick, and is applied to the glass by a process of chemical vapor deposition during the manufacturing process at temperatures of 600°C. This ensures that the self-cleaning functionality will last the lifetime of the glass.[48] Other more cost-effective and easily adapted manufacturing techniques, such as using modified sol-gel methods to apply the coating, are being developed to apply clear, hard coatings to glass, plastics, painted surfaces, metals, and ceramics.

7.4.3   Gecko-inspired Self-cleaning Dry Adhesives and Fabrication

Inspired by gecko feet, a variety of synthesis strategies have been applied to construct gecko-mimetic adhesives using a combination of high adhesion and dry self-cleaning. The self-cleaning is important because it is a prerequisite for a reusable tape, climbing robots, etc. The first gecko-like self-cleaning adhesive was fabricated with high-aspect-ratio polypropylene microfibers.[49] In contrast to the conventional pressure-sensitive adhesive (PSA), the contaminated gecko-like stiff polypropylene fibrillar adhesive recovered 25–33% of the original shear adhesion force after multiple contacts (approximately 30 simulated steps) with a dry clean surface, demonstrating the dry self-cleaning property with microspheres (radius $\leq 2.5\,\mu m$). However, the synthetic microfibrillar adhesives could not self-clean larger particles, which is consistent with the Johnson–Kendall–Roberts (JKR) pull-off force model.

The above adhesives consist of one-level polypropylene fibers; other gecko-inspired adhesives with more complex tips were fabricated and tested for adhesion and self-cleaning. Polyurethane elastomer microfibres with mushroom-shaped tips were fabricated using a lithographic technique combined with post-processing that involved dipping and soft mold casting.[50] Contamination of the samples was achieved by bringing each sample into contact with a monolayer of glass spheres until a predefined compressive load was reached. The sample was then retracted at the same speed as the loading speed. Under dry conditions, a significant adhesive recovery greater than that of previously reported synthetic fibrillar adhesive both in attachment strength (140 kPa) and the percentage of recovered adhesion (up to 100%) on a smooth substrate was observed. These observations show that synthetic gecko adhesives could recover from adhesion loss on contamination at a rate comparable to that of the gecko. The mechanics analysis on contact self-cleaning concluded that embedding contaminants between adjacent fibers or lamellae-inspired grooves, by particle rolling, is an important mechanism of contact self-cleaning for elastomeric gecko adhesives. Furthermore, it was observed that the relative size of the contaminants to the size of the microfiber tips in an array of synthetic adhesives strongly determined how and to what degree the adhesive could be cleaned.

In addition to dry self-cleaning, gecko-like adhesives also presented superhydrophobicity and wet self-cleaning with water droplets. Carbon nanotube-based adhesives were synthesized and their superhydrophobicity, self-cleaning ability, and shear resistance were optimized for length and pattern size.[51] Similar to the procedure used for gecko feet and synthetic mushroom adhesives described above, the synthetic tapes could be cleaned by contact mechanics. After several contacts with mica (or glass substrate), the majority of dirt particles (silica particles ranging from 1 to 100 μm in size) on the synthetic tapes were cleaned. Only a few small silica particles remained stuck to the surface and in between the pillars of the carbon nanotube. Beyond their dry self-cleaning ability, the synthetic carbon nanotube-based gecko tapes could also be cleaned by water, mimicking the leaves of lotus and lady's mantle plants in nature. When rinsed with water, the water droplets rolled off very easily and picked up dirt particles along the way. After mechanical cleaning, the shear strength recovered back to 90% (60% for water-cleaned samples) of the values measured before soiling, which is superior to such strength recovery in geckos. Similarly, the mushroom-shaped microfiber arrays exhibited both wet self-cleaning and dry adhesion. Almost 100% of silica particles (5–50 μm in diameter) on the elastomer arrays were cleaned, similar to the lotus effect found in nature.[52] This remarkable wet self-cleaning ability could be attributed to the mushroom-shaped tip geometry of the fiber array.[53]

Inspired by the unique spatular structure and its dynamic effect, artificial setae using synthetic polyester microfibers (10 μm in diameter) were fabricated.[21] Each fiber (150 μm in length) is cut to form a micro pad at its tip, and then bonded to an AFM cantilever (Figure 7.13a), followed by a layer-by-layer gluing of three layers of wrinkled graphene multilayers (~5 nm thick) on the micro pad (Figure 7.13b,c). The use of graphene layers can drastically enhance the adhesion capability of the artificial setae, as graphene has been previously reported to generate ultrastrong adhesion over various surfaces. The wrinkled graphene layers on a pad, mimicking gecko spatulae on a seta, could also increase the surface compliance and contact area to further enhance the adhesion. Unlike traditional adhesives (e.g., Scotch tape), the graphene layer can generate reversible/tunable adhesion, which is critical in a variety of applications, including self-cleaning and small object manipulation in air and under water. Using the artificial seta attached to an AFM probe as a micromanipulator, the researchers successfully manipulated microspheres on various substrates and were able to pick up, translate, and precisely assemble the micro-sized particles (d = 1–20 μm) in a patterned fashion (Figure 7.13d). The manipulation processes were demonstrated (pick-up at low pull-off speed, and drop-off at high pull-off speed). In addition to microspheres, the micromanipulator can also manipulate other types of small objects, such as debris and microfiber segments. Figure 7.13e shows the

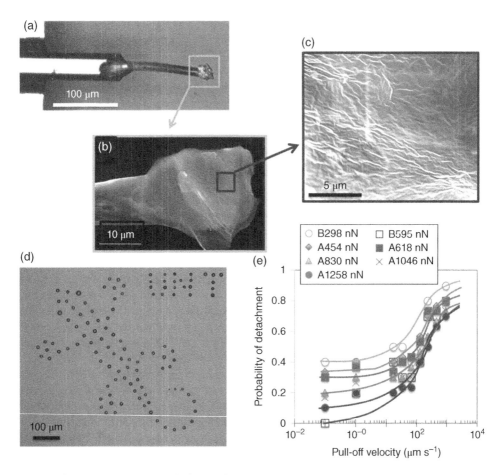

Figure 7.13 Micromanipulation of microparticles by artificial setae. (a) Optical image of the biomimetic micromanipulator made of a polyester microfiber ($d=10\,\mu m$, $L=150\,\mu m$) with a graphene layer-decorated micro-pad at its tip, glued on an AFM tipless cantilever (front view). (b) SEM image of the micro-pad with three layers of (c) wrinkled graphene (thickness ~5 nm) on the top. (d) Optical image of a gecko pattern and a logo precisely assembled with SiO_2 microparticles ($d=1-25\,\mu m$) on a glass slide by the biomimetic micromanipulator. (e) Probability of particle detachment as a function of pull-off velocity, measured with 30 trials, mimicking gecko walk (shear velocity $V_s=4\,\mu m\,s^{-1}$, drag distance $s=2\,\mu m$, and contact time $t_c=0$). A and B in the legend refer to two micromanipulators A and B, and the numbers are values of preloads. Scale bar in a–d: 100 μm, 10 μm, 5 μm, and 100 μm, respectively. Source: Xu *et al.* (2015).[24] Reproduced with permission of Nature Publishing Group.

probability of microsphere detachment events in a 30-trial experiment. Similar to the natural setae, the probability of particle detachment is within a range of 0–40% at low pull-off velocity, but rapidly ramps up and reaches ~80% at a normal pull-off velocity of ~1000 μm s^{-1}.

It is desirable that the micromanipulator can reliably and repetitively pick up and drop off particles or other small objects. The detachment probability is almost zero at a low speed of $V_n<1\,\mu m\,s^{-1}$ and a relatively high preload of 1.3 μN, suggesting that the micromanipulator can pick up the particles with nearly 100% success. In contrast, at a high pull-off speed

($V_n \sim 1000\,\mu m\,s^{-1}$) and low preload ($\sim 0.4\,\mu N$), the probability is close to 1, indicating that the micromanipulator can reliably drop off particles. Apart from the pull-off velocity and preload force, the shear velocity also affects the probability. If the pads are damaged, the micromanipulator can be repaired by gluing a new layer of graphene on the pads, indicating good reusability. These results demonstrate that the self-cleaning and micromanipulation capabilities based on the distinctive dynamic effect of gecko spatular nano pads are robust and efficient in synthetic bioinspired adhesives.

7.4.4 Underwater Organisms–inspired Self-cleaning Surfaces and Fabrication

Biofouling is a widespread and serious problem for both aquatic organisms and artificial underwater structures. In marine environments, fouling organisms adsorb and subsequently attach and grow on submerged surfaces. These adhering biomaterials significantly increase drag force on commercial vessels. As a result, biofouling results in an increase in fuel consumption and functional and monetary costs for cleaning, and loss of hull strength due to biocorrosion. Current antifouling agents (such as tributyltin and lead) usually have a negative impact on the environment, therefore it is necessary to design and manufacture universal, environmentally friendly coatings with both antifouling and fouling-release properties. In fact, nature exhibits various antibiofouling mechanisms to control fouling. Underwater organs, for example, some shark skin, whale skin, carp scales, mollusks, gorgonian corals, and others, possess natural antifouling defenses, which can inspire the design of artificial antifouling and fouling-release materials for various applications.

Shark skin is a typical biomimetic model for the design of antifouling surfaces with low adhesion/drag. Shark skin contains very small individual tooth-like scales called dermal denticles, which are covered by specially sized and spaced riblets oriented parallel to the swimming direction. The unique skin topological structures and a mucosal coating secreted by epidermal cells not only prevent sea plants and organisms from adhering to it (self-cleaning), but also enable sharks to reduce the drag force during swimming. Inspired by shark skin, a variety of antifouling self-cleaning surfaces have been developed using various methods,[30] for example poly(dimethylsiloxane) elastomer was used to fabricate artificial non-toxic shark skin. In the fabrication, the dimensions are reduced and the tips of the ribs are flattened, but the basic pattern of the placoids is maintained. Nevertheless, the artificial shark skin was found to effectively inhibit the settlement of fouling organisms. This bioinspired surface topography, Sharklet AF™ (Figure 7.14), reduced the attachment of zoospores of *Ulva* by 86% compared with smooth elastomer.[54]

Pilot whale skin also is clean and free of all fouling organisms due to the existence of nanoridge pores and zymogel on the skin surface (Figure 7.15a).[55] Whale skin is composed of tiny pores (0.1 µm) enclosed by raised nanoridges. Between the ridges there is a rubber-like gel containing enzymes that denature proteins and carbohydrates. The gel, which oozes out of the gaps between skin cells, is replenished as the whale sheds its skin. This structure reduces the surface area available for organisms such as bacteria and diatoms to adhere to the skin. If these organisms are successful in hanging onto the gel the enzymes will attack them. Thus, the pilot whale skin surface and the intercellular gel containing both polar and non-polar functional groups contribute to short- and long-term fouling reduction.[56] In addition, high shear water flow and the liquid–vapor interfaces of air-bubbles during jumping give the organisms little

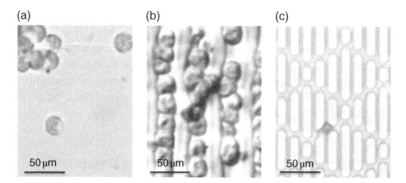

Figure 7.14   (a) Smooth surface 5 mm wide, 5 mm spaced, and 5 mm high, (b) channels, and (c) 4 mm high Sharklet AFTM in PDMSe. Images were taken via light microscopy. Scale bars.25 mm. Source: Carman *et al.* (2006).[54] Reproduced with permission of Taylor & Francis.

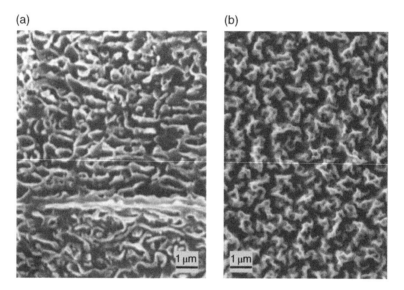

Figure 7.15   (a) SEM image of self-cleaning pilot whale skin. The pores enclosed by the nanoridges are displayed very clearly. (b) SEM image of pilot whale-inspired self-cleaning structures fabricated by polyelectrolyte self-assembly. Source: Cao *et al.* (2010).[57] Reproduced with permission of John Wiley and Sons.

time for adhesion. This special skin surface property, combined with the speed of movement, surfacing, and jumping, may contribute to removing weakly adhered epibionts. The self-cleaning and antifouling mechanism of pilot whale skin provides interesting biomimetic approaches for the design and fabrication of artificial whale skin. The whale skin structure was duplicated by templating methods. The complex skin morphology was replicated through layer-by-layer spray-coating deposition of poly(acrylic acid) and polyethylenimine polyelectrolytes (Figure 7.15b).[57] The resulting surfaces were modified with poly(ethylene glycol) and trideca-fluoroctyltriethoxysilane to obtain different surface morphologies. For the same materials, the morphology that most closely resembled the size and shape of features on the skin of *G. melas* resulted in the lowest fouling retention as measured by settlement of spores of *Ulva* regardless of surface chemistry,[57] therefore mimicking the hierarchical surfaces of whale shin should be able to create highly effective surfaces that encounter the different preferences of the target organisms.

Fish such as grass carp can freely swim and effectively resist oil pollution or biofouling underwater, demonstrating self-cleaning. The fish body is well protected from plankton and kept clean by multifunctional fish scales. Inspired by self-cleaning fish scales, the artificial bioinspired fish scale surface has been fabricated on a silicon wafer with a micro/nanostructure by lithography and chemical etching.[33] The resulting fish scale-mimicking surface exhibits extreme superoleophobicity with oil in water. The fish scale replicas have an extremely low oil contact angle in water ($163 \pm 2°$), resulting in low adhesion to the oil. To make the fish-inspired self-cleaning surfaces more robust, the hierarchical surfaces were constructed with a macromolecule–nanoclay hydrogel.[58] Because of its high compressive and tensile strength, the superoleophobicity and mechanical strength of the surface were simultaneously achieved with the interaction of flexible macromolecules and rigid nanoclays.

References

1. Bixler, G.D. & Bhushan, B. Rice and butterfly wing effect inspired low drag and antifouling surfaces: A review. *Critical Reviews in Solid State and Materials Sciences* **40**, 1–37 (2015).
2. Liu, K. & Jiang, L. Bio-inspired self-cleaning surfaces. *Annual Review of Materials Research* **42**, 231–263 (2012).
3. Darmanin, T. & Guittard, F. Recent advances in the potential applications of bioinspired superhydrophobic materials. *Journal of Materials Chemistry A* **2**, 16319–16359 (2014).
4. Young, T. An essay on the cohesion of fluids. *Philosophical Transactions of the Royal Society of London* **95**, 65–875 (1805).
5. Wenzel, R.N. Resistance of solid surfaces to wetting by water. *Industrial & Engineering Chemistry Research* **28**, 988–945 (1936).
6. Cassie, A.B.D. & Baxter, S. Wettability of porous surfaces. *Transactions of the Faraday Society* **40**, 546–551 (1944).
7. Lafuma, A. & Quere, D. Superhydrophobic states. *Nature Materials* **2**, 457–460 (2003).
8. Koishi, T., Yasuoka, K., Fujikawa, S., Ebisuzaki, T. & Zeng, X.C. Coexistence and transition between Cassie and Wenzel state on pillared hydrophobic surface. *Proceedings of the National Academy of Sciences of the United States of America* **106**, 8435–8440 (2009).
9. Cao, L.L., Hu, H.H. & Gao, D. Design and fabrication of micro-textures for inducing a superhydrophobic behavior on hydrophilic materials. *Langmuir* **23**, 4310–4314 (2007).
10. Extrand, C.W. & Kumagai, Y. Liquid-drops on an inclined plane – the relation between contact angles, drop shape, and retentive force. *Journal of Colloid and Interface Science* **170**, 515–521 (1995).
11. Yoshimitsu, Z., Nakajima, A., Watanabe, T. & Hashimoto, K. Effects of surface structure on the hydrophobicity and sliding behavior of water droplets. *Langmuir* **18**, 5818–5822 (2002).
12. Youngblood, J.P., Sottos, N.R. & Extrand, C. Bioinspired materials or self-cleaning and self-healing. *Materials Bulletin* **33**, 732–741 (2008).
13. Liu, K., Tian, Y. & Jiang, L. Bio-inspired superoleophobic and smart materials: Design, fabrication, and application. *Progress in Materials Science* **58**, 503–564 (2013).
14. Nishimoto, S. & Bhushan, B. Bioinspired self-cleaning surfaces with superhydrophobicity, superoleophobicity, and superhydrophilicity. *RSC Advances* **3**, 671–690 (2013).
15. Koch, K., Bhushan, B., Jung, Y.C. & Barthlott, W. Fabrication of artificial Lotus leaves and significance of hierarchical structure for superhydrophobicity and low adhesion. *Soft Matter* **5**, 1386–1393 (2009).

16. Samaha, M. & Gad-el-Hak, M. Polymeric slippery coatings: nature and applications. *Polymers* **6**, 1266–1311 (2014).

17. Neinhuis, C. & Barthlott, W. Characterization and distribution of water-repellent, self-cleaning plant surfaces. *Annals of Botany – London* **79**, 667–677 (1997).

18. Koch, K. & Barthlott, W. Superhydrophobic and superhydrophilic plant surfaces: an inspiration for biomimetic materials. *Philosophical Transactions Series A: Mathematical, Physical, and Engineering Sciences* **367**, 1487–1509 (2009).

19. Kudo, A. & Miseki, Y. Heterogeneous photocatalyst materials for water splitting. *Chemical Society Reviews* **38**, 253–278 (2009).

20. Federle, W., Riehle, M., Curtis, A.S.G. & Full, R.J. An integrative study of insect adhesion: Mechanics and wet adhesion of pretarsal pads in ants. *Integrative and Comparative Biology* **42**, 1100–1106 (2002).

21. Hansen, W.R. & Autumn, K. Evidence for self-cleaning in gecko setae. *Proceedings of the National Academy of Sciences of the United States of America* **102**, 385–389 (2005).

22. Hu, S.H., Lopez, S., Niewiarowski, P.H. & Xia, Z.H. Dynamic self-cleaning in gecko setae via digital hyperextension. *Journal of the Royal Society Interface* **9**, 2781–2790 (2012).

23. Autumn, K. & Hansen, W.R. Gecko setae: A self-cleaning adhesive nanostructure. *Integrative and Comparative Biology* **43**, 828–828 (2003).

24. Xu, Q., Wan, Y., Hu, T.S., Liu, T.X., Tao, D. *et al.* Robust self-cleaning and micromanipulation capabilities of gecko spatulae and their bio-mimics. *Nature Communications* **6**, 8949 (2015).

25. Russell, A.P. Integrative functional morphology of the gekkotan adhesive system (Reptilia: Gekkota). *Integrative and Comparative Biology* **42**, 1154–1163 (2002).

26. Clemente, C.J., Bullock, J.M., Beale, A. & Federle, W. Evidence for self-cleaning in fluid-based smooth and hairy adhesive systems of insects. *Journal of Experimental Biology* **213**, 635–642 (2010).

27. Clemente, C.J. & Federle, W. Mechanisms of self-cleaning in fluid-based smooth adhesive pads of insects. *Bioinspiration & Biomimetics* **7**, 046001 (2012).

28. Crawford, N., Endlein, T. & Barnes, W.J. Self-cleaning in tree frog toe pads; a mechanism for recovering from contamination without the need for grooming. *Journal of Experimental Biology* **215**, 3965–3972 (2012).

29. Federle, W., Barnes, W.J., Baumgartner, W., Drechsler, P. & Smith, J.M. Wet but not slippery: Boundary friction in tree frog adhesive toe pads. *Journal of the Royal Society Interface* **3**, 689–697 (2006).

30. Magin, C.M., Cooper, S.P. & Brennan, A.B. Non-toxic antifouling strategies. *Materials Today* **13**, 36–44 (2010).

31. Bers, A.V. & Wahl, M. The influence of natural surface microtopographies on fouling. *Biofouling* **20**, 43–51 (2004).

32. Scardino, A., De Nys, R., Ison, O., O'Connor, W. & Steinberg, P. Microtopography and antifouling properties of the shell surface of the bivalve molluscs *Mytilus galloprovincialis* and *Pinctada imbricata. Biofouling* **19**, 221–230 (2003).

33. Liu, M.J., Wang, S.T., Wei, Z.X., Song, Y.L. & Jiang, L. Bioinspired design of a superoleophobic and low adhesive water/solid interface. *Advanced Materials* **21**, 665–669 (2009).

34. Marmur, A. Super-hydrophobicity fundamentals: implications to biofouling prevention. *Biofouling* **22**, 107–115 (2006).

35. Latthe, S.S., Terashima, C., Nakata, K. & Fujishima, A. Superhydrophobic surfaces developed by mimicking hierarchical surface morphology of lotus leaf. *Molecules* **19**, 4256–4283 (2014).

36. Nystrom, D., Lindqvist, J., Ostmark, E., Hult, A. & Malmstrom, E. Superhydrophobic bio-fibre surfaces via tailored grafting architecture. *Chemical Communications*, 3594–3596 (2006).

37. Kim, T.-i., Tahk, D. & Lee, H.H. Wettability-controllable super water- and moderately oil-repellent surface fabricated by wet chemical etching. *Langmuir* **25**, 6576–6579 (2009).

38. Li, Y. *et al.* Superhydrophobic bionic surfaces with hierarchical microsphere/SWCNT composite arrays. *Langmuir* **23**, 2169–2174 (2007).

39. Zhu, J., Hsu, C.M., Yu, Z.F., Fan, S.H. & Cui, Y. nanodome solar cells with efficient light management and self-cleaning. *Nano Letters* **10**, 1979–1984 (2010).

40. Raut, H.K. *et al.* Multiscale ommatidial arrays with broadband and omnidirectional antireflection and antifogging properties by sacrificial layer mediated nanoimprinting. *ACS Nano* **9**, 1305–1314 (2015).

41. Wong, T.S. *et al.* Bioinspired self-repairing slippery surfaces with pressure-stable omniphobicity. *Nature* **477**, 443–447 (2011).

42. Li, Y., Sasaki, T., Shimizu, Y. & Koshizaki, N. hexagonal-close-packed, hierarchical amorphous TiO_2 nanocolumn arrays: transferability, enhanced photocatalytic activity, and superamphiphilicity without UV irradiation. *Journal of the American Chemical Society* **130**, 1475514762 (2008).

43. Abidi, N., Cabrales, L. & Hequet, E. Functionalization of a cotton fabric surface with titania nanosols: Applications for self-cleaning and UV-protection properties. *ACS Applied Materials and Interfaces* **1**, 2141–2146 (2009).

44. Bedford, N.M. & Steckl, A.J. photocatalytic self cleaning textile fibers by coaxial electrospinning. *ACS Applied Materials and Interfaces* **2**, 2448–2455 (2010).

45. Miyauchi, M., Nakajima, A., Hashimoto, K. & Watanabe, T. A highly hydrophilic thin film under 1 mu W/cm(2) UV illumination. *Advanced Materials* **12**, 1923 (2000).

46. Zhang, X.T. et al. Self-cleaning particle coating with antireflection properties. *Chemistry of Materials* **17**, 696–700 (2005).

47. Wang, R.H., Wang, X.W. & Xin, J.H. Advanced visible-light-driven self-cleaning cotton by Au/TiO$_2$/SiO$_2$ photocatalysts. *ACS Applied Materials and Interfaces* **2**, 82–85 (2010).

48. Gould, P. Smart, clean surfaces. *Materials Today* **6**, 44–48 (2003).

49. Lee, J. & Fearing, R.S. Contact self-cleaning of synthetic gecko adhesive from polymer microfibers. *Langmuir* **24**, 10587–10591 (2008).

50. Menguc, Y., Rohrig, M., Abusomwan, U., Holscher, H. & Sitti, M. Staying sticky: contact self-cleaning of gecko-inspired adhesives. *Journal of the Royal Society Interface* **11** (2014).

51. Sethi, S., Ge, L., Ci, L., Ajayan, P.M. & Dhinojwala, A. Gecko-inspired carbon nanotube-based self-cleaning adhesives. *Nano Letters* **8**, 822–825 (2008).

52. Kim, S., Cheung, E. & Sitti, M. Wet self-cleaning of biologically inspired elastomer mushroom shaped microfibrillar adhesives. *Langmuir* **25**, 7196–7199 (2009).

53. Davies, J., Haq, S., Hawke, T. & Sargent, J.P. A practical approach to the development of a synthetic Gecko tape. *International Journal of Adhesion and Adhesives* **29**, 380–390 (2009).

54. Carman, M.L. et al. Engineered antifouling microtopographies – correlating wettability with cell attachment. *Biofouling* **22**, 11–21 (2006).

55. Baum, C., Meyer, W., Stelzer, R., Fleischer, L.G. & Siebers, D. Average nanorough skin surface of the pilot whale (*Globicephala melas*, Delphinidae): considerations on the self-cleaning abilities based on nanoroughness. *Marine Biology* **140**, 653–657 (2002).

56. Baum, C. et al. Surface properties of the skin of the pilot whale *Globicephala melas*. *Biofouling* **19**, 181–186 (2003).

57. Cao, X. et al. Interaction of zoospores of the green alga ulva with bioinspired micro- and nanostructured surfaces prepared by polyelectrolyte layer-by-layer self-assembly. *Advanced Functional Materials* **20**, 1984–1993 (2010).

58. Lin, L. et al. Bio-inspired hierarchical macromolecule-nanoclay hydrogels for robust underwater superoleophobicity. *Advanced Materials* **22**, 4826–4830 (2010).

8

Stimuli-responsive Materials

8.1 Introduction

Stimuli-responsive materials, or smart materials, are those that sense their environments and selectively respond to external stimuli via an active control mechanism. Because of their intrinsic ability to change their physical or chemical properties on command, smart materials react to their environment by themselves. The change is inherent to the material and not a result of electronics. The properties of these materials, such as volume, color, or viscosity, may change in response to external stimuli, that is, a change in temperature, stress, electrical current, or magnetic field. In many cases this reaction is reversible, for example the coating on spectacles that reacts to the level of UV light, turning ordinary glasses into sunglasses when one goes outside and back again when returning inside. There are many groups of smart engineering materials, each exhibiting particular properties that can be harnessed in a variety of applications. These include shape memory alloys, piezoelectric materials, magneto-rheological and electro-rheological materials, magnetostrictive materials, and chromic materials.

Although various smart materials have been developed, and these materials combine sensing and actuation, none have integrated any method used by plants for sensing and precise control. To realize such capabilities in conventional engineering systems requires a control step to be inserted into a closed-loop sensing and response system. These systems are almost exclusively electronic. Nature provides another way to achieve this control that has yet to be exploited in engineering systems: chemical sensing, signal transmission, and control.[1] In contrast to "smart" engineering material systems, nature creates truly smart materials utilizing completely different strategies to accomplish shape changes upon external stimuli. Muscle, for example, is a masterpiece of biological actuators. Movement and force in muscle are generated by a band of sarcomere, a basic molecular machine, in which the myosin cross-bridges interact cyclically with thin (actin) filaments and transport them past the myosin thick filaments.[2]

Plants can also generate a variety of movements, including nastic (movements independent of the spatial direction of a stimulus) or tropic (movements influenced by the direction of a stimulus), active (movements produced by live plants that activate and control their responses by moving ions and by changing the permeability of membranes based on potential actions) or passive (movements mainly based on dead tissue that can undergo predetermined modifications when the environmental conditions change), and reversible (movements specifically based on reversible variations in the turgor pressure).[3,4] These movements are driven by an osmotic pump, another molecular machine based on the sensing and osmotic actuation mechanism.[5] There are many other stimuli-responsive materials, for example hair cells for mechanosensory purposes in many living organisms across the animal kingdom, ranging from insects and spiders to fish and mammals,[6] and putative magnetic receptor and multimeric magnetosensing rod-like protein complex that allow animals to detect the Earth's magnetic field for the purpose of orientation and/or navigation.[7] In these cases, unlike "smart" engineering materials that rely on atomic structures and composition, the stimuli-responsive units are designed at the cellular or larger length scale. These basic molecular machines and working principles could be considered as role models for the biomimetic design and fabrication of smart materials.

The replication of the stimuli-responsive mechanisms found in nature will lead to the development of highly efficient solutions in advanced engineering materials design and processing. Intelligent, yet highly robust, multifunctional structures can be produced by applying such principles to the wider sensing, signal transmission, and control systems found in animals and plants. Such efficient and elegant structures could find application in a variety of fields.[1] There is a large body of work focused on biomimetic animal and plant stimuli-responsive capabilities. In this chapter, actuation mechanisms in animals and plants are selected and discussed, followed by an explanation of biomimetic design strategies and resultant synthetic materials with sensing and actuation capability. Particular attention is paid to microstructural mimics and the design principles of the stimuli-responsive materials.

8.2   The Biological Models for Stimuli-responsive Materials

8.2.1   Actuation Mechanism in Muscles

Muscles are natural actuators that contract to generate force under the control of the nervous system. There are three types of muscles, skeletal, cardiac, and smooth, that are primarily responsible for maintaining and changing body posture and locomotion, the contraction of the heart, and the movement of internal organs, respectively. Among them, skeletal muscle is a form of striated muscle tissue that it is voluntarily controlled via electrical signals. Upon receiving an electrical pulse from the nervous system, these biological actuating materials, which have hierarchic fiber units and a highly ordered fibril network, can drastically change their stiffness while remaining strong and tough. For example, the asynchronous flight muscle of some inserts, one of the most powerful muscles in nature, exhibits high passive stiffness and pronounced stretch activation, and can operate at very high frequencies (100–1000 Hz).[8] The rapid oscillatory contraction and high stroke are derived from a distinct highly-ordered fibril structure, in which highly crystalline proteins form a strong mechanical link and the thick filaments slide stepwise along single crystal-type myofibrils. Mimicking such structures in biological systems could lead to the development of a new class of advanced adaptive materials capable of actuation and changes in mechanical properties.

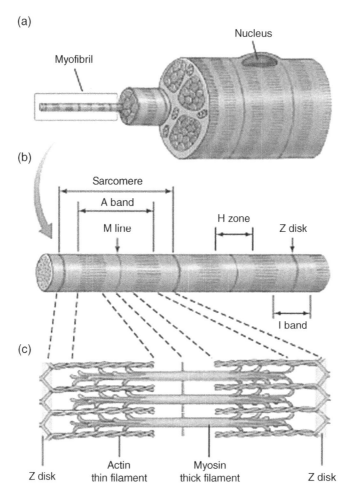

Figure 8.1 Structure of vertebrate skeletal muscle (striated muscle). (a) Muscle fiber with a diameter of 50–100 μm and consisting of many myofibrils (diameter 1–2 μm). In insect flight muscle, both muscle fibers and myofibrils are thicker than these. (b) A magnified view of a myofibril, made of 2–2.5 μm long sarcomeres connected in series. (c) In a sarcomere, the myofilaments, made of the contractile proteins actin and myosin, are arranged in a hexagonal lattice. Source: http://sportsnscience.utah.edu/tag/body/.

Skeletal muscle has a multiscale structure and is highly compartmentalized (Figure 8.1). At the macroscale, a tough layer of connective tissue sheaths and anchors muscle tissue to tendons at each end, and also protects muscles from friction against other muscles and bones. At the microscale, the muscle comprises fascicles in bundles, each of which contains 10–100 or more microscale muscle fibers (Figure 8.1a) collectively sheathed by a perimysium, which is also a pathway for nerves and the flow of blood within the muscle. Within the cells of the muscle fibers are myofibrils, which themselves are bundles of protein filaments organized into repeating units called sarcomeres (Figure 8.1c). The collagenous membrane at each length scale supports muscle function both by resisting passive stretching of the tissue and distributing forces applied to the muscle. Scattered throughout the muscles are muscle spindles, which provide sensory feedback information to the central nervous system.

Sarcomeres are the fundamental units performing contractile work and are the origin of the contraction force. At this level, the sarcomeres are butted up end to end, running the length of each myofibril. A given myofibril contains approximately 10,000 sarcomeres, each of which is about 3 μm in length. While each sarcomere is small, several sarcomeres added together span the length of the muscle fiber. Each sarcomere consists of thick and thin bundles of proteins referred to as myofilaments. The thick and thin filaments overlap but are distributed alternatively in parallel. Thick filaments contain myosin, while thin filaments contain actin. Actin and myosin collectively are referred to as contractile proteins, which cause muscle shortening when they interact with each other. Additionally, thin filaments contain the regulatory proteins troponin and tropomyosin, which regulate interaction between the contractile proteins. The I band is the part of the sarcomere that contains thin filaments, while the A band contains an area of overlap between the thin and the thick filaments. A single I band spans two neighboring sarcomeres. A Z line attaches neighboring sarcomeres. The thin filaments are attached to the Z lines on each end of the sarcomere, while the thick filaments reside in the middle of the sarcomere.[9]

When the muscle is stimulated to contract by nerve impulse, calcium channels open in the sarcoplasmic reticulum (which is effectively a storage house for calcium within the muscle) and release calcium into the sarcoplasm (fluid within the muscle cell). Some of this calcium attaches to troponin, which causes a change in the muscle cell that moves tropomyosin out of the way so that the cross-bridges can attach and produce muscle contraction. During this process calcium fills the binding sites in the troponin molecules. This alters the shape and position of the troponin, which in turn causes movement of the attached tropomyosin molecule. Movement of tropomyosin permits the myosin head to contact actin, which causes the myosin head to swivel. During the swivel, the myosin head is firmly attached to actin, so when the head swivels it pulls the actin (and therefore the entire thin myofilament) forward. Obviously, one myosin head cannot pull the entire thin myofilament. Many myosin heads swivel simultaneously, or nearly so, and their collective efforts are enough to pull the entire thin myofilament. As a result, the head is once again bound firmly to actin. However, because the head was not attached to actin when it swiveled back, the head will bind to a different actin molecule (i.e., one further back on the thin myofilament). Once the head is attached to actin, the cross-bridge again swivels. As long as calcium is present, the above process will continue, and the thin myofilament is "pulled" by the myosin heads of the thick myofilament. Thus, the thick and thin myofilaments are actually sliding past each other. As this occurs, the distance between the Z lines of the sarcomere decreases. As the sarcomeres get shorter, the entire muscle contracts. Relaxation occurs when stimulation of the nerve stops. Calcium is then pumped back into the sarcoplasmic reticulum, breaking the link between actin and myosin. Actin and myosin return to their unbound state, causing the muscle to relax. Relaxation (failure) also occurs when ATP is no longer available.

8.2.2 Mechanically Stimulated Morphing Structures of Venus Flytraps

Apart from truly smart materials – muscles with morphing/actuation ability exhibited by animals –plants show an amazing variety of movements and shape changes. Unlike the actuation in animals, shape changes in plants do not require control by nervous systems; they depend on processes within living cells. Plant movements can be reversible or irreversible; reversible turgor movements are more interesting as biological prototypes for the biomimetic design of materials with actuation/morphing capability. Reversible turgor movements can be nastic or tropic. Nastic movements are independent of the direction of stimuli (e.g., stomata opening in photosynthesis, circadian sleep movements, influenced by incident light, but not controlled by light angle). Tropic movements are induced by and in a relative direction to a stimulus (e.g., heliotropism in sun-tracking plants).[1]

Among nastic movement plants, the Venus flytrap has an ingenious solution to scaling up movements in non-muscular engines. The Venus flytrap (*Dionaea muscipula*) is a carnivorous plant that catches and digests small insects using an active trapping mechanism with one of the most rapid movements in the plant kingdom, first described by Darwin in 1875. This small plant consists of five to seven leaves with a maximum size of 3–7 cm. Each leaf has two parts: the upper leaf and the lower leaf (Figure 8.2a–c). The upper leaf is composed of a pair of trapezoidal lobes that are held together by a blade (midrib). At the center of each lobe, there are three sensitive trigger hairs and a red anthocynanin pigment that attracts insects, while at the edge of each lobe hair-like projections (cilia) stick out. The lower leaf, also known as the footstalk, has an expanded leaf-like structure. The trapping action is triggered by a series of tiny hairs at the crease where two leaves join. When a fly or spider walk across these hairs, touching two or more of them in succession, the two leaves quickly close to capture the prey

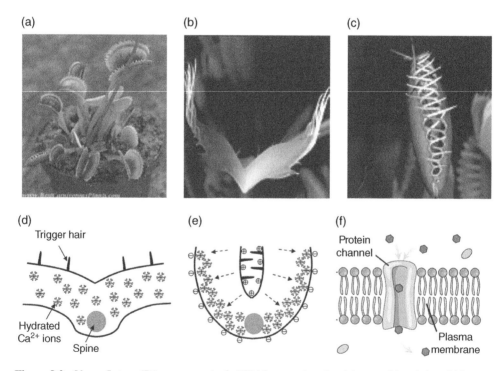

Figure 8.2   Venus fly trap (*Dionaea muscipula Ellis*) in natural setting (a), open (b) and closed (c) states of the snaps (folds, lobes). (d) Cross-section view before the leaf motion and (e) after the leaf motion. (f) Principle of ATP-powered transport for the actuation. Source: Shahinpoor (2011).[11] Reproduced with permission of IOP Publishing.

within hundreds of milliseconds. After the trap closes, the cilia slowly mesh and lock the trap. It takes 4–5 days for the leaf to digest the prey with secreted digestive enzymes. After the digestion, the trap reopens, restoring the convex shape of each lobe.[10]

The rapid movements of Venus flytrap leaves are the result of osmotic movement at the cellular level. In many plants there are special cells, called biological motor cells, that can drive ion diffusion to create osmotic pressure. Biological motor cells have a membrane that contains embedded biological proteins (Figure 8.2f). These special proteins that act as electro-chemical transporters (pumps, cotransporters, and channels) and can selectively transport small species across the membrane to establish an ion gradient or osmotic gradient. Driven by the osmotic gradient, water diffuses into cells, expanding the cellular volume. If expansion of the cells in the arrays occurs non-uniformly, an overall shape changes on the macro scale. In Venus flytrap leaves similar motor cells can drive the macroscale movements of the leaf. Before the leaf is triggered, it has a flat shape since the resistance in the upper and lower epidermis layers is equivalent (Figure 8.2d), and pressure uniformly distributes within the internal layers. Upon triggering, ion fluxes through the cell membranes, creating osmotic movement of water molecules toward the outermost surface of the leaf. The leaves rapidly fold when the outer epidermal cell layer suddenly expands (Figure 8.2e). This ability to rapidly expand is attributed to the osmotic pressure created by an ATP-driven proton pump located on the plasma membrane, which expels protons, creating both a very negative membrane potential (–120 to –250 mV) and an acidic external pH.[10]

The trigger hairs are critical to the trap closure. When insects touch trigger hairs in the open trap, the mechanical stimuli generates a receptor potential that causes a propagation reaction. In order to trigger the reaction, the hairs must be touched twice before the trap is closed *in vivo*, but at high temperatures (36–40 °C) only one stimulus is required for trap closure. Receptor potentials generate adenosine phosphates (APs), which can propagate through the plasmodesmata of the plant to the midrib. Through propagation, the APs deliver the electrical charge that accumulates in the trap. Once a threshold value of the charge is reached, ATP hydrolysis and fast proton transport start through protein channels, and aquaporin opening is initiated (Figure 8.2f).[12,13] Fast proton transport induces transport of water and a change in turgor. As the process is controlled by electrical charge accumulation, electrical stimulation between the lobes and midrib of the Venus flytrap can also cause closure of the trap.[14] The type of stimuli does not determine the speed of closing. Venus flytraps close in 0.3 s when triggered either mechanically or electrically.[10] The trap has curvature elasticity and is composed of outer and inner hydraulic layers where different hydrostatic pressures can build up. Because there is a hydrostatic pressure difference between the outer and inner layers of the lobe in its open state, the trap pre-stores high elastic energy. Once applied stimuli open channels connecting the two layers, water rushes from one hydraulic layer to another through the channels, resulting in a rapid closure due to trap relaxation to its equilibrium configuration.[15]

The rapid movements of Venus flytraps are also related to its unique shape change. The lobes of the flytraps can be considered as shell structures. When the inner layers of the lobes contract while the outer ones expand during closure, the shape of the leaf changes from convex to concave. The leaf closure movement was analyzed using biomechanics. It was proposed from elasticity theory that a dimensionless parameter, $\alpha = W^4\kappa^2/h^2$ (where W is the size of the lobe, h is the thickness, and κ is the curvature), controls bistability.[16] This hypothesis was tested by carefully examining the flytrap closure process. In the experiment, arrays of submillimetric UV-fluorescent dots were drawn on the surface of the leaves. During the plant movement, the dots were irradiated and recorded by high-speed camera during trap closure. It was found that the sudden change in the intrinsic curvature along the direction perpendicular to the midrib was the main source of the driving force behind trap closure, verified by measuring the strain field on closure. The experiment also showed that the speed of the lobe during closure increases with increasing α.

8.2.3   Sun Tracking: Heliotropic Plant Movements Induced by Photo Stimuli

Another type of active motions/actuation is tropic, paratonic movements induced by different external stimuli, for example sunlight. In many cases the movements are also related to the relative direction of the light. Heliotropic plants, for example, move in response to, and in a direction relative to the angle of, incident sunlight.[17] Unlike phototropic growth movements, which are also made in response to light, heliotropic movements are reversible. There are a wide variety of forms of plant movement within heliotropism, for example heliotropic movements observed in flowers and leaves. It is believed that solar-tracking raises the air temperature in the flower, increasing its attractiveness to pollinating insects, while the tracking movements of leaves serve to maximize photosynthetic light absorption and minimize water loss.[18] Similar to Venus flytrap leaves, the movement of heliotropic leaves is also driven by a specific motor organ called the pulvinus, located in the petiole (stem),[17] whilst in the flowers the entire stem is involved in the motion.[18,19] The perception of the light stimulus for leaf heliotropism can either take place in the pulvinus or distant to the motor organ on the leaf itself. Leaf heliotropism can be either diaheliotropic or paraheliotropic. In diaheliotropic movement the plane of the lamina (leaf) is maintained perpendicular to the incident sunlight throughout the day, while in paraheliotropic movement the lamina remains edge on to the incident light. Some plants combine these two types of sun-tracking throughout the day, balancing the need for absorption of oblique light early and late in the day. This combination can reduce water stress in the middle of the day.[20]

Cornish Mallow (*Lavatera cretica* L.) is one of the most fascinating sun-tracking plants, and performs diaheliotropic leaf movements. During the daytime, the plane of the lamina (leaf) is perpendicular to the incident sunlight. This sun-tracking action is driven by bending of the pulvinus, the specialized motor organ located at the connection between the leaf and the stem (3–4 mm in length). Light perception primarily occurs distant from the pulvinus in the leaf veins, as shown in Figure 8.3a.

The sun-tracking action of the Cornish Mallow involves sensing of light stimuli, control and filtering of stimuli signals, actuation in response to the stimuli signal, and memory of an aspect of the stimulus signal to allow for orientation towards the position of sunrise. The Cornish Mallow is sensitive to blue light (410–500 nm) and detects this light in the five to seven major leaf veins that radiate out from the end of the stem (Figure 8.3a).[22] When the leaf is exposed to oblique light (45° from above), all of the veins appear to contribute to the overall response, but their individual contribution varies (Figure 8.3c).[22] As shown in Figure 8.3b, the subcellular component of the cells along the vein are anisotropically ordered with respect to the direction of the vein. This anisotropic orientation of photo receptor molecules makes the leaves preferentially respond to light incident at 45–55° movement (response dependent on the polarization of light).[21] The combination of light perception in variably oriented veins, and the anisotropic arrangement of photoreceptor molecules within these veins make it possible for light to be perceived as a vector, and its angle of incidence upon the leaf reacted to.

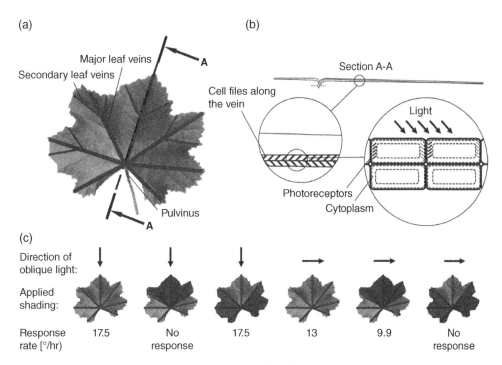

Figure 8.3   Sun-tracking mechanism of Cornish mallow.[1,21] (a) Leaf of the Cornish mallow showing major and secondary veins, and the pulvinus. (b) Sectional diagram showing light perception primarily taking place in the cell files that form these veins. A possible arrangement of the photoreceptors for vectorial sensing of the light is shown in the large detail, where the photoreceptors (indicated by a vertical row of diagonal lines) are located within the layer of the cytoplasm (indicated by dotted lines) adjacent to the transverse cell walls. Their orientation at an angle to the adjacent wall, and antagonistically arranged at opposite ends of the cells, is expected to lead to differences in the level of excitation from the displayed angle of incident light. This is indicated by the varied line thickness in the diagram. (c) The rate of reorientation (degree per hour) of the Cornish mallow leaf with differential shading when exposed to a light beam at 45° from above. Arrows indicate direction of beam, no response indicates that the leaf remained within 3° of the initial position. Source: Dicker et al. (2014).[1] Reproduced with permission of IOP Publishing.

Similar to the closing mechanism of a Venus flytrap, the movement of Cornish mallow leaves is driven by osmotic pressure that is induced by the movement of solvent to a region of higher solute concentration. After the light direction is transferred to a chemical signal by the photoreceptors, the signal will travel through the vascular xylem and phloem tissue and arrive at the pulvinus. The chemical signal (AP) will cause osmotic pressure and thus water uptake in the pulvinus. As uptake of water by cells varies on either side of the longitudinal pulvinar axis, this causes different internal cell pressure on the two sides. This differential internal cell pressure (turgor) leads to pulvinus bending, adjusting the orientation of the leaves.[23] The amplitude, speed, and frequency of the pulvinar movement is defined by the number of cells involved, as well as their capacity, rate of uptake, and loss of water.[24]

As discussed above, the multifunctional structural features of the Cornish Mallow allow for orientation towards the sunrise. Although various smart materials have been developed, and these materials combine sensing and actuation, including photo-actuators, none have integrated any method for eliminating signal conflict or ensuring elimination of control objective error quickly with stability and minimal overshoot. To realize such capabilities in conventional engineering systems, a control step has to be inserted into a closed-loop sensing and response system, but these systems are almost exclusively electronic, resulting in complicated structures and control systems. The Cornish Mallow leaf provides a biological strategy that achieves this control in different way that has yet to be exploited in engineering systems: chemical sensing, signal transmission, and control.[1]

8.2.4 Biomimetic Design Strategies for Stimuli-responsive Materials

As discussed above, there are various strategies that can generate movement in materials in response to external stimuli, including electrical, mechanical, or photo input. Table 8.1 lists some of the extensively studied actuation mechanisms employed in nature.

Table 8.1 Biomimetic design strategies for smart materials.

Functions	Actuation/morphing mechanisms	Biological prototypes	Illustration	Examples of possible applications
Contraction/ stiffness change	Filament sliding	Muscles		Artificial muscle, actuators
Nastic motion	Mechanically induced osmatic pressure	Venus flytrap (*Dionaea muscipula*) Mimosa (*Mimosa pudica*)		Sensors, actuators, water purification
Snap-buckling	Mechanically induced shell buckling	Venus flytrap, bladderwort (*Utricularia*)		Morphing structures
Sun tracking	Photo-induced osmatic pressure	Sunflower, leaves		Solar tracking for solar cells

Living muscles have evolved over millions of years as premier living generators of force, work, and power with unique characteristics compared with standard artificial materials. The muscle functions are controlled by a complex excitation–contraction coupling, which involves communication between muscles and nerves, and a series of chemical, biological, mechanical, and electrical cascades in an extremely complicated manner. It is therefore unlikely that a muscle–nerve system that allows force generation in a similar manner can be reproduced artificially. Nevertheless, the stimuli-responsive characteristics of muscle provide an inspiration for the design of novel electromechanical materials with force, work, and power generation functions similar to natural muscle. Alternatively, these characteristics could be achieved with known mechanical components. In addition, the mechanical behavior and force generation of biological skeletal muscle are determined by many other important factors, including muscle fiber type, motor unit recruitment, architecture, structure, and morphology of skeletal muscle. These structural factors may inspire new ideas in artificial muscle design for better control of the mechanical behavior and active force generation of the artificial muscle.[25]

In the design of synthetic muscle, two strategies are usually considered. One is a bottom-up approach for the collection of individual molecular units, starting from the molecular level. In this approach, the unique molecular structures of muscle are mimicked to achieve actuation. These molecular structures should contain actuation units that can contract and generate force under external electric signals. The ordered fibril network of natural muscle is fabricated through the chemical synthesis of artificial molecular machines at the molecular level. These new materials would allow the development of a new generation of machines with life-like movements and outstanding performance. The second strategy is to use a top-down approach that relies primarily upon the response of a bulk material. Instead of mimicking the detailed structures of natural muscle, this approach uses materials with actuation capability similar to natural muscle under electric stimuli. There are a variety of materials that have such properties, including electroactive polymers (EAPs), ion-based EPAs, shape-memory polymers, alloys, and liquid crystalline elastomers, etc.

Osmotic actuation is a ubiquitous plant-inspired actuation strategy that has a very low power consumption but is capable of generating effective movements in a wide variety of environmental conditions. Osmosis represents one of the fundamental natural processes, for example cellular chemical exchanges, and provides the basis for some high-performance natural movements (e.g., Venus flytrap leaf closure, cnidocysts exocytosis in Cnidarian, etc.). Osmosis is the spontaneous flow of a solvent across a selectively permeable membrane from a region of higher water chemical potential (i.e., lower solute concentration) to a region of lower water chemical potential (i.e., higher solute concentration). Forward osmosis is a powerless process based on the chemical potential non-equilibrium of two chemicals separated by an osmotic membrane; its major technological challenge is to control the solute concentrations between two separated solutions. The actuation performances directly depend on the solute and its solubility in the selected solvent and the dissociation coefficient. Moreover, the efficiency of the osmotic actuation system is strongly correlated with the features of the osmotic membrane. In the design of osmotic devices the membrane should have (1) a high mechanical strength, in order to resist the high pressure obtained and limit its deflection, (2) a high water permeability coefficient, which is directly correlated to the actuation velocity, and (3) a very low solute permeability, to avoid the loss of actuation reversibility.

8.3   Biomimetic Synthetic Stimuli-responsive Materials and Processes

Skeletal muscle, which has hierarchic fiber units and a highly ordered fibril network, can drastically change its stiffness upon receiving an electrical pulse from the nervous system while remaining strong and tough. By analogy, an artificial muscle is defined as material that can reversibly contract, expand, or rotate under an external stimulus such as voltage, current, pressure, or temperature. With their high flexibility, versatility, and power-to-weight ratio compared with traditional rigid actuators, artificial muscles are a highly disruptive emerging technology.

A large body of artificial muscles has been developed, which can be divided into two types: a collection of individual molecular units, starting from the molecular level, or a bulk material. In the first group, the ordered fibril network of natural muscle is mimicked through chemical synthesis of artificial molecular machines at the molecular level. In the second group, the materials used for artificial muscle are broad, including those that can generate mechanical forces or deformation under external stimuli, such as shape-memory polymers and alloys, and liquid crystalline elastomers.

8.3.1   Motor Molecules as Artificial Muscle: Bottom-up Approach

As discussed in Section 8.2.1, in muscular tissues the coordinate movements of thousands of myosin heads lead to the gliding of thick myosin filaments along thin actin filaments, resulting in a cooperative contraction of the entire sarcomere. In particular, although the individual shifts of the proteins take place in a range of 10 nm, their collective translation produces a 1 μm contraction of the sarcomeres. Coupling of these micrometric contractions leads to macroscopic motions. The chemical synthesis of artificial molecular machines has carried out to mimic their biological counterparts with the aim of engineering dynamic microscopic devices and macroscopic functional materials by bottom-up approaches.[26]

To mimic sarcomere units of muscle at the molecular level it is important that the molecular motion is coupled in a cooperative and coherent manner within an ordered mechanical setting so that the muscles can coherently push and pull much larger objects, as the muscles in our bodies do. To this end, the artificial muscle units should be designed to be able to produce internal motions such as (1) translation in bistable single rotaxanes or on molecular tracks, (2) scissor-like motions, and (3) unidirectional circumrotation (in multistation catenanes) or rotation (around a central bond) in molecular motors that are driven far from equilibrium by external stimuli (e.g., light).[26] Furthermore, linear motor-molecules could be used to provide a range of properties and enable a variety of auxiliary functions that will ultimately facilitate their inclusion in an integrated system. These integrated systems could be obtained by organization at three hierarchical levels:[27] (1) in the lowest level, single linear motor-molecules are synthesized by self-organizing protocols, (2) in the second level, motor-molecules are self-assembled into a molecular ensemble, and (3) the force and strain can be amplified in the highest level of organization by stacking the self-assembled monolayers on top of each other.

Progress has recently been made in the design of synthetic molecular muscles. In particular, the linear amplification of muscle-like translational molecular motions by orders of magnitude has been demonstrated experimentally. A pH-triggered muscle-like system in which tailored bistable daisy chain rotaxanes are combined linearly was synthesized by taking advantage of a metallosupramolecular polymerization process (Figure 8.4).[26] The material can contract and

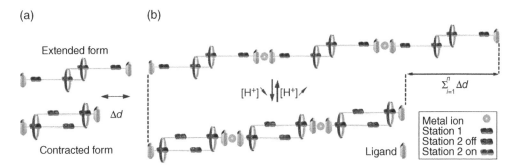

Figure 8.4 General principle of (a) bistable [c2]daisy chain rotaxane and (b) the supramolecular polymer in this study. Here, the stoppers are ligands that can also bind to metal ions to produce coordination polymers. The integrated translational motion of the supramolecular polymer chain is the product of the individual contractions and extensions of each molecular machine by the degree of polymerization. Source: Du et al. (2012).[26] Reproduced with permission of John Wiley and Sons.

expand by controlling the pH of the solution. Under the stimulation of pH, polymer units can simultaneously move, allowing the whole polymer chain to contract or extend over about 10 μm. This simultaneous unit movement amplifies the muscle movement by a factor of 10,000, along the same principles as those used by muscular tissues. These results, obtained using a biomimetic approach, could lead to numerous applications for the design of artificial muscles, micro-robots, or the development of new materials incorporating nano-machines endowed with novel multiscale mechanical properties.

8.3.2 Electroactive Polymers as Artificial Muscle: Top-down Approach

The second type of artificial muscles uses more traditional materials and can be divided into three major groups based on the actuation mechanism: pneumatic, thermal, and electric field actuations. These artificial muscles mimic the actuation function of natural muscles instead of their molecular structures. Pneumatic artificial muscles (PAMs) are made with an internal bladder covered by a braided mesh shell that encircles the bladder, the two ends of which are attached to tendon-like structures. Upon applying gas pressure to the bladder, the isotropic volume expands in the radial direction but contracts in the axial direction, driving the braided wires to translate the volume expansion to a linear contraction along the axis of the actuator. The force-length properties of PAMs are reasonably close to those of biological muscle, but the force-velocity is not comparable. Among the most commonly used PAMs is a cylindrically braided muscle known as the McKibben muscle.[28]

Shape-memory materials, such as shape-memory alloys (SMAs) and liquid crystalline elastomers, are thermally-driven actuators. These materials can be deformed and then returned to their original shape when exposed to heat, functioning like artificial muscles. Although the stimuli-responsive speed of the thermally-driven actuator is relatively low, this type of artificial material offers heat resistance, impact resistance, low density, high fatigue strength, and large force generation during shape changes.[29]

EAPs are materials that can be actuated through the application of electric fields, so they are more like natural muscles in terms of stimuli signals, forces, strains, and responsive times. Mechanically, the actuation process is like the osmotic movement found in plant motor cells;

both are related to the diffusion of ions in an electrolyte solution. EAPs can be classified into two categories: (1) ionic polymers, including polymer gels, ionic polymer–metal composites, conjugated polymers, and carbon nanotubes (CNTs),[30] and (2) field-responsive polymers, such as electrostrictive polymers, dielectric elastomers, liquid crystal elastomers, and CNT aerogels. EAP-based artificial muscles offer a combination of light weight, low power requirements, resilience, and agility for locomotion and manipulation. Two examples of prominent EAPs are discussed in detail below.

8.3.2.1   Ionic EAPs as Artificial Muscle

Ionic EAPs are materials that can be actuated through the diffusion of ions in an electrolyte solution (in addition to the application of electric fields).[31] In ionic polymer–metal composites, ion-exchange membranes made from nafion (perfluoronated polysulphonic or carboxylic acids) are coated with platinum or gold electrodes on each side of the membrane by chemical reduction of a metal salt dissolved in the swollen membrane. The membranes have a complicated chemical structure and a morphology displaying hydrophilic clusters surrounded by the hydrophobic backbone of the perfluoronated polymer. When a potential (e.g., 1 V) is applied over the membrane, an initial fast bending occurs due to fast cation movements toward cathode (Figure 8.5a). This process may also include water migration and electrostatic interactions under the applied potentials.

Conjugated or conducting polymers derived from pyrrole, aniline, thiophene, acetylene, and ethyldioxithiophene can also perform actuation when a potential is applied. These materials are doped by an oxidation process during their electrochemical synthesis, which makes them conductors.[32] To make them work as actuators, the materials are coated with two electrodes and put in an electrolyte to form an electrolytic cell (Figure 8.5b). When an electrical potential is applied between the two electrodes of the electrolytic cell, electrons are removed (if anion-driven) from or added (if cation-driven) to the backbone of the conjugated polymer, depending on the type of electrolyte used. The gained or lost electrons in the polymers are balanced by the addition and removal of electrolyte ions to or from the polymer chains, which will cause volume expansion or contraction. The polymers are deformed due to ion diffusion, and the electrical energy is thus transferred into mechanical energy. Different actuation effects can be achieved by constructing the polymers in different structures, such as thick film, which shows the volume change in multiple directions, stand-alone thin film, which expands or contracts in a single direction, and multi-layer film, which bends like a cantilevered composite beam. For example, a bilayer structure consists of an active conjugated polymer layer and a passive layer. The active layer causes bending of the bilayer and the passive one prevents the linear strain of the active layer.

Similar design principles can be applied to construction nanotube actuators: two CNT electrodes as the working and counter electrodes immersed in an electrolyte (Figure 8.5c).[33] However, the actuation principle of nanotubes is not based on ion intercalation. Upon applying a potential difference between the two electrodes, the ionic charges start accumulating on the surface of each nanotube electrode, which leads to the redistribution of the electronic charge in the nanotube electrodes and the formation of a double layer at the nanotube–electrolyte interface. This charge injection and electrostatic effects cause dimensional changes (i.e., strain) in the nanotube electrodes.[33] Different shapes of electrodes can be made to generate

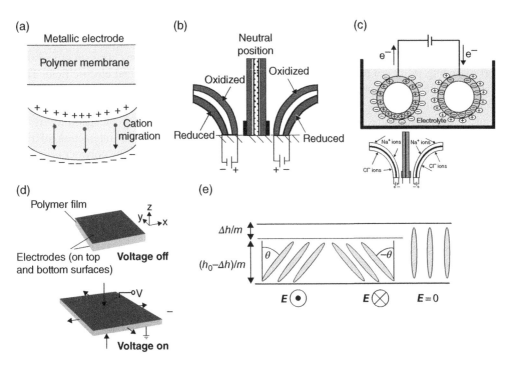

Figure 8.5 Schematic representations of the actuation principle of EAP actuators. (a) Natural state: the IPMC actuator is flat and application of a potential difference between the electrodes causes cations to migrate towards the cathode and the IPMC actuator bends. (b) The actuation principle for a multi-layer conjugated polymer actuator: this actuator is anion-driven and the anions in the salt move into the positively charged electrode to cause a volume expansion. (c) The actuation principle for a CNT actuator, where the opposite signed ionic species accumulate on the surface of each nanotube electrode to establish a supercapacitor filled with an electrolyte (NaCl) and two nanotubes as its electrodes, causing a trilayer cantilever actuator bending under applied potentials. (d) Functional element and typical response of a field-responsive EAP. Electrostrictive polymers and dielectric elastomers are functionally similar. Source: Carpi *et al.* (2011).[34] Reproduced with permission of IOP Publishing. (e) a liquid crystal elastomer actuator. Source: Lehmann *et al.* (2001).[35] Reproduced with permission of Nature Publishing Group.

different actuation. Bending-type CNT actuators, for example, are wired like the electrodes of an electrolytic cell and are immersed in an electrolyte such as NaCl (Figure 8.5c).

8.3.2.2   Field-responsive EAPs as Artificial Muscle

Field-responsive EAPs are based on materials that respond to an applied electric field. The basic functional element of an electronic EAP is shown in Figure 8.5d. The structure of field-responsive EAP muscles is very simple, consisting of field-responsive EAPs between two electrodes. When a potential is applied to the electrodes, the materials (as well as the electrodes) expand in area and come closer together as the film thickness decreases, which converts electrical to mechanical energy by bringing opposite charges closer. When the potential is removed, the materials recover their original shape using elastic energy stored in the materials. This type of artificial muscle can achieve high actuation strains and stresses, fast responses and long lifetimes, but requires large driving electric fields (up to the order of 100 V/μm) due to the electrostatic nature of their activation mechanisms. The stress and the corresponding strain are roughly proportional to the square of the applied electric field. The response to the applied potential depends on the material structures, and mechanic and electric properties. If the response is dominated by the field-induced reorientation of a crystalline or semicrystalline structure, the polymer is electrostrictive (in this case, the permittivity is dependent on the electric field). If the response is dominated by the interaction of the electrostatic charges on the electrodes (often called the Maxwell stress), the polymer is called a dielectric elastomer.[34]

PVDF-based materials are electrostrictive polymers, which have specific energy density greater than that of the best piezoelectric ceramics. These materials produce moderate amounts of force and strain (up to 5%), therefore they are especially suitable for diaphragm actuators (such as actuators for pumps), bending beam actuators, and directly linear actuators as well as sonar transducers due to their high efficiency and very fast speed of response. Compared with PVDF-based materials, elastomers usually are soft and good insulators, and can be good dielectric elastomer actuators. When a voltage is applied across a dielectric elastomer film, the unlike charges on the opposing highly compliant electrodes will attract each other, generating an attractive force to squeeze the film thickness, while the like charges on each electrode will produce a repelling force to expand the film thickness. The actuation pressure, p, for dielectric elastomers can be written as[36] $p = \varepsilon\varepsilon_o E^2 = \varepsilon\varepsilon_o (V/t)^2$, where ε_o is the permittivity of free space, ε is the dielectric constant of the material, E is the imposed electric field, V is the applied voltage, and t is the polymer thickness. The strain exhibited by a dielectric elastomer depends on its modulus, boundary conditions, and external loading. For small strains (e.g., <10%) under free boundary conditions, the thickness strain, s_z, can be written as $s_z = -p/Y = -\varepsilon\varepsilon_o V^2/(Yt^2)$, where Y is the modulus of elasticity of the polymer–electrode composite film. According to these equations, the most successful materials should have a relatively low modulus of elasticity and can sustain high electric fields. Many commercially available materials, including silicone rubber (polydimethyl siloxane) and acrylic elastomers, meet such requirements and can generate more than 100% strain.

Liquid crystal elastomers consist of liquid crystal groups attached to one another via a polymer network, which combine the mechanical properties of the elastomer network with the electric properties of the liquid crystals. In particular, an electric field is able to realign intrinsically polarized liquid crystal groups (Figure 8.5e), generating active strains typically of a few per cent. Apart from electric field, heat (either directly or via optical, electrical, or magnetic sources) can induce much higher strains (up to the order of 100%) in liquid crystal elastomers, although in this case the response speed is significantly limited by heat diffusion.

8.3.3   Venus Flytrap Mimicking Nastic Materials

Plants such as the Venus flytrap (*Dionaea muscipula*) and Mimosa (*Mimosa pudica*) can quickly close their leaves, driven by a biological process called nastic motion in specialized motor cells.[10,16] These nastic movements occur when biological reactions drive water to flow into or out of the motor cells. By analogy, this system can be considered to be the muscles of plants. With this inspiration, a novel nastic material has been created based on the principles of biological nastic motion, which allows for significant deformation due to internal micro-actuations. This type of actuator is considered to be a high-energy density-smart material.[37]

The biomimetic osmotic actuator consists of four main elements: a reservoir chamber (RC), an osmotic membrane (OM), an actuation chamber (AC) composed of rigid walls, and a deformable part for actuation and force output. The AC also has a compliant wall that deforms under pressure to transduce the actuation force. Due to the difference in concentration of osmolytes in the AC and RC, a driving osmotic potential is established, drawing water from the RC to the AC. The increasing volume of water presses the compliant wall to convert chemical energy to mechanical force. Figure 8.5a shows the biomimetic design of the actuator. Solvent flows from the RC to the AC through the OM, and actuation work is gathered/transduced through the bulging deformation of a disk-shaped portion of the actuation chamber boundary.

A forward osmosis-based actuator with a typical size of 10 mm and a characteristic response time of 2–5 minutes (inset of Figure 8.6b) has been successfully designed and fabricated.[38] This biomimetic actuator can produce forces above 20 N, exhibiting low power consumption, high energy density, and remarkable efficiency. This actuator is the fastest osmotic actuator developed so far. As shown in Figure 8.6b, the achieved timescale is very close to the relaxation time of an ideal giant plant cell with the same typical size as the actuator.[38] More interestingly, the actuator timescale is not far from that of typical plant cells, such as the stomata shown in Figure 8.6b. This comparable timescale between device and actual plant cells indicates that the plant osmotic pump and related actuation could be fully achieved from different materials and design. It should be noted that the resulting actuator volumetric stiffness is very close to a typical cell bulk modulus, e.g. 30 MPa.[39] These results demonstrate that a bioinspired approach has great potential in developing actuation materials and devices. Such an approach has proved to be successful in extracting smart cues from the plant kingdom.

The fast motion of the Venus flytrap also involves a snap-buckling instability of the shell-like geometry of the lobes at the macroscale.[16] This mechanical buckling of plates (in a two-dimensional analogue of Euler buckling) and biaxial compressive loads mimic the fabrication of an array of lens-like bistable shells.[40] A layer of biaxially pre-stretched polydimethylsiloxane (PDMS) is patterned with periodic arrays of holes and then coated with a thin film of uncured PDMS by spin coating. The two layers of PDMS are then bonded by cross-linking to form micro-lens arrays on thin films. The micro-lenses can transfer from concave to convex shape by a snap-through movement, similar to that of the Venus flytrap. The focal point of each micro-lens can changed from above the structure surface (when the micro-lens is convex) to below the structure surface (when the micro-lens is concave) (Figure 8.7). The micro-lens can dynamically and reversibly change shape and aspect ratio by changing the strain applied to the shell structures.

8.3.4 Biomimetic Light-tracking Materials

Inspired by sun-tracking plants, a photochemically activated actuator was developed involving photochemical sensing, signal transmission, and control. Figure 8.8 shows the configuration of a simplified biomimetic analogue of the sun-tracking Cornish mallow.[1] This device consists of two connected stimuli-responsive materials: a reversible photobase that changes pH when exposed to light, and pH-sensitive hydrogels that swell under different pH environments.

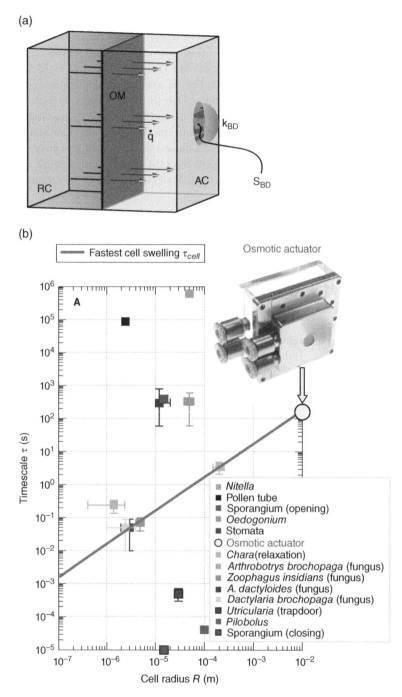

Figure 8.6   (a) Schematic of biomimetic osmotic actuator. (b) Osmotic actuator timescale and computed performance. The osmotic actuator timescale matches with the relaxation time of an ideal, giant plant cell with the same typical size, i.e. cell radius. Source: Sinibaldi *et al.* (2014).[38] Reproduced with permission of Oxford University Press.

(a)

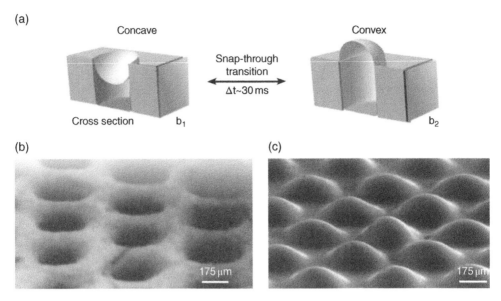

Figure 8.7 (a) Schematic of the transition between concave and convex (transition time 30 ms). SEM images of "snapping" microlens shells in (b) concave and (c) convex states. Source: Holmes *et al.* (2008).[40] Reproduced with permission of the Royal Society of Chemistry.

The two materials are coupled with each other to form an artificial leaf vein that mimics the ability of plants to sense light direction. In particular, this artificial vein comprises two clear tubes filled with a light sensitive chemical solution. One tube is covered in shades orientated at 45° to the tube surface, while the other has shades in opposed orientation. With these two tubes arranged in symmetry, the amount of light reaching and reacting with the light-responsive contents is gradually varied between a maximum in one tube when light is orientated at 45° from above, to a minimum in both tubes when the light source is positioned directly (90°) above the device, to 135°, when the light falling on the opposing tube is maximized. In this way, the orientation of the light source is converted to chemical signals.

The selection of a suitable light-sensitive chemical solution is important to accurately transform the chemical signals. 4,4-bis(dimethylamino)triphenylmethane leucohydroxide, known as malachite green carbinol base (MGCB), can reversibly change pH when exposed to light. The pH change results from the formation of hydroxide ions upon irradiation. A short (3 minute) exposure to high-intensity UV light (280–410 nm) results in the pH of a solution of MGCB rapidly changing from pH 5.4 to 10. After removal of the light radiation the pH returns to the initial value after 15 minutes. The diffusion of these ions, and the resulting pH change, provides both chemical signaling and energy transfer from the photosensing component to the actuator.

After the light direction information is converted into a chemical signal, it can be used to stimulate the actuation to make the artificial vein respond to the light direction change. This achieved by a novel hydrogel core, which is a flexible matrix composite (H-FMC) that reversibly swells when exposed to differing stimuli in solution.[41] Hydrogels are three-dimensional polymeric networks; on absorbing water they can swell to many times their dry weight.[42]

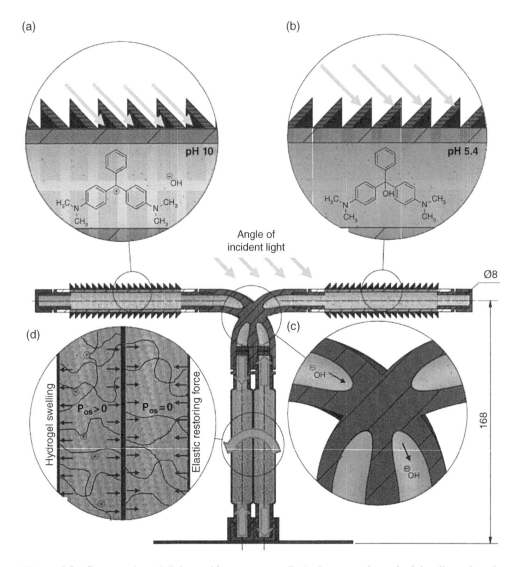

Figure 8.8 Cross-section of light tracking prototype displaying operating principle, dimensions in millimeters. (a) Angled shades allow light exposure of reversible photobase solution, such as malachite green carbinol base (MGCB), shown here in light grey. Irradiation of the solution with UV light splits the molecule into an MG cation and hydroxide anion, increasing the solution pH to 10. (b) On the opposing sensor tubes the shades shadow the MGCB solution from light exposure, hydroxide ions remain attached, and the pH remains at 5.4. (c) Hydroxide ions from the illuminated sensor tube diffuse to the opposing hydrogel core (such as epoxy hydrogel) FMC actuator. (d) The normally ionized epoxy hydrogel is neutralized by the arrival of hydroxide ions, stopping the swelling-inducing osmotic pressure that is present at lower pH. The elastic restoring force from the FMC then becomes dominant and the actuator extends. Meanwhile the opposing actuator is still exposed to pH 5.4 and remains in its contracted state. The differential force from the two actuators generates a bending moment that acts to bend the device to minimize the error in angle of incidence of incoming UV light with sensor tube normal. This effect becomes diminished as this error is reduced, resulting in automatic control of the device. (e) Unactuated device with quarter section removed to reveal internal structure. (f) Full device in actuated state showing orientation towards the light source. Source: Dicker et al. (2014).[1] Reproduced with permission of IOP Publishing.

(e)   (f)

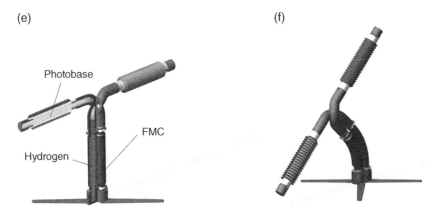

Figure 8.8 *(Continued)*

By doing so, hydrogels directly convert chemical energy into mechanical work, for example generating force and producing actuation. The volume change is caused by environmental factors such as pH because the polymer chains in the hydrogels contain either acidic or basic groups, which can be ionized. When the photosensing component generates protons and changes pH, it switches the swelling of the hydrogel on and off. Just as in the plant cells that form the pulvinus, swelling of the hydrogel is driven by osmotic pressure that further depends on pH. The swelling continues until it reaches an equilibrium between the ionic pressure and the elastic restoring force of the polymer.

Although the swelling of the hydrogel is controlled by the chemical signal, to make it work for direction adjustment of the artificial veins it is important to partially constrain the three-dimensional swelling of the hydrogel to control it to do useful work. To this end, the hydrogel is stored within an FMC tube (a cylindrical fiber-reinforced composite laminate with an elastomeric matrix). In this way, the partial constraint increases the fixed charge density from the ionized groups and resulting osmotic pressure. Moreover, hydrogel swelling in the radial and circumferential directions is converted into high axial strains for actuation due to the Poisson's ratio of the anisotropic FMC laminate.

The materials device mimicking the sun-tracking function of the Cornish Mallow leaf can be used for solar power generation. This biomimetic solar tracking system can change the orientation of photovoltaic (PV) panels according to the angle of incidence of incoming light such that the efficiency of the panels is maximized. In addition, sun tracking in this way does not consume power. It is estimated that this process could typically increase the annual energy yield of PV solar panels by 30–40% compared to optimally oriented fixed panels.[43] In addition, chemical sensing, control, and actuation could have a wider field of applications.[1]

References

1. Dicker, M.P., Rossiter, J.M., Bond, I.P. & Weaver, P.M. Biomimetic photo-actuation: sensing, control and actuation in sun-tracking plants. *Bioinspiration & Biomimetics* **9**, 036015 (2014).
2. Geeves, M.A. & Holmes, K.C. Structural mechanism of muscle contraction. *Annual Review of Biochemistry* **68**, 687–728 (1999).

3. Joyeux, M. At the conjunction of biology, chemistry and physics: the fast movements of Dionaea, Aldrovanda, Utricularia and Stylidium. *Frontiers in Life Science* **5**, 71–79 (2011).

4. Guo, Q. *et al.* Fast nastic motion of plants and bioinspired structures. *Journal of the Royal Society Interface* **12**, 0598 (2015).

5. Raven, J.A. & Doblin, M.A. Active water transport in unicellular algae: where, why, and how. *Journal of Experimental Botany* **65**, 6279–6292 (2014).

6. Studart, A.R. Biologically inspired dynamic material systems. *Angewandte Chemie International Edition* **54**, 3400–3416 (2015).

7. Qin, S.Y. *et al.* A magnetic protein biocompass. *Nature Materials* **14** (2015).

8. Iwamoto, H., Inoue, K. & Yagi, N. Evolution of long-range myofibrillar crystallinity in insect flight muscle as examined by X-ray cryomicrodiffraction. *Proceedings of the Royal Society B: Biological Sciences* **273**, 677–685 (2006).

9. Davies, K.E. & Nowak, K.J. Molecular mechanisms of muscular dystrophies: old and new players. *Nature Reviews Molecular Cell Biology* **7**, 762–773 (2006).

10. Taya, M., Stahlberg, R., Li, F.H. & Zhao, Y. Sensors and actuators inherent in biological species. *Proceedings of SPIE* **6529**, U16–U26 (2007).

11. Shahinpoor, M. Biomimetic robotic Venus flytrap (*Dionaea muscipula Ellis*) made with ionic polymer metal composites. *Bioinspiration & Biomimetics* **6**, 046004 (2011).

12. Rea, P.A. The dynamics of H+ efflux from the trap lobes of *Dionaea muscipula Ellis* (Venus-flytrap). *Plant Cell Environment* **6**, 125–134 (1983).

13. Williams, S.E. & Bennett, A.B. Leaf closure in the Venus flytrap – an acid growth-response. *Science* **218**, 1120–1122 (1982).

14. Volkov, A.G., Adesina, T., Volkova-Gugeshashvili, M.I., Williams, J. & Jovanov, E. Electrophysiology of venus flytrap (*Dionaea muscipula Ellis*). *Biophysical Journal*, **111a** (2007).

15. Volkov, A.G., Adesina, T., Markin, V.S. & Jovanov, E. Kinetics and mechanism of *Dionaea muscipula* trap closing. *Plant Physiology* **146**, 694–702 (2008).

16. Forterre, Y., Skotheim, J.M., Dumais, J. & Mahadevan, L. How the Venus flytrap snaps. *Nature* **433**, 421–425 (2005).

17. Ehleringer, J. & Forseth, I. Solar tracking by plants. *Science* **210**, 1094–1098 (1980).

18. Sherry, R.A. & Galen, C. The mechanism of floral heliotropism in the snow buttercup, *Ranunculus adoneus*. *Plant Cell Environment* **21**, 983–993 (1998).

19. Vandenbrink, J.P., Brown, E.A., Harmer, S.L. & Blackman, B.K. Turning heads: the biology of solar tracking in sunflower. *Plant Science* **224**, 20–26 (2014).

20. Donahue, R.A., Berg, V.S. & Vogelmann, T.C. Assessment of the potential of the blue-light gradient in soybean pulvini as a leaf orientation signal. *Physiologia Plantarum* **79**, 593–598 (1990).

21. Koller, D., Ritter, S., Briggs, W.R. & Schafer, E. Action dichroism in perception of vectorial photoexcitation in the solar-tracking leaf of *Lavatera cretica* L. *Planta* **181**, 184–190 (1990).

22. Schwartz, A. & Koller, D. Phototropic response to vectorial light in leaves of *Lavatera cretica* L. *Plant Physiology* **61**, 924–928 (1978).

23. Martone, P.T. *et al.* Mechanics without muscle: Biomechanical inspiration from the plant world. *Integrative and Comparative Biology* **50**, 888–907 (2010).

24. Morillon, R., Lienard, D., Chrispeels, M.J. & Lassalles, J.P. Rapid movements of plants organs require solute-water cotransporters or contractile proteins. *Plant Physiology* **127**, 720–723 (2001).

25. Gao, Y. & Zhang, C. Structure–function relationship of skeletal muscle provides inspiration for design of new artificial muscle. *Smart Materials and Structures* **24**, 033002 (2015).

26. Du, G.Y., Moulin, E., Jouault, N., Buhler, E. & Giuseppone, N. Muscle-like supramolecular polymers: Integrated motion from thousands of molecular machines. *Angewandte Chemie International Edition* **51**, 12504–12508 (2012).

27. Flood, A.H. *et al.* Biomimetic molecules as building blocks for synthetic muscles – A proposal. *NSTI Nanotechnology Conference and Trade Show* **1**, 114–117 (2004).

28. Krishna, S., Nagarajan, T. & Rani, A.M.A. Review of current development of pneumatic artificial muscle. *Proceedings of SPIE* **11**, 1749–1755 (2011).

29. Ariano, P. *et al.* Polymeric materials as artificial muscles: an overview. *Journal of Applied Biomaterials & Functional Materials* **13**, 1–9 (2015).

30. Foroughi, J. *et al.* Torsional carbon nanotube artificial muscles. *Science* **334**, 494–497 (2011).

31. Bar-Cohen, Y. *Electroactive Polymer (EAP) Actuators as Artificial Muscles: Reality, Potential, and Challenges* (SPIE Press, Bellingham, WA; 2001).
32. Smela, E. Conjugated polymer actuators. *MRS Bulletin* **33**, 197–204 (2008).
33. Baughman, R.H. *et al.* Carbon nanotube actuators. *Science* **284**, 1340–1344 (1999).
34. Carpi, F., Kornbluh, R., Sommer-Larsen, P. & Alici, G. Electroactive polymer actuators as artificial muscles: are they ready for bioinspired applications? *Bioinspiration & Biomimetics* **6**, 045006 (2011).
35. Lehmann, W. *et al.* Giant lateral electrostriction in ferroelectric liquid-crystalline elastomers. *Nature* **410**, 447–450 (2001).
36. Pelrine, R., Kornbluh, R., Pei, Q.B. & Joseph, J. High-speed electrically actuated elastomers with strain greater than 100%. *Science* **287**, 836–839 (2000).
37. Freeman, E. & Weiland, L.M. High energy density nastic materials: Parameters for tailoring active response. *Journal of Intelligent Material Systems and Structures* **20**, 233–243 (2008).
38. Sinibaldi, E., Argiolas, A., Puleo, G.L. & Mazzolai, B. Another lesson from plants: The forward osmosis-based actuator. *Plos One* **9** (2014).
39. Forterre, Y. Slow, fast and furious: understanding the physics of plant movements. *Journal of Experimental Botony* **64**, 4745–4760 (2013).
40. Holmes, D.P., Ursiny, M. & Crosby, A.J. Crumpled surface structures. *Soft Matter* **4**, 82–85 (2008).
41. Dicker, M.P.M., Weaver, P.M., Rossiter, J.M. & Bond, I.P. Hydrogel core flexible matrix composite (H-FMC) actuators: theory and preliminary modelling. *Smart Materials and Structures* **23** (2014).
42. Meenach, S.A., Anderson, A.A., Suthar, M., Anderson, K.W. & Hilt, J.Z. Biocompatibility analysis of magnetic hydrogel nanocomposites based on poly(N-isopropylacrylamide) and iron oxide. *Journal of Biomedical Materials Research Part A* **91a**, 903–909 (2009).
43. Mousazadeh, H. *et al.* A review of principle and sun-tracking methods for maximizing solar systems output. *Renewable & Sustainable Energy Reviews* **13**, 1800–1818 (2009).

9

Photonic Materials

9.1 Introduction

Photonic materials are periodic optical nanostructures that affect the motion of photons, resulting in enhanced reflection, diffraction, interference, or transmission. These materials can exhibit brilliant structural colors through their unique nanostructures, which confine light with particular wavelengths and control the direction of light propagation. This leads to interesting optical phenomena such as inhibition of spontaneous emission, high-reflecting omni-directional mirrors, and low-loss waveguiders. With these distinct functions, the photonic materials, especially those with tunable photonic band gap, promise to be useful in different forms in a range of applications such as efficient microwave antennas, zero-threshold or microlasers, low-loss resonators, optical fibers, switches, sensors, display devices, and so on.[1] However, because of the complex nanoarchitectures required for creating distinct functions, it is challenging to efficiently generate photonic structures even with the assistance of powerful computers and advanced nanotechnologies.

Nature has created various excellent photonic materials that generate brilliant colors, providing bioinspiration for the design of artificial photonic materials. Many living creatures make use of coloration to adapt to their surrounding environments, warn their predators through mimicry, or mislead their natural predators through camouflage. With millions years of evolution, the structures and functions of the natural photonic nanoarchi-tectures are almost perfect for helping natural creatures survive.[2-4] Among the many mechanisms for producing colors, pigment color and structural color are the most frequently used to generate colors in natural creatures. Pigment color is generated by chemical chromophores through the selective absorption of light. In contrast, structural color usually relies on periodical photonic nanostructure, through which the transportation of light is controlled. Compared with pigment color, structural color is much more efficient in energy consumption and the use of light. It is the most vivid and brightest in nature.[1] In addition, living creatures can also reversibly and actively change their structural colors. This property of tunable structural color materials is desirable for many novel optical technologies.

A completely opposing function to structural color or reflector is also widely distributed in the animal world. Unlike structural colorations that enhance color reflection, thus brightness, these nanoarchitectures work as antireflectors to suppress light reflection, and therefore enhances the optical transparency of the surface. For example, some nocturnal insects, such as moths and some butterflies, have hexagonal arrays of non-close-packed cylindrical nipples with sub-300 nm size on the surface of their compound eyes.[4] These two-dimensional arrays of photonic crystals generate a gradient in refractive index between the air and the surface of the cornea. Because of this refractive index gradient, the structure can dramatically reduce the reflection losses and increase transmission of light at the interface simultaneously over a wide range of wavelengths and a large field of view.

Photonic materials have potential for industrial, commercial, and military application, with biomimetic surfaces that could provide brilliant colors, adaptive camouflage, efficient optical switches, and low-reflectance glass. Inspired by these biological photonic nanostructures, researchers have created artificial photonic materials and improved the performance of many optical devices.[5,6] In this chapter, the microstructures and mechanisms of natural and bioinspired photonic materials will be analyzed, and then artificial structural color materials will be introduced to demonstrate bioinspired design principles, fabrications, and applications. Examples of typical photonic materials will be presented with an emphasis on the photonic mechanisms of natural creatures and bioinspired photonic materials.

9.2   Structural Colors in Nature

Structural colors can be generated by many kinds of photonic nanostructures on the surfaces of plants, insects, and fishes. As shown in Figure 9.1, the structures range from one-dimensional (1D) nanostructured photonic grating or multilayers to two-dimensional (2D) and three-dimensional (3D) colloidal crystals.[7] 1D photonic nanostructures are diffraction gratings found in some insects, birds, fish, plant leaves, berries, and algae (Figure 9.1a,b). 3D photonic solids exist in the scales of pachyrrhynchine weevils that have a close-packed hexagonal arrangement analogous to (mineral) opal. 2D photonic materials are most commonly seen in beetles, feathers, and fish, but occur less frequently in the color generation than their 1D and 3D counterparts. These structures can create structural colors by mechanisms including diffraction gratings, selective mirrors, photonic crystals, crystal fibers, and deformed matrices.[7] Each mechanism creates a specific color or combination of colors, providing a specific solution to the problem. In addition, a variety of materials used in structural colors are found in living creatures; the type of photonic material depends on many factors, including the physiology, ecology or environment with which the natural creatures are associated. The following sections describe the structures, functions, and mechanisms of these biological photonic materials in more detail.

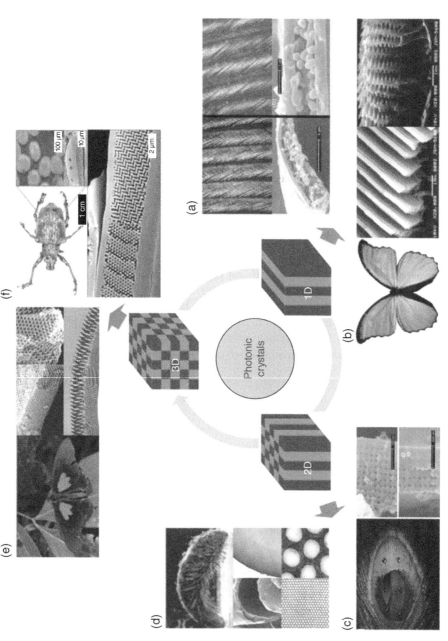

Figure 9.1 Typical natural photonic structures with various structural colors. (a) 1D periodicity in the form of multilayers existing in neck feathers of domestic pigeons. (b) Some discrete 1D periodicity found in *Morpho* butterflies. (c) 2D photonic crystal structure in the barbules of male peacocks with intricate, colorful eye patterns. (d) 2D periodicity of cylindrical voids embedded in a high refractive index solid medium in iridescent setae from polychaete worms. (e) 3D inverse opal structures appearing in the green color of *Parides sesostris*. (f) 3D diamond-based photonic crystal structure in the beetle *L. augustus*. Source: Wang & Zhang (2013)[7] (http://www.mdpi.com/1424-8220/13/4/4192). Used under CC by 3.0 http://creativecommons.org/licenses/by/3.0/.

9.2.1   One-dimensional Diffraction Gratings

A diffraction grating is a nanoscale array of parallel ridges or slits that disperses white light into its constituent wavelengths (Figure 9.2).[8] When white light beams on the surface with diffraction gratings, it diffracts into full spectra via reflection, creating a rainbow-like reflectance. Beetles have many types of diffraction gratings, such as strigulose microsculpture, microtrichiae, or modified setae. Iridescence resulting from diffraction gratings always takes the form of one or more ordered spectra, that is, colors are ordered in the same sequence as the colors in the spectrum of visible light, red–orange–yellow–green–blue–violet. These colors are different from the unordered "faux spectra" produced in some beetles by variable thickness multilayer reflectors (Figure 9.1d). The spectra produced by diffraction gratings are also "ordered" in a different sense: when compared with a standard birefringence chart used in mineral analysis, they follow the same sequence of spectral orders, from zero-order (pure specular reflectance), to first-order (saturated red/yellow/blue) to high-order spectra of secondary and tertiary colors.[8]

In principle, light diffraction is the result of interaction between the incident light and structure of an object. To generate colors by diffraction the grating spacing d must be wider than the wavelength of the color to cause diffraction. When a plane wave of incident monochromatic light of wavelength λ hits a grating in the normal direction (perpendicular to the grating), each slit in the grating acts as a quasi point-source from which light is reflected in all directions. The diffracted light containing all the interfering wave components emanates from each slit in the grating. At any given point over the grating surface, the distance the light travels from each slit in the grating to this point will vary, and the phases of the waves will be also different. The waves with different phases at this point will add or subtract from one another to create peaks and valleys by constructive and destructive interference. When the path difference between the light from adjacent slits is equal to half the wavelength, $\lambda/2$, the waves will experience destructive interference and thus will cancel each other to create points of minimum intensity. Similarly, when the path difference is λ, constructive interference occurs

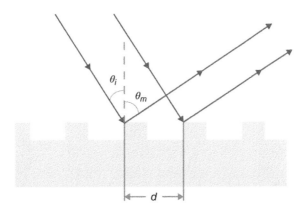

Figure 9.2   Schematic of cuticular grating. When white light beams on the surface with diffraction gratings, it diffracts into full spectra via reflection, creating the rainbow-like reflectance through interference.

and the phases will add together to form maxima. According to Bragg's laws, the maxima occur at angles θ_m, which satisfy the relationship

$$d\left(\sin\theta_i + \sin\theta_m\right) = m\lambda \qquad (9.1)$$

where θ_i and θ_m are the angles between the incident light and the grating's normal vector, and the diffracted ray and the grating's normal vector, respectively, d is the distance from the center of one slit to the center of the adjacent slit, and m is an integer representing the propagation mode of interest. For the incident light hitting a grating in the normal direction, the equation becomes $d(\sin\theta_m) = m\lambda$.

Although the grating equation (Equation 9.1) is derived based on an idealized grating, the relationship can be applied to any regular structure of the same spacing. The detailed materials structure of the grating determines the distribution of the diffracted light, but it will always give maxima in the directions given by the grating equation. When the maxima occur, the grating surfaces will appear a certain color corresponding with the wavelength.

9.2.2   Multilayer Reflectors

A multilayer is composed of alternatively distributed parallel layers with different refractive indices (Figure 9.3a). Multilayer reflectors are the most common iridescence mechanism in beetles. Many beetles have multilayer structures on their body skins that generate structural colors (Figure 9.3b). If the spacing of these layers approaches one-quarter the wavelength of visible light (approx. 380–750 nm), one or more colors will be produced by constructive interference (Figure 9.3c).[9] Multilayer structures can be far more elaborate and efficient than single-layer film and can generate stronger iridescence, combine two colors, or balance out the inevitable change of color with angle to give a more diffuse, less iridescent effect.[10] Multilayer structures are formed during insect integument. Thin parallel layers of chitin

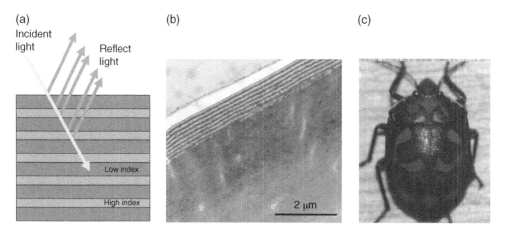

Figure 9.3 Multilayer reflectors. (a) A simple cuticular multilayer reflector. (b) Transmission electron micrograph (TEM) of epicticle and exocuticle. (c) Simple multilayer color in *T. diopthalmus*. Source: Fabricant *et al.* (2013).[11]

(sometimes interspersed with other materials) that differ in refractive index are secreted by the epidermis and later harden during sclerotization.[11]

The colorful appearances of multilayered photonic materials can be ascribed to interference and reflection, which can be described by Bragg's and Snell's laws,[1] given by:

$$\lambda = 2D\left(n_{eff}^2 - \cos^2 \theta\right)^{1/2} \tag{9.2}$$

where λ is the wavelength of the reflected light, n_{eff} is the volume-weighted average of the refractive index of the constituent photonic materials, D is diffracting plane spacing, and θ is the Bragg angle of incidence of the light falling on the nanostructures (the angle between the light and multilayer surface). The color reflected by a multilayer structure depends on the refractive index of the component layers and their periodicity. Layers with a greater optical thickness reflect longer wavelengths than thinner layers. The peak wavelength λ_{max} ($\theta=0$) can be expressed as

$$\lambda_{max} = 2\left(n_a d_a + n_b d_b\right) \tag{9.3}$$

where n is the refractive index, d is the actual layer thickness ($D = d_a + d_b$), and a and b are the alternating layers in the reflector ($n_{eff} = (n_a d_a + n_b d_b)/D$).[9] When reflector thickness varies between body regions, the surface of the body will generate multiple colors of different hue. It was shown that differences in the periodicity of reflecting layers could be the result of polymorphism and geographical variations in metallic color within species.[12] It is estimated that the average refractive index of individual reflecting layers in beetles ranges between 1.4 and 1.73.[12,13]

According to Equation 9.2, the *apparent* color of a simple multilayer reflector also varies with the angle of observation. If angle of illumination is constant with increasing angle of observation, that is, deviating from normal, the *apparent* color will undergo a blue shift to a shorter wavelength.[14] The color shift occurs because reflected light rays are traveling a shorter distance through each layer when increasing Bragg angle, leading to constructive interference occurring at shorter wavelengths. This relationship between viewing angle and apparent hue results in a multicoloured appearance in convex beetles. When viewed from above, the lateral regions of the pronotum and elytra will be blue shifted with respect to the top of the segment, as shown in Figure 9.3c.[8]

9.2.3 Two-dimensional Photonic Materials

2D photonic structures are cylindrical arrays embedded in a medium, for example cylindrical voids embedded in a high-refractive-index solid medium in iridescent setae from polychaete worms. This type of structure is less seen for generating structural color, but more often acts as an antireflection and self-cleaning layer. In a broader sense, several photonic structures, such as butterfly scales, can be considered as quasi-2D structures that can be equally regarded as 2D photonic materials or 2D diffraction gratings (bigratings).[2] *Morpho didius* butterfly wings, for example, have a ridge lamellar structure; each ridge consists of 10–12 alternate layers of cuticle and air (Figure 9.4). The spacing between the ridges is in the range of 0.5–5 µm, while the gaps in the lamellae are nanoscale. Like the multilayer reflectors, the lamellae have the gaps measured crest to crest, which matches the visible color wavelength from the red wavelength of 700 nm to the violet wavelength of 400 nm. The cuticle layer runs obliquely with respect to the plane of the scale shown in Figure 9.4c. The upper ends of the cuticle layers are distributed randomly over a whole area of the scale. As a result, the heights of the ridges are not so regular and are distributed within a layer interval of 0.2 μm. With this irregular distribution, no interference should occur because the light diffracted at each ridge is randomly superimposed. Thus, the properties of interference and diffraction essentially depend on a single ridge. When natural white light hits this nanostack of shelves or cuticles at different depths, its different component colors are bounced back by the shelves at different distances, and therefore arrive at a point at different times, resulting in interference and so brighter color. Because of the unique nanostructures, *Morpho* wings generate vivid blue light by this quasi-2D diffraction grating.[2]

In addition to the elaborate structure, the material properties also play an important role in generating the color. As can be seen in Figure 9.4c, the width of the layer is so small that the

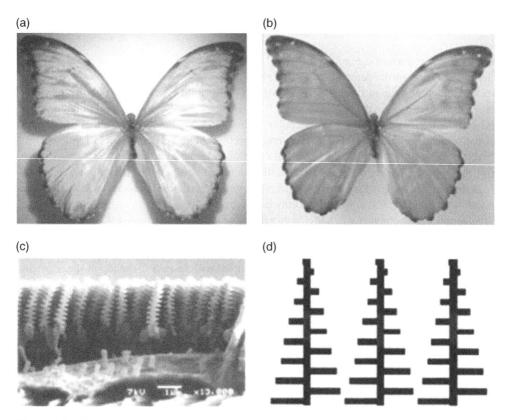

Figure 9.4 *Morpho didius* butterfly (a) close to normal artificial illumination and (b) diffuse, natural illumination. (c) SEM image of a broken scale of *M. rhetenor* showing the Christmas tree-like cross-section of the ridges. (d) Transverse cross-section of the photonic crystal constituted by chitinous material in parallel ridges and air. Source: Biro & Vigneron (2011).[2] Reproduced with permission of John Wiley and Sons.

multiple reflections at each layer do not occur owing to the diffraction effect. There is a large difference in the refractive indices between cuticle ($n = 1.56$) and air ($n = 1$), and the layer thickness nearly satisfies an ideal multilayer, causing high reflectivity with a large bandwidth. Thus, the *Morpho* butterfly realizes simultaneously the high reflectivity in a wide wavelength range and the generation of diffusive light in a wide angular range.[2] Although these two characteristics usually contradict each other, the combination of the regular and irregular nanostructures of *Morpho* butterflies meet these seemingly conflicting requirements.

9.2.4 Three-dimensional Photonic Crystals

3D photonic materials are those that have periodically packed (crystalline) structures, as shown in Figure 9.5a. For example, the photonic crystals in the scales of pachyrrhynchine weevils (*Pachyrrhynchus* and *Metapocyrtus*) have a close-packed hexagonal arrangement analogous to (mineral) opal, producing scintillating, gem-like reflectance, while those in *Lamprocyphus* have a diamond-based lattice (i.e., a face-centered cubic (FCC) system rather than a hexagonal one). In contrast to the close-packed colloidal crystals, in which microbeads occupy the crystal lattice, the inverse opal structures in the butterflies are composed of lattices of hollow air-filled voids within a network of interconnecting cuticle (Figure 9.5b). This photonic nanostructure appears to be inverse to the diamond-like tetrahedral structure, which could offer excellent reflectivity over a broad angle range.[1]

3D photonic crystals generate colors in a similar way to broadband multilayer reflectors. Both work through an "iridescence-reducing" color mechanism, that is, they not only reflect vivid, saturated interference colors but also reduce the angle dependency of the chromatic effect. However, for 3D photonic crystals, color is more related to the highly ordered lattice of nanoscale spheres. Figure 9.5c shows 3D photonic crystals in weevils and longhorn beetles. The inverse of this structure, spherical lacunae in a chitin matrix, produces the same optical effect.[15,16] Another example is peacock feathers that have brilliant color. A peacock feather consists of many barbs, each of which has a lot of branches called barbules. Each barbule has the shape of connected segments, typically 20–30 mm; its transverse cross-section is crescent-shaped and the brilliant color comes from these segments in the barbule. In fact, beneath the surface layer of the barbule there are several layers consisting of periodically arrayed particles in a lattice structure, whose diameters range from 110 to 130 nm. The layer intervals are dependent on the color of the feather: 140–150, 150, and 165–190 nm for blue, green, and yellow feathers, respectively.[17] Below the lattice structure, the particles are rather randomly distributed. In contrast to the transverse direction, the particles have an elongated shape of length 0.7 mm with a random arrangement in the longitudinal direction.[17] The slender particles are melanin granules and cause the barbule to be dark brown. Thus, the photonic crystal in the peacock forms a square lattice nearly in 2D. This peacock photonic crystal gives the peacock its brilliant color.

Compared with mineral crystals, which are usually highly ordered, photonic crystals in living organisms are not "perfect". Interestingly, it is this inherent imperfection that partially results in optimized diffuse reflection. For example, the 3D photonic structures in weevils usually contain an irregular assemblage in regular domains. These natural photonic materials are usually polycrystal instead of single crystal. Each grain in the polycrystal forms a highly ordered photonic crystal in short-range order (grain size); the orientation of the different

(a)   (b)

(c)   (d)

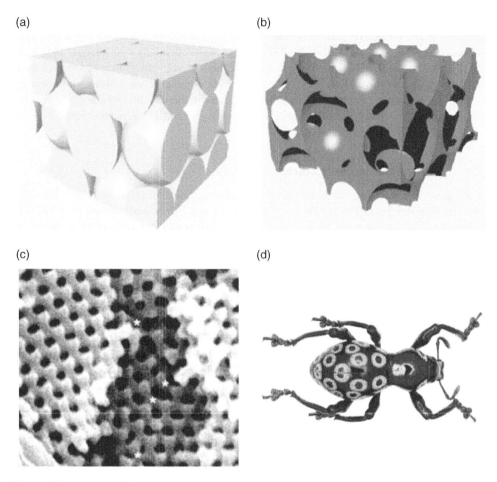

Figure 9.5 Graphical illustration of (a) an opal and (b) an inverse opal. Source: Ozin & Arsenault (2008).[18] Reproduced with permission of Elsevier. (c, d) Three-dimensional photonic crystals in weevils and longhorn beetles: (c) *Pachyrrhynchus congestus pavonius* (Curculionidae: Entiminae), SEM of crystal structure from scale interior and (d) *P. gemmatus*, habitus view. Source: Welch & Vigneron (2007)[19] and Welch *et al.* (2007).[20] Reproduced with permission of Springer.

grains varies across the scale (long-range disorder). The size of the domains ranges from single crystal to polycrystal, to amorphous structure. These various levels of disorder have strong optical effects; different coherence lengths can produce different visual effects, including iridescence (Figure 9.5d), dull metallic colors, and whites.[8]

9.2.5   Tunable Structural Color in Organisms

In nature, many animals can reversibly change their structural colors to match their surrounding environment. These animals form a large family, including *Paracheirodon innesi*, paradise whiptail, blue damselfish, tortoise beetle, hercules beetle, and so on.[1,21,22] In most cases, the structural color changes are achieved by adjusting the structures or properties of the 1D photonic multilayers. According to Equation 9.2, the structural color of multilayer structures can be tuned by changing (1) the diffracting plane spacing D, (2) the average refractive index n_{eff}, (3) the Bragg glancing angle θ, and (4) the n_{eff} and D simultaneously. Figure 9.6c shows three mechanisms of tuning structural color in nature where the key factors are refractive index, periodic nanostructure, and incident light angle.

Changing the diffracting plane spacing D is a typical way of tuning color. *Paracheirodon innesi*, for instance, can reversibly change skin color to evade its pursuers. This fish normally displays a structural color of cyan, produced by interference of light on periodically stacked microstructures of reflecting platelets in skin cells. Under stressful conditions, the fish rapidly changes color from blue to yellow. This rapid color shift is triggered by the simultaneous spacing variation of adjoining reflecting plates (Figure 9.6a). This space

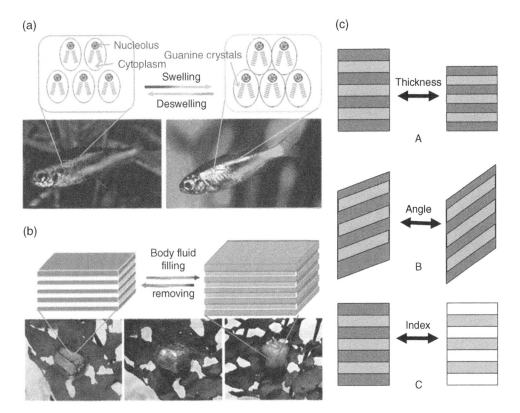

Figure 9.6 Natural creatures with variable structural colors. (a) *Paracheirodon innesi* normally displays a structural color of cyan. After the swelling of the oval iridophores, its stripe changes to yellow. (b) *Charidotella egregia* switches its structural color from gold within 1.5 minutes. This transformation is triggered by touching its wings, plausibly simulating a missed attempt at predation. Source: Zhao *et al.* (2012).[1] Reproduced with permission of the Royal Society of Chemistry. (c) Color tuning mechanisms in living creatures: (A) expansion and compression of extracellular space between protein platelets in cephalopods, (B) tilting protein platelet in the iridophore of neon tetra fish, (C) changing refractive index of a porous layer by absorbing liquid onto the beetle shell. Source: Zhao *et al.* (2012)[1] (http://www.ncbi.nlm.nih.gov/pmc/articles/PMC3673079/).

increase could be caused by swelling of the fish skin cells due to the inflow of water and a consequent increase in spacing among the platelets.[23]

Some insets and fish change their color by tuning glancing angle. The South American tropical fish *Neon tetra* changes its color from green in daytime to violet–blue at night.[24] The color change is achieved by tilting the platelets in the iridophore, as shown in Figure 9.6c(B).[25] Jellyfish have a similar color tuning mechanism based on a deformable and transparent body that iridesces across the whole visible spectrum due to their combs beating. This phenomenon originates from the angle-dependent reflection from a 2D photonic crystal formed in the combs.[26]

Some beetles can change color by varying the refractive index of the epidermis, as shown in Figure 9.6c(C). For example, the tortoise beetle and hercules beetle alter their structural colors by varying the amount of water in the cuticle and thus both the thickness and the average reflective index variations of their thin skin films (Figure 9.6b).[22] The *Hoplia coerulea* beetle changes its color from bright blue to emerald green.[27] Its skin film consists of a hygrochromic structure, a light-scattering, weakly ordered 3D lattice with air-filled porous regions. As moisture infiltrates the spaces, the difference in refractive index between chitin and lacunae reduces dramatically, resulting in more light absorption, and the elytron thus looks black in color.[28] This reversible color, also known as hygrochromic color, is induced by changing the refractive index of porous layers because of partially hydrating or dehydrating multilayer structures. The mechanisms behind this phenomenon should be useful for designing color-tunable materials.

9.3   Natural Antireflective Structures and Microlenses

While reflectors are often seen for structural colors, antireflective surfaces are also widely distributed in the animal world. Antireflectors eliminate reflection over broadband and large view angles, and enhance the optical transparency of the surface. For example, moths' eyes are covered with a natural nanostructured film. The reflectivity at the surface was found to decrease by two orders of magnitude compared with that of a flat cornea. This water-repellent, antireflective coating makes moths' eyes among the least reflective surfaces in nature. Similar protuberances have been found on the transparent wing of a hawk moth, *Cephonodes hylas*,[29] the corneal surfaces of butterflies,[30] and the eyes of some species of fly and mosquito.[31]

9.3.1   Moth-eye Antireflective Structures

The antireflective layer on the surface of the compound eye of a night moth (Figure 9.7a,b) consists of a hexagonal array of nanoscale cone-shaped protuberances, corneal nipples, with a spacing and height of 170 and 200 nm, respectively.[32] This type of structure reduces the reflectance of the lens surface from approximately 4% to less than 1%. Because of the strong antireflection of nipple arrays on the corneal surfaces of moths, the reduction in reflection increases transmission of light into the eye, making its visual system more sensitive to the surroundings, especially at night. In addition, reflection of sunlight from the eyes or wings could expose the location of the moth to passing predators. With antireflection properties, the moth-eye structure reduces the visibility of the moth, making it less vulnerable to predation.[33]

The nipple arrays found on winged surfaces appear to serve only as camouflage, making the moths less visible. In fact, for some species of moth, for example *Cryptotympana aquila* and

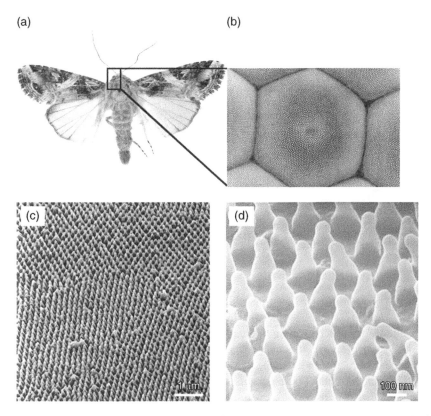

Figure 9.7   (a) Moth and (b) moth eye structure. Source: www.phys.org. (c) Moth-eye antireflective structures where nipple arrays can be found and (d) images of natural moth-eye structures found on the transparent wing of *Cephonodes hylas*. Source: Asadollahbaik *et al.* (2014).[35]

Cephonodes hylas (Figure 9.7c,d), nipple arrays can be found on both the dorsal and ventral surfaces of transparent sections of wing, increasing optical transmission to above 90% over a broad band of wavelengths, with a peak of ~98%. This makes the creature practically invisible to any passing predators.[33] In addition to antireflection, the protuberance arrays possess multifunctionalities, including superhydrophobicity to a surface, which leads to the antifogging, anti-adhesive, and self-cleaning properties exhibited by a range of species.[34]

The key for moth-eye structures in reducing reflection is their dimensions, which are less than the wavelength range of incident light. The wavelength of visible light ranges from 400 nm to 700 nm. When light interacts with nipple arrays with much smaller values (170 and 200 nm), it behaves as if it were encountering a continuous refractive index gradient between the air and the medium. This regular modulation of the surface is considered as to be a refraction matching; reflection is reduced by effectively removing the air–lens interface. In principle, if the light hits a plane surface, part or all of the light will be reflected by the surface (Figure 9.8a). However, if the surface has subwavelength features that are regular in size and arrangement (Figure 9.8b), they are known as a zero-order diffraction grating because all the higher orders are evanescent and only the zero order propagates. Consequently, when the wavelength of light is larger than the size of surface ridges it would see the surface as if it had

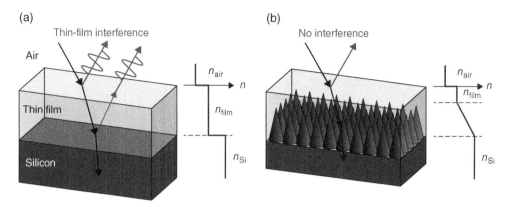

Figure 9.8 Moth-eye antireflective structures showing subwavelength structures and their analogous refractive index profiles, as experienced by incident light: (a) ridged profile and (b) tapered profile. Source: Yang *et al.* (2013).[37] Reproduced with permission of IOP Publishing.

a single layer of antireflective coating (ARC) of refractive index governed by the ratio between the ridges and channels. Likewise, a tapered profile will introduce a gradual change in refractive index from one medium to the other (Figure 9.8b).

The moth-eye nipple array effectively acts as an impedance-matching device, avoiding the sudden change in refractive index in the light transition between two media, thus reducing the reflectance of incident light. This mechanism can be understood by imagining the subwavelength-structured surface as a stack of thin layers, and so destructive interference between reflections from each layer cancel out all reflected light, maximizing the proportion of light transmitted. Apart from the nipple arrays, which show an apparently graded refractive index, squid eyes employ a "real" graded refractive index lens to reduce the reflection. This optical design entails a radial gradient of protein density, with low density in external layers and high density in internal layers.[36] This system also avoids spherical aberration, which is a serious technological problem in lens fabrication and causes considerable image blurring. Squid has found a clever solution to this problem.

9.3.2 Brittlestar Microlens with Double-facet Lens

Brittlestar, *Ophiocoma wendtii*, possesses a remarkable microlens array. These creatures belong to the group of marine invertebrates known as echinoderms. In this light-sensitive species, a calcitic skeleton forms a hexagonal array of microlenses integrated with pores (Figure 9.9a–c).[38] These arrays are located in the skeletal plates that protect the upper-arm joints on each of the brittlestar's five arms. Each microlens is about one-twentieth of a millimetre in diameter and is linked to six neighbours (Figure 9.9c). The cross-section image in Figure 9.9d shows that each protuberance has the appearance of a double lens with about 20–30 µm in radius of curvature of the upper face and rather less in that of the lower face. This double lens system gives a focal length of about 10 µm, with a focal-spot size of less than 3 µm. These focal spots are connected with bundles of nerve fibers of about that size, which are responsible for the sensitivity to light stimuli.

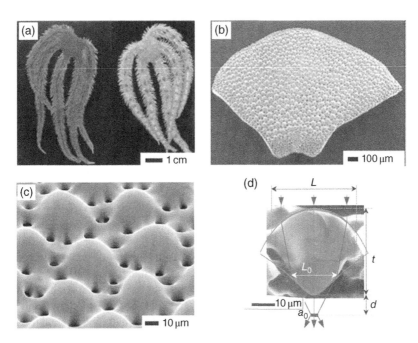

Figure 9.9 (a) Light-sensitive brittlestar *O. wendtii* changes color markedly from day (left) to night (right). (b) Scanning electron micrograph (SEM) of a dorsal arm plate (DAP) of *O. wendtii* cleansed of organic tissue. (c) SEM of the peripheral layer of a DAP from *O. wendtii* with enlarged lens structures. (d) High-magnification SEM of the cross-section of an individual lens in *O. wendtii*. Curved lines represent the calculated profile of a lens compensated for spherical aberration. The operational part of the calcitic lens (L_0) closely matches the profile of the compensated lens (bold lines). The light paths are shown in blue. L_0 is the operationa. Source: Yu *et al.* (2013).[40] Reproduced with permission of Elsevier.

There are strict requirements for the construction and operation of microlenses. First, there has to be exquisite control of calcite growth to form the lens structures. Second, each microlens should ideally have minimal optical aberration, and that seems to be the case. This last point both has been checked in experiments using an extracted array of lenses as the focusing elements and by modeling the optical response of such structures.

The microlens array that brittlestars possess is birefringence-free and aberration-free. These lens structures are formed by exquisite control of calcite growth. It is known that calcite is optically anisotropic, with different refractive indices for light polarized in different directions. Brittlestar has cleverly avoided birefringence effects by growing single crystals with the optical axis parallel to the axis of the double lens. More importantly, the top surface of the microlens in the skeleton of brittlestar is spherical while the bottom surface has a characteristic aspherical form, as seen in Figure 9.9d. The double-facet lens design renders the microlens totally compensated for spherical aberration. The micropores surrounding the lens act as light-intensity adjusting system by transporting the pigment that regulates the illumination dose reaching the lens. With such structural features, microlenses in brittlestars exhibit a unique focusing effect, signal enhancement, intensity adjustment, angular selectivity, and photochromic activity.[39] Such a design gives the living organisms quick and accurate detection and imaging of the surrounding environment and provides a solution to microlens design and fabrication.

9.3.3   Biomimetic Strategies for Structural Colors and Antireflection

Nature has developed optical structures employing a variety of strategies. These strategies include five fundamental optical processes and their combinations: (1) thin-film interference, (2) multilayer interference, (3) diffraction grating effect, (4) photonic crystals, and (5) light scattering.[41] In terms of optical functions, the natural photonic materials can be divided into two opposite groups: (1) materials for generating structural color with enhanced reflection, and (2) materials for reducing reflection and enhancing transmission. According to variations in the refractive index and period in space, photonic materials can be classified as 1D, 2D, or 3D. These intricate and elaborate structures generate a great variety of precise and subtle optical effects that can be used for optical materials and device development. Typical optical microstructures in nature, their optical mechanisms, and corresponding biomimetic strategies are summarized in Table 9.1.

These strategies provide an inspiration for the biomimetic design and fabrication of various types of artificial photonic materials and devices with the potential for applications ranging from everyday life to advanced optical technology. Based on natural fundamental structures, different optical structures can be constructed to generate much more diverse optical structures with excellent optical effects. For example, multilayer-based structures have many variations demonstrating the flexibility of nature's design skills. The ridge structures in butterfly wing scales are a modification of multilayer structure. In addition to fundamental interference, they even induce two optical effects in contrast: broad angular *Morpho* blue versus limited-view iridescence, and antireflection versus metallic white. Moreover, the same optical effect can be realized through very different structures. For example, an antireflection effect can be achieved by two distinguished structures: nipple array and inverse-V type ridges. Sculpted multilayers and 3D photonic crystal domains can produce the same color-mixing as causes green, and both chirped stacks and ridge structures can act as broadband reflectors.[40] These biological prototypes can be translated into advanced optical materials that are useful in the area of optical fibers, photovoltaic devices, Bragg mirrors, displays, display technologies, colorimetric sensors, and so on. All these natural photonic nanostructures, including their special biological functions and design strategies, offer an enormous number of blueprints for artificial photonic materials design, fabrication, and applications.

9.4   Bioinspired Structural Coloring Materials and Processes

With the inspiration of natural photonic structures, efforts have been made for the fabrication of artificial photonic materials with desired nanostructures. Similar to other nanomaterial fabrication, the fabrication methods for photonic materials can be classified into top-down and bottom-up approaches. In the top-down strategies, macroscopic tools, such as lithography and focused ion beam, are used to transfer a computer-generated pattern onto a larger piece of bulk material, and then a nanostructure is "sculpted" by physically removing materials or directly adding/rearranging ("writing") materials on a substrate. By contrast, bottom-up approaches take advantage of physicochemical interactions for the hierarchical synthesis of ordered nanostructures through the self-assembly of basic building blocks.[1] Compared with the top-down fabrication processes, bottom-up approaches are low cost, time-effective, and not restricted by nanoscales. These processes have been used to fabricate various artificial optical materials and devices, including structural coloration materials, antireflection coatings, and microlens arrays.

Table 9.1 Typical optical microstructures in nature exhibiting diverse optical effects through smart optical mechanisms and corresponding biomimetic strategies.

Photonic structures	Biological prototypes	Mechanisms	Illustrations	Biomimetic applications
1D	Diffraction grating	Diffraction		Film with wide angular structural color
	Multilayer	Multilayer interference and diffraction		Film with wide angular structural color
	Sculpted multilayer	Concurrent multilayer interference in different directions		Security labeling
2D	Quasi-2D multilayer ridges	Multilayer interference in different directions		Film with wide angular structural color
	Photonic crystal in bird feather, butterfly wing scale	Forbidding/guiding light via stopbands		Photonic crystal fiber, surface of LED
3D	Photonic crystal	Differently oriented crystal domains		Omnidirectional structural color

(Continued)

Table 9.1 (Continued)

Photonic structures	Biological prototypes	Mechanisms	Illustrations	Biomimetic applications
	Inverse opal	Differently oriented crystal domains		Omnidirectional structural color
Tunable structures	Multilayer on tropical fish neon tetra	Angle change		Sensors, displays
	Multilayer in *Paracheirodon innesi*	Dimensional change		Sensors, displays
	Multilayer in beetle scale	Refractive index change		Humidity sensors
Antireflective structures	Moth eye	Continuous index fitting		Antireflection coatings/surfaces
	Inverse type	Continuous index fitting		Antireflection coatings/surfaces
Microlens with double-facet lens	Brittlestar	Birefringence and aberration correction, intensity adjustment		Sensors, microfluid systems

9.4.1   Grating Nanostructures: Lamellar Ridge Arrays

Simple diffraction grating can generate brilliant colors, as demonstrated by beetles and butterfly wings. Inspired by nature, researchers have employed a variety of nanofabrication approaches to mimic smartly designed microstructures in nature. These methods can be generally classified as template-aided and non-template-replicated. In the templating approach, the biological prototypes (e.g., butterfly wing scales) are copied to form biomimetic structures by using atom layer deposition, a conformal evaporated-film-by-rotation technique, a computer-controlled surface sol–gel process, and molding lithography. For instance, the alumina replica shown in Figure 9.10a possesses exactly the lamellar microstructure of *Morpho* butterfly wing scales.[42] In this templating process, butterfly wing scales were coated with Al_2O_3 at $100\,°C$ and then annealed under $800\,°C$ in air to remove the organic template. The colors of the alumina replica change with different thicknesses of the alumina coating. A carbon replica of *Morpho* scales (Figure 9.10b) was also fabricated using the focused-ion-beam chemical-vapor-deposition (FIB-CVD) method without using templates.[43] In this process, a precursor of phenanthrene was utilized, resulting in a diamond-like carbon replica. This replica shows high reflectivity under illumination light of $440\,nm$ and brilliant blue reflection over a wide-angle range under optical microscope (Figure 9.10c).[43]

Although the detailed structure of butterfly wing scales can be duplicated, it is desirable to simplify the biomimetic models while keeping the major structural feature of the scales, avoiding the structural complexity of *Morpho* scales so that the broad angular structural blue can be created through mass production. This was achieved by using nanostructured multilayer materials to replace the alternating solid cuticle and air layers in *Morpho* scales. In this way, the authentic optical performance was realized by a much simpler structural design with appropriate materials. In this process, electron beam (EB) lithography and dry etching were used to fabricate the shelf box structure (Figure 9.10d). Comparable to dimensional parameters of natural scale, on a quartz plate, EB deposition was applied to deposit seven alternative layers of TiO_2 (about $40\,nm$ thick) and SiO_2 (about $75\,nm$ thick) onto the shelf box structure (Figure 9.10e).[44] The resulting film exhibits structural blue (Figure 9.10f). Nanoimprinting is a simple and versatile method of fabricating nanostructures on surfaces, and this process has been introduced to replace the EB lithography step. In the first step of the process a master mold was fabricated by EB lithography and dry etching, then a UV curable resin was utilized to act as a substrate for the shelf box structure. The artificial *Morpho* film shows a vivid blue color similar to butterfly wings (Figure 9.10f), therefore structural colors could be duplicated by the replicas and simplified biomimetic models. The future of the study of structural colors is extremely promising because the optical and physical investigation of their mechanisms can be immediately confirmed by fabricating reproductions directly from first principles.[40]

Perhaps, the most interesting biomimetic example in tunable structural color is Qualcomm's Mirasol display, which is the first full color, video-capable display on a prototype e-reader, built on the concept of the iridescence of butterfly's wings.[45] Mimicking the reflective properties of biological multilayers to create its color, two layers structures are constructed, consisting of electrically charged, tiny flexible membranes overlaid onto a mirrored surface, between which are sandwiched with a tiny separation (Figure 9.11). A thin-film stack on a glass substrate is suspended above a reflective membrane. By adjusting the spacing between the plates, the color wavelengths can be precisely adjusted to make the reflection light for interference to generate colors (e.g., blue, red, green as pixels). When an electrical potential is

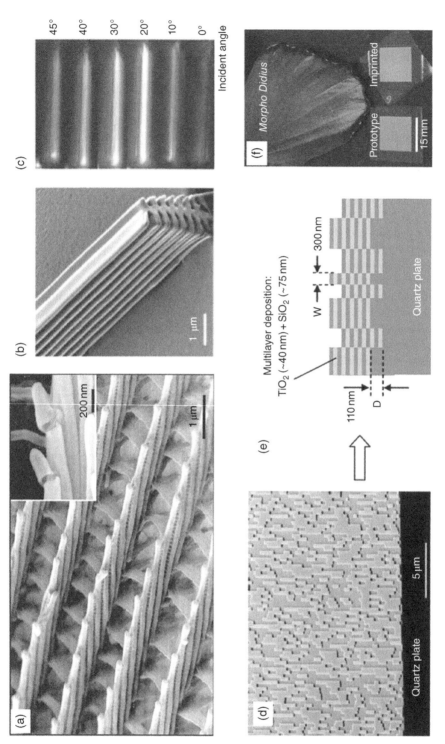

Figure 9.10 (a) Alumina replica of exactly the same fine multilayer ridge structure of *Morpho* scale fabricated through atomic deposition.[42] The inset shows two broken rib tips. (b) Pine-shaped multilayer structure fabricated through FIB-CVD. (c) Optical microscope images of *Morpho* butterfly-scale quasi-structure observed with a 5–45° incidence angle of white light. (d, e) Schematic of the fabrication artificial analogue combining electrobeam lithography and dry etching.[44] (f) Photograph of *Morpho didius* (upper) and replicated *Morpho*-blue (left: prototype, right: plate fabricated by nanoimprinting method for massproduction).[40,42–44] Source: Huang *et al.* (2006).[42] Reproduced with permission of Elsevier.

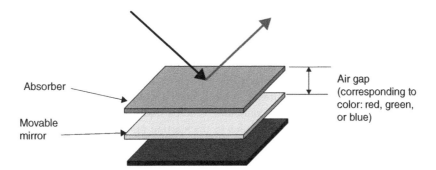

Figure 9.11   Qualcomm's Mirasol display working principle. In each pixel there are absorber and movable mirrors. The color is generated by adjusting the air gap between the absorber and the mirror. Source: https://www.qualcomm.com/products/mirasol/technology.

applied to the array of plates, electrostatic attraction brings the plates together, all the light colors except UV are absorbed, and one sees black. When the potential is released, however, one of three width gaps results, corresponding to the red, green or blue wavelengths. These colors in a pixel array are combined to create any color desired. The plate gaps can be adjusted 1000 times a second. The movable plates are bistable and therefore consume low power; energy is used to make changes but not maintain states. The displays require virtually no illumination or backlighting, a power-sucking requirement of LCD screens, resulting in significant energy savings. Though simple in structure, the sandwich plate units provide modulation, color selection, and memory while eliminating active matrices, color filters, and polarizers.

The Mirasol display technology is one of the most visible examples of bioinspired design in the technology sector. Based on the principles of structural color that are demonstrated by nature, the reflective display has been marketed as an energy saver that will yield a rich color display when used in full sunlight. This display technology comes from the adaptation of silicon chip manufacturing technology to MEMS devices. Although the butterfly's stacked shelves are not mimicked exactly, nor is its hierarchy of scale, it is a role model that shows that the same color effect can be achieved if the major feature is abstracted in biomimetics.

9.4.2   Multilayer Photonic Nanostructures and Fabrication Approaches

Synthetic multilayer photonic nanostructures mimicking the biological multilayer structures seen in the scales of beetles can generate vivid color with a broader view angle. These photonic materials consist of periodic stacks of layers with alternating high and low refractive indices. These stacks can be simply fabricated by layer-by-layer deposition techniques, such as electro-deposition of charged nanoparticles, the Langmuir–Blodgett technique, sol–gel chemistry-based deposition, and so on. The materials used in the fabrication are usually inorganic dielectric materials or inorganic nanoparticle/polymer composites. Porous structures or stimuli-responsive polymer materials were introduced into the layers of the multilayers to enhance variable stop bands.[1]

The multilayer photonic nanostructures were synthesized by self-assembly of organic dielectric materials, such as block copolymers.[46] The block copolymers are composed of two or more covalently linked chemically different polymers (or blocks). Multilayer films with

PBGs can be formed by assembling the block copolymers under certain conditions of temperature, external fields, solvent and its evaporation rate, and so on.[1] Multilayer photonic nanostructures have also been fabricated by soft lithography with holographic patterning.[47] With a photopolymer as the starting material, holographic lithography produces periodic index modulations using the interference pattern of coherent light beams. A relatively large area of multilayer periodic structures can be achieved with this single-step fabrication approach. In addition, polymer films containing ordered liquid crystals can be produced by adding non-reactive liquid crystal molecules to the starting photopolymer.[48] Other kinds of additives, such as nanoparticles and non-reactive solvent, could also be mixed with the photopolymer to make multilayer photonic nanostructures. The interference pattern of coherent light beams was fabricated by exposing photosensitive polymer-silver halide photographic emulsions in a holographic approach.[49] The virtual pattern can comprise fringes of ultra-fine metallic silver grains embedded within the thickness of the polymer film. These silver fringes lie in planes parallel to the substrate surface and are separated by a distance of approximately half the wavelength of the laser beam used to generate them.[1] These synthetic multilayer photonic nanostructures mimicking biological multilayer structures can generate vivid structural colors.

9.4.3   Three-dimensional Photonic Crystals and Fabrication

Nanoparticles can be assembled into colloidal crystals with highly ordered arrays, analogous to standard crystals whose repeating subunits are atoms or molecules. These arrays are found in many biological photonic materials and can display shiny structural colors. Gem opal, for example, is composed of close-packed domains of colloidal nanoparticles, which generate brilliant colors.[50] The 3D photonic crystals in the colloidal nanoparticles have been fabricated by several methods. The simplest approach is natural sedimentation.[51] To improve the efficiency, higher temperatures, external electrical fields, enhanced gravitational fields in centrifugation and suspension flow filtration, as well as oscillatory shear and sonic fields, have been applied in sedimentation. Among these approaches, the vertical deposition method is one of the most successful in the fabrication of high-quality colloidal crystal films. A method based on vertical deposition, the lifting substrate, operates much faster (Figure 9.12a). Here the concentration of particles essentially remains unchanged during film formation, so the thickness of the films is relatively uniform.[52] The thicknesses of the films can be tailored from just a single layer to several tens of layers by precisely controlling the concentrations of the nanoparticles and the lifting speed (Figure 9.12b). The structural color of the colloidal crystal film can also be tailored from red to blue by using different sizes of nanoparticles (Figure 9.12c).

Biomimetic inverse opal films have also been synthesized using various methods. Inverse opal is the interestial space between the particles in colloidal crystals. The common fabrication method for inverse colloidal crystals (inverse opal) involves replication of colloidal crystal templates. In this process, the materials are filled into the free voids of the colloidal crystal templates and then the templates are removed. There are two infiltration processes: non-reactive and reactive. In non-reactive infiltration, ultrafine nanoparticles much smaller than the channels of the free voids, such as silica, gold, or titania nanoparticles, are infiltrated with dispersion liquid into the voids.[1] After the dispersion liquid evaporation or electrophoresis treatment, the ultrafine nanoparticles can fully fill the void of the colloidal crystal templates, and the low-shrinkage inverse opals are obtained after removing the templates. In reactive

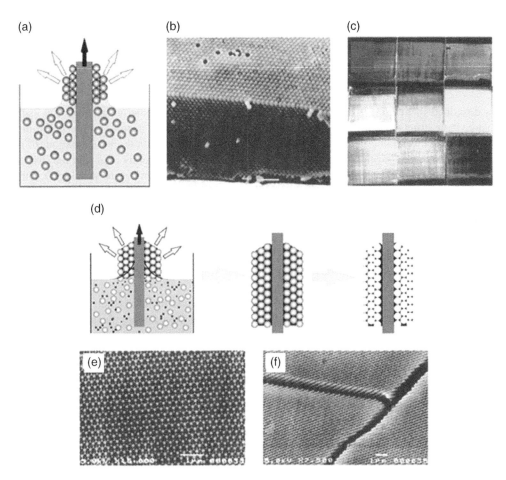

Figure 9.12 Schematic representation of various methods for close-packed colloidal crystal formation: (a) lifting substrate, (b) cross-sectional images of colloidal crystals with the thickness of 15 layers and 31 layers, and (c) the photos of colloidal crystal films composed of different sizes of silica nanoparticles.[52] (d) Scheme of co-deposition preparation routes to inverse opals, and SEM image of the silica inverse opal film: (e) plan-view image and (f) cross-sectional image Source: Zhao *et al.* (2012).[1] Reproduced with permission of the Royal Society of Chemistry.

infiltration, monomer–initiator mixture or pre-polymer solution is infiltrated into the void of the colloidal crystal, then polymerized or UV cured to form polymeric inverse opals. Metal oxide or silica inverse opals could also be fabricated by infiltrating the precursor structure with the corresponding sol–gel solution, followed by hydrolyzation and solidification in the void of the colloidal crystal templates.[1] In addition to those processes, inverse opals were fabricated by co-deposition (Figure 9.12d) in which colloidal particles as opal template and the solution/materials in the interstitial sites simultaneously deposit to form a composite film.[53] Usually, the deposition is carried out using a mixture of the templating colloidal nanoparticles with the matrix material precursor. In order to fill the interstitial space between colloidal nanoparticles, the matrix material precursor should be smaller than the space without interfering with the template crystal formation. Silica and titania inverse opals have been formed by this method (Figure 9.12e,f).

9.4.4 Tunable Structural Colors of Bioinspired Photonic Materials

As described in Section 9.2.2, the scales of some fish and the epidermises of insects can change their structural color. These examples may provide valuable inspiration for the bioinspired design of variable photonic materials. According to diffraction theory (Equation 9.2), the structural colors of photonic materials can be tuned by independently adjusting the Bragg glancing angle θ, the average refractive index n_{eff}, and the diffracting plane spacing D, and simultaneously changing n_{eff} and D. However, a change in one parameter usually causes a change in others. For example, when changing D, the photonic materials should be swollen or shrunk, and this process is always accompanied by a change in n_{eff}, therefore there is no pure protocol for tuning the structural colors of the photonic materials based on changing the diffracting plane spacing. In general, the diffracting plane spacing is a more dominate factor in tuning the structural colors. Various structural color tuning technologies have been developed to design and fabricate bioinspired tunable optical materials.

Many photonic nanostructured materials are anisotropic and therefore their color is angle dependent. This feature can be used to tune the apparent color by rotating the photonic materials themselves under external triggers. A photonic ball with a tunable glancing angle by a magnetic field, for example, can turn their color by applying a magnetic field.[54,55] To fabricate magnetochromatic microspheres, uniform superparamagnetic colloidal particles were selected and dispersed in a liquid. The color of the magnetochromatic microspheres can then be conveniently switched between the "on" and "off" states by using an external magnetic field.[54] Upon applying a vertical magnetic field, the particle chains stand straight, turning "on" their diffraction, and therefore the corresponding color that could be observed from the top. If the field is switched horizontally, the microspheres are forced to rotate 90° to lay down the particle chains, thus turning "off" the diffraction, and the microspheres show their original color. These orientation controllable magnetochromatic microspheres could be applied in structural color displays as well as forgery protection and other light-mediated authentication.[1]

For the photonic materials with multiple components or composite structure, the refractive indices can be described by the effective medium theory. According to this theory, the refractive index of photonic materials takes the average value of the refractive indices of all their components,[56] which can be approximately calculated by:

$$n_{eff}^2 = \sum n_i^2 V_i \qquad (9.4)$$

where n_i and V_i are the refractive index and the volume fraction, respectively, of the individual components. The apparent color of the photonic composites can therefore be tuned by varying the components or tuning the refractive index of the composed materials. Various materials and processes have been used to tune the color of photonic materials, including solvent infiltration,[57] phase change,[58] molecular reaction,[59] etc. The close-packed colloidal crystals, for example, were fabricated by the self-assembly of monodisperse nanoparticles, during which free voids are formed among these nano-components. The average refractive index of these structures with voids can be adjusted by filling the free voids (air) with solutions with a high refractive index, thus leading to a red shift of the stop-band. Apart from the opal structural colloidal crystals, the inverse opals are more effective in varying their average refractive index in a wider range, and show a more manifest shift in their stop-bands since they have more void percentage (about 74%) than the opal structural colloidal crystals (26%). The solution infiltrating-induced shift of the stop-bands could be further enhanced by using low refractive index materials such as gold, silver, or titania as the structural materials of the colloidal crystals.[57]

9.4.5 Electrically and Mechanically Tunable Opals

Photonic ink, or P-Ink, is a composite comprising an opal embedded in a matrix of a specialized redox-active polyferrocenylsilane gel.[60] When this metallopolymer gel is incorporated into an electrochemical cell, it swells or shrinks reversibly upon varying the applied voltage, and concomitantly the color of the P-Ink material can be shifted as depicted in Figure 9.13. In this way, the color of the device can be electrically tuned to any wavelength across the whole visible spectrum.[18] This color tuning method does not need expensive color filters or the subpixellation of red–green–blue elements, thereby wasting much less of the accessible light output. Obviously, this technology could form the basis of full-color opal coatings for a wide range of applications.

Color change under pressure in a controllable way has also been demonstrated, for example the color of an elastomeric inverse opal dubbed Elast-Ink can respond to mechanical pressure.[61] This inverse opal material is fabricated by mixing opal with synthetic rubber to form a composite opal. After the rubber is cross-linked, the opal in the composite is dissolved away, leaving an interconnected network of air voids embedded in a rubbery matrix, as shown in Figure 9.13b. The elastic inverse opal has a highly porous architecture and can be easily compressed, and its color gradually shifts across the entire visible spectrum as increasing pressure is applied. This pressure-tunable optical material can be used to develop a highly sensitive and accurate fingerprint sensor. When pressing on this material, the elastic inverse opal captures in full color the topography of the ridges and valleys on a person's finger. With potential applications in forensics, biometrics, security, and authentification devices, this material could provide viable solutions in an increasingly security-conscious world.[18,62]

9.5 Bioinspired Antireflective Surfaces and Microlenses

The tapered pillar arrays found on the surface of moth eyes minimize the reflection over broadband and large view angles. Inspired by the effectiveness of natural moth-eye structures for reducing surface reflection, efforts have been made to reproduce such structures for technological applications. Moth-eye structures designed for visible wavelengths require pillar spacings on the scale of several hundred nanometers, too small to be fabricated using standard photolithography techniques. A range of other nanoscale fabrication techniques, such as EB lithography, nanoimprinting, colloidal lithography, self-masked dry etching, and interference lithography, have been employed to create artificial moth-eye antireflective structures in a variety of materials.[63] The process usually creates a pattern in a polymer film on top of a hard substrate, followed by etching to transfer the pattern into the hard substrate. Polymer moth-eye structures are often used as an antireflective surface for glass by careful choice of the polymer so that its refractive index closely matches that of the substrate.

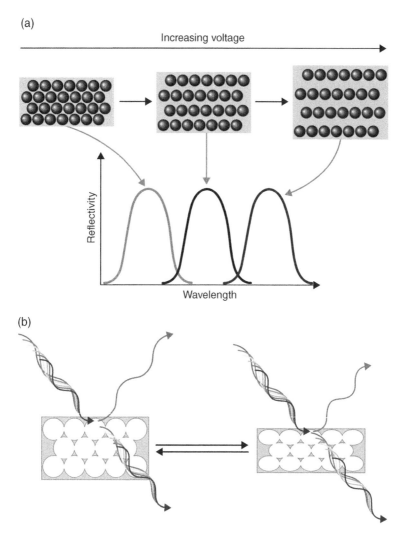

Figure 9.13 (a) Illustration of the operation of P-Ink voltage-tuneable full-color opal coatings. (b) Elastic inverse opal, Elast-Ink, where compression reduces the lattice constant and blue-shifts the color. Source: Ozin & Arsenault (2008).[18] Reproduced with permission of Elsevier.

Moth-eye antireflective films can be applied to solar cells, displays, and other optical devices. These structures have been formed on the encapsulant surface of organic solar cells (OSCs), reducing the reflectance at the air–glass interface and thereby increasing the performance of the cell by 2.5–3.5%.[64] For light-emitting devices (LEDs), light extraction is important to the efficiency of the devices. Because of the total internal reflection and the waveguiding modes in the glass substrate, only about 20% of the generated light can irradiate from the LEDs. By applying the biomimetic concept of antireflection, the perceived brightness of LED/LCD screens could be enhanced to reduce their power consumption. To enhance the light extraction, silica cone arrays were fabricated on the surfaces of the indium tin oxide

(ITO) glass substrate. These arrays modulate the above two bottleneck factors: the internal reflection and the waveguiding, and as a result the light luminance efficiency of white LEDs is significantly improved by a factor of 1.4 in the normal direction with an even larger enhancement for large viewing angles.[49,65]

Another fascinating application of antireflection structures is the use in the micro Sun sensor for Mars rovers.[66] The location coordinates of the rover are calculated on the basis of the recorded image by an active pixel sensor. However, the accuracy of the location coordinates is

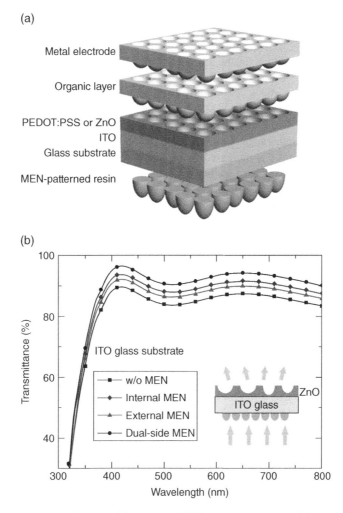

Figure 9.14 (a) Schematic diagram of OLED and OSC integrated with dual-side MENs. Internal MEN is first patterned on a PEDOT:PSS or ZnO layer on ITO-coated glass substrate for OLED or OSC, respectively. The organic active layers and metal electrode are then deposited on the corrugated PEDOT:PSS or ZnO layer. The diagram is not to scale. (b) Transmittance of the incident light from ITO side to PEDOT:PSS-coated OLED substrates patterned without MEN (squares) and with internal (diamond), external (triangles), and dual-side (circles) MEN. Inset depicts the schematic of the measurement procedure. Source: Zhou *et al.* (2014).[67] Reproduced with permission of Nature Publishing Group.

severely affected by the ghost image resulting from the multiple internal reflections of the optical system. This problem was solved by fabricating dense nanotip arrays on the surfaces of the sensor. The nanotip arrays reduced internal reflection by nearly three orders of magnitude compared with no treatment, resulting in a more reliable three-axis attitude.

As an example of moth-eye structure application to LED, a dual-side moth-eye-inspired coating with broadband antinanoreflective (MEN) and quasi-omnidirectional properties has been developed to improve the efficiency of LEDs.[67] Figure 9.14a shows the device structure constructed with dual-side MEN. The structure was fabricated by an imprinting technique using a perfluoropolyether (PFPE) mold containing the MEN. The MEN structure was then transferred by applying compressive stress on the surface of poly(3,4-ethylene dioxythiophene):polystyrene sulfonic acid, ZnO, or UV-curable resin films supported on ITO glass substrates. The structure on the surface of the PFPE molds contains two important features that a moth's eye possesses that are critical to produce the desired functionality: sub-wavelength structures and a continuously tapered morphology.[68] With these features of moth's eyes, the LEDs and OSCs with MEN show an effective gradient refractive index profile at the interface, and the light is therefore manipulated in all azimuthal directions over the entire emission wavelength range. Overall, the light out-coupling efficiency of LEDs with stacked triple emission units is increased by bioinspired nanostructures more than two-fold compared to a conventional device, resulting in a drastic increase in external quantum efficiency and current efficiency to 119.7% and 366 cd/A without introducing spectral distortion and directionality. Similarly, OSCs show a 20% increase in light in-coupling efficiency, and consequently an enhanced power conversion efficiency of 9.33%.[67]

Double-facet microlens structures are also found in nature; in the brittlestar light-sensitive system they help to effectively eliminate the aberration and birefringence that spherical lenses usually have. Inspired by the unique lens design and the outstanding optical properties of microlenses in brittlestars, researchers have successfully fabricated the biomimetic analogue through three-beam interference lithography from a negative tone photoresist, SU8.[69] The symmetry and connectivity of the patterns were controlled by adjusting the beam wave vectors and their polarization. The pore size and porosity are determined by the laser intensity and exposure time. After the photoresists were exposed and dissolved, microlens arrays were synthesized. These synthetic double-facet microlens structures are strikingly similar to their biological prototype and capable of focusing light. The artificial microlens arrays can be used as photomasks. A variety of different microscale patterns can be created by using the same photomask and simply adjusting the illumination dose and the distance between the mask and the photoresist film.[69]

References

1. Zhao, Y.J., Xie, Z.Y., Gu, H.C., Zhu, C. & Gu, Z.Z. Bio-inspired variable structural color materials. *Chemical Society Reviews* **41**, 3297–3317 (2012).
2. Biro, L.P. & Vigneron, J.P. Photonic nanoarchitectures in butterflies and beetles: valuable sources for bioinspiration. *Laser Photonics Review* **5**, 27–51 (2011).
3. Parker, A.R. & Townley, H.E. Biomimetics of photonic nanostructures. *Nature Nanotechnology* **2**, 347–353 (2007).
4. Vukusic, P. & Sambles, J.R. Photonic structures in biology. *Nature* **424**, 852–855 (2003).
5. Takeoka, Y. Fusion materials for biomimetic structurally colored materials. *Polymer Journal* **47**, 106–113 (2015).
6. Zhao, N. *et al.* Bioinspired materials: from low to high dimensional structure. *Advanced Materials* **26**, 6994–7017 (2014).

7. Wang, H. & Zhang, K.Q. Photonic crystal structures with tunable structure color as colorimetric sensors. *Sensors (Basel)* **13**, 4192–4213 (2013).
8. Seago, A.E., Brady, P., Vigneron, J.P. & Schultz, T.D. Gold bugs and beyond: a review of iridescence and structural colour mechanisms in beetles (Coleoptera). *Journal of the Royal Society Interface* **6** (Suppl 2), S165–S184 (2009).
9. Land, M.F. The physics and biology of animal reflectors. *Progress in Biophysics & Molecular Biology* **24**, 75–106 (1972).
10. Ball, P. Nature's color tricks. *Scientific American* **306**, 74–79 (2012).
11. Fabricant, S.A., Kemp, D.J., Krajicek, J., Bosakova, Z. & Herberstein, M.E. Mechanisms of color production in a highly variable shield-back stinkbug, *Tectocoris diopthalmus* (Heteroptera: Scutelleridae), and why it matters. *Plos One* **8**, UNSP e64082 (2013).
12. Kurachi, M., Takaku, Y., Komiya, Y. & Hariyama, T. The origin of extensive colour polymorphism in *Plateumaris sericea* (Chrysomelidae, Coleoptera). *Naturwissenschaften* **89**, 295–298 (2002).
13. Noyes, J.A., Vukusic, P. & Hooper, I.R. Experimental method for reliably establishing the refractive index of buprestid beetle exocuticle. *Optical Express* **15**, 4351–4358 (2007).
14. Kinoshita, S. & Yoshioka, S. Structural colors in nature: the role of regularity and irregularity in the structure. *ChemPhysChem* **6**, 1442–1459 (2005).
15. Galusha, J.W., Richey, L.R., Gardner, J.S., Cha, J.N. & Bartl, M.H. Discovery of a diamond-based photonic crystal structure in beetle scales. *Physical Review E* **77** (2008).
16. Parker, A.R. & Martini, N. Structural colour in animals – simple to complex optics. *Optics & Laser Technology* **38**, 315–322 (2006).
17. Zi, J. *et al.* Coloration strategies in peacock feathers. *Proceedings of the National Academy of Sciences of the United States of America* **100**, 12576–12578 (2003).
18. Ozin, G.A. & Arsenault, A.C. P-Ink and Elast-Ink from lab to market. *Materials Today* **11**, 44–51 (2008).
19. Welch, V.L. & Vigneron, J.P. Beyond butterflies – the diversity of biological photonic crystals. *Optical and Quantum Electronics* **39**, 295–303 (2007).
20. Welch, V., Lousse, V., Deparis, O., Parker, A. & Vigneron, J.P. Orange reflection from a three-dimensional photonic crystal in the scales of the weevil *Pachyrrhynchus congestus pavonius* (Curculionidae). *Physical Reveiw E* **75** (2007).
21. Fudouzi, H. Tunable structural color in organisms and photonic materials for design of bioinspired materials. *Science and Technology of Advanced Materials* **12**, 064704 (2011).
22. Vigneron, J.P. et al. Switchable reflector in the Panamanian tortoise beetle *Charidotella egregia* (Chrysomelidae: Cassidinae). *Physical Review E* **76** (2007).
23. Lythgoe, J.N. & Shand, J. The structural basis for iridescent color changes in dermal and corneal iridophores in fish. *Journal of Experimental Biology* **141**, 313–325 (1989).
24. Lythgoe, J.N. & Shand, J. Changes in spectral reflections from the iridophores of the neon tetra. *Journal of Physiology – London* **325**, 23–& (1982).
25. Yoshioka, S. et al. Mechanism of variable structural colour in the neon tetra: quantitative evaluation of the Venetian blind model. *Journal of the Royal Society Interface* **8**, 56–66 (2011).
26. Welch, V., Vigneron, J.P., Lousse, V. & Parker, A. Optical properties of the iridescent organ of the comb-jellyfish *Beroe cucumis* (Ctenophora). *Physical Review E* **73** (2006).
27. Rassart, M., Simonis, P., Bay, A., Deparis, O. & Vigneron, J.P. Scale coloration change following water absorption in the beetle *Hoplia coerulea* (Coleoptera). *Physical Review E* **80** (2009).
28. Rassart, M., Colomer, J.F., Tabarrant, T. & Vigneron, J.P. Diffractive hygrochromic effect in the cuticle of the hercules beetle *Dynastes hercules*. *New Journal of Physics* **10** (2008).
29. Yoshida, A., Motoyama, M., Kosaku, A. & Miyamoto, K. Antireflective nanoprotuberance array in the transparent wing of a hawkmoth, *Cephonodes hylas*. *Zoological Science* **14**, 737–741 (1997).
30. Stavenga, D.G., Foletti, S., Palasantzas, G. & Arikawa, K. Light on the moth-eye corneal nipple array of butterflies. *Proceedings of the Royal Society B: Biological Sciences* **273**, 661–667 (2006).
31. Liu, K.S., Yao, X. & Jiang, L. Recent developments in bio-inspired special wettability. *Chemical Society Reviews* **39**, 3240–3255 (2010).
32. Bernhard, C.G. Structural and functional adaptation in a visual system. *Endeavour* **26**, 79–84 (1967).
33. Boden, S.A. & Bagnall, D.M. Moth-eye antireflective structures. *Encyclopedia Nanotechnology* 1467–1477 (2012).
34. Peisker, H. & Gorb, S.N. Always on the bright side of life: anti-adhesive properties of insect ommatidia grating. *Journal of Experimental Biology* **213**, 3457–3462 (2010).

35. Asadollahbaik, A. *et al.* Reflectance properties of silicon moth-eyes in response to variations in angle of incidence, polarisation and azimuth orientation. *Optical Express* **22**, A402–A415 (2014).

36. Sweeney, A.M., Des Marais, D.L., Ban, Y.-E.A. & Johnsen, S. Evolution of graded refractive index in squid lenses. *Journal of the Royal Society Interface* **4**, 685–698 (2007).

37. Yang, Q., Zhang, X.A., Bagal, A., Guo, W. & Chang, C.-H. Antireflection effects at nanostructured material interfaces and the suppression of thin-film interference. *Nanotechnology* **24**, 235202 (2013).

38. Aizenberg, J., Tkachenko, A., Weiner, S., Addadi, L. & Hendler, G. Calcitic microlenses as part of the photoreceptor system in brittlestars. *Nature* **412**, 819–822 (2001).

39. Aizenberg, J. & Hendler, G. Designing efficient microlens arrays: lessons from Nature. *Journal of Materials Chemistry* **14**, 2066–2072 (2004).

40. Yu, K., Fan, T., Lou, S. & Zhang, D. Biomimetic optical materials: Integration of nature's design for manipulation of light. *Progress in Materials Science* **58**, 825–873 (2013).

41. Kinoshita, S. & Yoshioka, S. Structural colors in nature: The role of regularity and irregularity in the structure. *ChemPhysChem* **6**, 1442–1459 (2005).

42. Huang, J., Wang, X. & Wang, Z.L. Controlled replication of butterfly wings for achieving tunable photonic properties. *Nano Letters* **6**, 2325–2331 (2006).

43. Watanabe, K., Hoshino, T., Kanda, K., Haruyama, Y. & Matsui, S. Brilliant blue observation from a *Morpho*-butterfly-scale quasi-structure. *Japanese Journal of Applied Physics Part 2 – Letters & Express Letters* **44**, L48–L50 (2005).

44. Saito, A. *et al.* Reproduction, mass-production, and control of the *Morpho*-butterfly's blue. *Proceedings of SPIE* **7205**, Advanced Fabrication Technologies for Micro/Nano Optics and Photonics II **720506** (2009).

45. Mitchell, M.W. Brilliant displays. *Scientific American* 94–97 (2007).

46. Kim, S.O. *et al.* Epitaxial self-assembly of block copolymers on lithographically defined nanopatterned substrates. *Nature* **424**, 411–414 (2003).

47. Campbell, M., Sharp, D.N., Harrison, M.T., Denning, R.G. & Turberfield, A.J. Fabrication of photonic crystals for the visible spectrum by holographic lithography. *Nature* **404**, 53–56 (2000).

48. Tondiglia, V.P., Natarajan, L.V., Sutherland, R.L., Tomlin, D. & Bunning, T.J. Holographic formation of electro-optical polymer-liquid crystal photonic crystals. *Advanced Materials* **14**, 187–191 (2002).

49. Song, Y.M. *et al.* Disordered antireflective nanostructures on GaN-based light-emitting diodes using Ag nanoparticles for improved light extraction efficiency. *Applied Physical Letters* **97**, 093110 (2010).

50. Marlow, F., Muldarisnur, Sharifi, P., Brinkmann, R. & Mendive, C. Opals: Status and prospects. *Angewandte Chemie International Edition* **48**, 6212–6233 (2009).

51. Davis, K.E., Russel, W.B. & Glantschnig, W.J. Settling suspensions of colloidal silica – observations and X-ray measurements. *Journal of the Chemical Society – Faraday Transactions* **87**, 411–424 (1991).

52. Gu, Z.Z., Fujishima, A. & Sato, O. Fabrication of high-quality opal films with controllable thickness. *Chemistry of Materials* **14**, 760–765 (2002).

53. Subramanian, G., Manoharan, V.N., Thorne, J.D. & Pine, D.J. Ordered macroporous materials by colloidal assembly: A possible route to photonic bandgap materials. *Advanced Materials* **11**, 1261–1265 (1999).

54. Kim, J. *et al.* Real-time optofluidic synthesis of magnetochromatic microspheres for reversible structural color patterning. *Small* **7**, 1163–1168 (2011).

55. Ge, J. *et al.* Magnetochromatic microspheres: Rotating photonic crystals. *Journal of the American Chemical Society* **131**, 15687–15694 (2009).

56. Gu, Z.Z. *et al.* Varying the optical stop band of a three-dimensional photonic crystal by refractive index control. *Langmuir* **17**, 6751–6753 (2001).

57. Kuo, C.-Y., Lu, S.-Y., Chen, S., Bernards, M. & Jiang, S. Stop band shift based chemical sensing with three-dimensional opal and inverse opal structures. *Sensors and Actuators B – Chemical* **124**, 452–458 (2007).

58. Han, G.-Z., Xie, Z.-Y., Zheng, D., Sun, L.-G. & Gu, Z.-Z. Phototunable photonic crystals with reversible wavelength choice. *Applied Physics Letters* **91** 141114 (2007).

59. Zhao, Y. *et al.* Encoded porous beads for label-free multiplex detection of tumor markers. *Advanced Materials* **21**, 569–572 (2009).

60. Arsenault, A.C. *et al.* P-Ink displays: Flexible, low power, reflective color. *Advanced Fabrication Technologies for Micro/Nano Optics and Photonics Vi* **8613** (2013).

61. Arsenault, A.C. *et al.* From colour fingerprinting to the control of photoluminescence in elastic photonic crystals. *Nature Materials* **5**, 179–184 (2006).

62. Arsenault, A.C., Puzzo, D.P., Manners, I. & Ozin, G.A. Photonic-crystal full-colour displays. *Nature Photonics* **1**, 468–472 (2007).
63. Li, Y., Zhang, J. & Yang, B. Antireflective surfaces based on biomimetic nanopillared arrays. *Nano Today* **5**, 117–127 (2010).
64. Forberich, K. *et al.* Performance improvement of organic solar cells with moth eye anti-reflection coating. *Thin Solid Films* **516**, 7167–7170 (2008).
65. Li, Y. *et al.* Improved light extraction efficiency of white organic light-emitting devices by biomimetic anti-reflective surfaces. *Applied Physical Letters* **96**, 153305 (2010).
66. Lee, C., Bae, S.Y., Mobasser, S. & Manohara, H. A novel silicon nanotips antireflection surface for the micro sun sensor. *Nano Letters* **5**, 2438–2442 (2005).
67. Zhou, L. *et al.* Light manipulation for organic optoelectronics using bio-inspired moth's eye nanostructures. *Scientific Reports* **4** (2014).
68. Lee, L.P. & Szema, R. Inspirations from biological, optics for advanced phtonic systems. *Science* **310**, 1148–1150 (2005).
69. Yang, S., Ullal, C.K., Thomas, E.L., Chen, G. & Aizenberg, J. Microlens arrays with integrated pores as a multi-pattern photomask. *Applied Physics Letters* **86** (2005).

10

Catalysts for Renewable Energy

10.1 Introduction

Fossil fuels currently provide more than 80% of global energy needs. Such large-scale burning of fossil fuels has posed serious problems in energy sustainability, human health, and environmental pollution. It is therefore critical to develop alternative or additional energy sources via renewable and clean energy pathways ensuring environmental protection and harmony. The solar radiation reaching the Earth's surface every day is phenomenal, in the range of terawatts. If even a few per cent of this abundant, clean, and free resource can be converted and stored, global energy needs will be met. To decrease our dependency on fossil fuels, it is necessary to develop clean and sustainable energy sources from solar radiation.

Renewable energy technologies, such as photochemical water splitting for hydrogen fuel generation, fuel cells for direct conversion from fuels to electricity, and batteries for energy storage, are promising for future energy sources, including portable power sources, engines, and power plants. These emerging energy conversion and storage technologies are currently under intensive research and development because of their extremely high energy density and low pollution. At the heart of these renewable energy technologies, there are three key chemical reactions: the oxygen reduction reaction (ORR) in fuel cells and metal-air batteries, the oxygen evolution reaction (OER) in water splitting and the charge process of metal-air batteries, and the hydrogen evolution reaction (HER) in water splitting. These reactions determine the efficiency of the devices. Since these photochemical and/or electrochemical reactions are sluggish, catalysts must be used to promote them. Noble metals and metal oxides are the most widely used catalysts in catalytic processes. Platinum group metal catalysts are used in fuel cells and metal-air batteries to accelerate the electrochemical processes,[1,2] while many metal oxides are used as catalysts and electrolytes (e.g., vanadium oxides in batteries) for the OER

Biomimetic Principles and Design of Advanced Engineering Materials, First Edition. Zhenhai Xia.
© 2016 John Wiley & Sons, Ltd. Published 2016 by John Wiley & Sons, Ltd.

and HER.[3,4] However, these metal-based catalysts often suffer from their high cost, low selectivity, poor durability, and sometimes detrimental environmental effects, which hamper the applications of the renewable technologies.[5,6] It is highly desirable to develop alternative materials that are earth-abundant, cost-effective, stable, and active in catalyzing ORR/OER/HER.

Nature has created efficient renewable energy machinery using Earth-abundant elements. Photosynthesis, for example, is an amazing energy converter evolved over billions of years. In photosynthesis, plants use sunlight as the energy source to carry out two important reactions: decomposition of water to molecular oxygen accompanied by reduction of CO_2 to carbohydrates and other carbon-rich products.[7,8] Although the overall energy conversion efficiency of a plant is low, typically much less than 1%, which is a result of energy losses associated with growth and other life processes, the primary photosynthetic energy conversion processes – light capture, charge separation, and water splitting – are extremely efficient. This efficient biological water oxidation is catalyzed by a metal–protein cluster $(CaMn_4O_5(H_2O)_4)$ in photosystem II (PSII), which controls reaction coordinates, proton movement, and water access.[9] This cluster is the only biological catalyst that can oxidize water to molecular oxygen, and it appears that it has remained basically unchanged during two billion years of evolution. The Mn_4/Ca cluster is the most efficient anodic "electrolysis" system known. In certain types of photosynthetic organisms, a specific enzyme, a hydrogenase enzyme, either iron nickel (Fe–Ni) or all iron (Fe–Fe), can bypass the CO_2 fixation process and lead to the generation of H_2.[10,11] The Fe-only hydrogenases are extremely efficient in hydrogen production, using a catalyst based on the cheap metal iron.

Nature also provides an alternative and efficient solution to ORR catalysis: oxygen-metabolizing enzymes containing an active site of iron(II)-porphyrin structure, which promote oxygen reduction.[12] These naturally occurring oxygen activation catalysts are considered as viable substitutes for precious metals in ORR catalysts for fuel cells. Efficient and rapid reduction of oxygen is carried out without the use of noble metals in nature by cytochrome c oxidase enzymes to power the cellular proton pump during the final stage of respiration. The enzymes display high selectivity for the four-electron reduction to water. Mimicking these biological molecular structures could create new artificial catalysts with the desired properties for energy conversion and storage.

There is a strong interest in replicating biological molecular structures in artificial systems for technological applications. Great progress has been made in biomimetic catalysts for energy conversion and storage toward renewable energy technologies. Inspired by nature, breakthroughs have been made in some key areas in clean energy generation, including ORR catalysts mimicking the hemeprotein core structure in the cytochrome complex, water splitting by an artificial metal–protein cluster in Photosystem II, hydrogen evolution using hydrogenase enzyme-inspired catalysts, artificial leaves, and nitrogen fixation. These bioinspired catalysts based on Earth-abundant, cost-effective materials significantly enhance the capability and efficiency of renewable energy technologies in practical applications.

The structures and mechanisms of biological and synthetic catalysts are focused on oxygen evolution, hydrogen generation, and oxygen reduction. The design principles, catalytic activities, and processes of these catalysts are discussed in this chapter. Examples are given of how to mimic nature to create catalytic materials for water splitting, fuel cells, metal–air batteries, and artificial photosynthesis.

10.2 Catalysts for Energy Conversion in Biological Systems

10.2.1 Biological Catalysts in Biological "Fuel Cells"

Oxygen reduction is one of the essential reactions pertinent to the majority of living beings that is slow in nature and requires enzymes and/or catalyst assistance.[13] The cytochrome complex, cyt$c$, is a small hemeprotein that is loosely associated with the inner membrane of the mitochondrion in most eukaryotic cells.[14] This hemeprotein contains an active site of iron(II)-porphyrin (Figure 10.1a) that efficiently catalyzes the ORR in mitochondria.[15] It is the fourth complex of the respiratory electron transport chain, which catalyzes the respiratory reduction reaction of O_2 to water. The reaction catalyzed is the oxidation of cytochrome c and the reduction of oxygen:[15]

$$4cytc_{red} + 4H^+ + O_2 \rightarrow 4cytc_{ox} + 2H_2O \qquad (10.1)$$

In the process, the enzyme binds four protons from the inner aqueous phase to make water, and in addition translocates four protons across the membrane. Oxygen is reduced by four electrons to water at the active site of the Fe(II)-porphyrin structure, but each electron is delivered one at a time from external cytochrome c. This strategy for the safe reduction of O_2 completely avoids the release of partly reduced intermediates that are quite harmful to a variety of cell components. Cytochrome c oxidase meets this crucial criterion by holding O_2 tightly between Fe and Cu ions.[15]

Vitamin B_{12} (V_{B12}) has a cobalt atom bonding with four nitrogen atoms and two ligands, similar to the Fe(II)-porphyrin structure.[16] V_{B12} also contains a phosphorus atom, and the dual-doping synergetic effect of both nitrogen and phosphorus into carbon can reportedly promote ORR activity better than the single doping of nitrogen.[17–19] In photosynthesis, the first event is the absorption of light by a photoreceptor molecule. The principal photoreceptor in the chloroplasts of most green plants is *chlorophyll a*, a substituted tetrapyrrole (Mg(II)-porphyrin) (Figure 10.1c). The four nitrogen atoms of the pyrroles are coordinated to a magnesium ion. Unlike a porphyrin such as heme, chlorophyll has a reduced pyrrole ring. Another distinctive feature of chlorophyll is the presence of *phytol*, a highly hydrophobic 20-carbon alcohol, esterified to an acid side chain.[15]

As discussed above, many biological catalysts contain metal–protein clusters that play a key role in catalyzing the ORR, OER, and other reactions. Among them is the metal-perpherin clusters (metal = Fe, Mg, Co, Zn, etc.) in chlorophyll, blood cells, and V_{B12}, etc. (Figure 10.1d). These metal–perpherin clusters have diverse biological functions, including the transportation of diatomic gases, chemical catalysis, diatomic gas detection, and electron transfer. The original metal–nitrogen bond in the metal–perpherin clusters is considered to be a promising ORR active site,[20] indicating the potential of the metal–perpherin as an ORR catalyst. Incorporating these key genetic units into proper media could create new multifunctional materials with desired properties for energy conversion applications.

10.2.2 Oxygen Evolution Catalyzed by Water-oxidizing Complex

Plants use photosynthesis to convert carbon dioxide and water into sugars and oxygen. The process starts in a cluster of manganese, calcium, and oxygen atoms at the heart of a protein

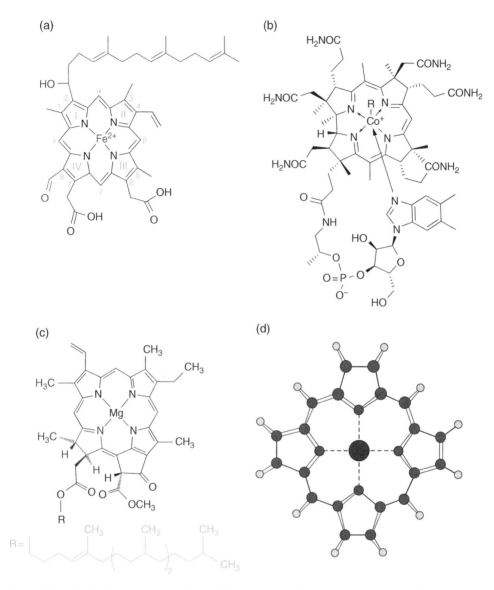

Figure 10.1 Biological structures of (a) Fe(II)-porphyrin with the inner membrane of the mitochondrion, (b) Co(II)-porphyrin in vitamin B_{12}, (c) tetrapyrrole (Mg(II)-porphyrin) in chlorophyll, and (d) their basic units to be considered for bioinspired catalysts.

complex, PSII, which splits water to form oxygen gas, protons, and electrons. In *light-dependent reactions* a water-oxidizing complex (WOC) performs one of the most important reactions in nature: water oxidation, the source of nearly all the atmosphere's oxygen, resulting in the development of modern organisms using respiration on our planet.[21] The oxygen evolution center (OEC) includes an oxo-bridged Mn_4/Ca cluster, surrounded by special proteins. This metal-containing protein cluster controls reaction coordinates, proton movement, and water

(a)

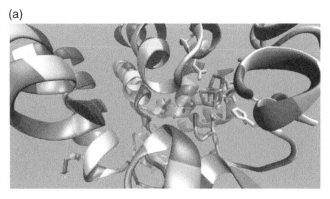

(b)

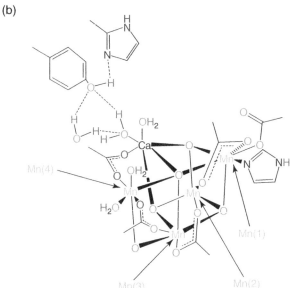

Figure 10.2 (a) The whole structure of the $CaMn_4O_5(H_2O)_4$ cluster resembles a distorted chair, with the asymmetric cubane. There is only a small fraction of the residues that come into direct contact with the manganese–calcium cluster. In other words, the structure could be considered as a nano-sized manganese–calcium oxide in protein environments. (b) $CaMn_4O_5(H_2O)_4$ cluster and the surrounding amino acids. Source: Najafpour *et al.* (2012).[21] Reproduced with permission of Elsevier.

access. All oxygenic photosynthetic organisms share a common set of "core" membrane proteins in PSII, which include the reaction centre proteins and inner chlorophyll a binding proteins, as well as the pigment components associated with the primary light-driven charge separation across the thylakoid membrane (Figure 10.2).[21] These proteins principally ligate the catalytic Mn cluster responsible for water oxidation, and stabilize/regulate the OEC. The OEC Mn_4/Ca cluster is the most efficient anodic "electrolysis" system known for catalyzing the oxidation of water:

$$2H_2O \rightarrow 4H^+ + O_2 + 4e^- \tag{10.2}$$

In effect, H_2O is split into O_2 and H_2, where the hydrogen is not in the gaseous form but bound by carbon. The OEC Mn_4/Ca cluster Mn_3Ca-cubane with the fourth Mn attached further away and can be considered to be $CaMn_4O_5(H_2O)_4$.[21]

The biological key units for catalysis, such as $CaMn_4O_5(H_2O)_4$, are nano-sized metal oxides (e.g., manganese oxide) with a dimension of about ~0.5 nm in a protein environment.[22] From biomimicking principles, nano-sized metal oxides are considered to be good structural and functional models for the cluster. The nanometer-sized particles have larger surface areas than bulk materials and this ensures that most of the active sites are at the surface, where they function as a water-oxidizing catalyst. In addition, nanoparticles are more active in catalyzing reactions due to the quantum effect.

10.2.3   Biological Hydrogen Production with Hydrogenase Enzymes

In certain photosynthetic organisms, a specific type of enzyme, hydrogenase enzymes, can by-pass the CO_2 fixation process and lead to non-negligible H_2 formation, for example algal [FeFe] hydrogenases generate hydrogen through the oxidation of reduced ferredoxin, an electron mediator that is initially reduced by photosystem I (PSI).[23] Hydrogenase enzymes catalyze the reaction:

$$2H^+ + 2e^- \leftrightarrows H_2 \tag{10.3}$$

in both directions, that is, for hydrogen production and consumption. The reaction takes place close to the thermodynamic potential.

Hydrogenases are classified as one of three types based on the metal atoms comprising the active site: [NiFe], [FeFe], or [Fe] only. As shown in Figure 10.3, hydrogenases have some common features in their structures. Each enzyme has an active site and a few Fe–S clusters that are buried in protein.[24,28] The active site, which is believed to be the place where catalysis takes place, is also a metallocluster, and each metal is coordinated by carbon monoxide (CO) and cyanide (CN^-) ligands.[25] In general, [FeFe] hydrogenases are considered to be strong candidates for an integral part of the solar H_2 production system because of their lower over-potential and higher catalytic activity.[26] However, [FeFe] hydrogenases are highly sensitive to O_2. [NiFe] hydrogenase enzymes consume molecular hydrogen as a fuel source, while Fe-only hydrogense enzymes produce molecular hydrogen. Of these three classes, the [Fe]-only hydrogenases ($[Fe]_H$) are highly evolved for catalysis and are able to produce 9000 molecules of H_2 per second and per enzyme molecule at 30 °C.[27] Because of this outstanding catalytic rate, the [Fe]-only hydrogenases are of special interest for industrial applications.

10.2.4   Natural Photosynthesis and Enzymes

Natural photosynthesis is a complex chemical process involving in many subreactions that need catalysts or enzymes. In this process, plants use sunlight as the energy source to carry out two important reactions: decomposition of water to molecular oxygen accompanied by reduction of CO_2 to carbohydrates and other carbon-rich products.[7,8]

$$6H_2O + 6CO_2 \xrightarrow{\text{(sunlight)}} C_6H_{12}O_6 + 6O_2 \tag{10.4}$$

Figure 10.4 illustrates the key processes. Photosynthesis occurs in two stages: *light-dependent reactions* (PSII) and *light-independent reactions* (PSI). In the first stage, an important process

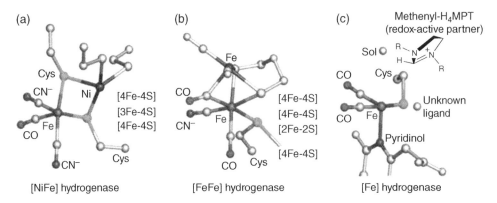

Figure 10.3 Superimposed active-site structure of the three phylogenetically unrelated hydrogenases: (a) [NiFe]-hydrogenase from *Desulfovibrio gigas*, (b) [FeFe]-hydrogenase from *Clostridium pasteurianum* and *Desulfovibrio desulfuricans*, (c) [Fe]-hydrogenase from *Methanocaldococcus jannaschii*. In [Fe]-hydrogenase, the fifth and sixth ligation sites are marked by the six spheres (upper right) connected to Ni. All three hydrogenase types have in common a low-spin iron ligated by thiolate(s), CO, and cyanide or pyridinol (considered as cyanide functional analog), which acts together with a redox-active partner (upper right in (a), upper part in (b) and lower part in (c)). Source: Allakhverdiev *et al.* (2010).[28] Reproduced with permission of Elsevier.

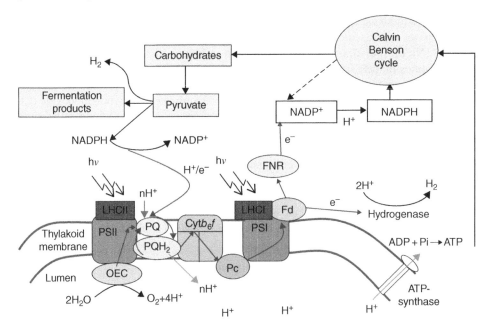

Figure 10.4 Carbon fixation and oxygen evolution take place in two distinct spaces in oxygenic photosynthesis: light-powered hydrogen production during oxygenic photosynthesis, as well as carbohydrate synthesis that can also be followed by hydrogen production. The photosynthetic processes are driven by the light energy captured by the light-harvesting complexes of PSII and PSI. Electrons are derived from H_2O by water oxidation at the WOC, also known as oxygen evolving complex (OEC), of PSII. The ATP and NADPH generated during the primary photosynthetic processes are consumed for CO_2 fixation in the Calvin–Benson cycle, which produces sugars and ultimately starch. Source: Najafpour *et al.* (2012).[21] Reproduced with permission of Elsevier.

is water oxidation to molecular oxygen. These reactions, which depend directly on light, take place in specific pigment–protein complexes in the thylakoid membranes. In these reactions light energy is converted into chemical energy. The end product of this set of reactions is the production of oxygen, the reducing power, nicotinamide adenine dinucleotide phosphate (NADPH), and adenosine triphosphate (ATP). In the second stage, chemical reactions, so called "dark reactions" that do not depend directly on light, take place in the stroma region. The electrons are taken through a series of uphill and downhill steps to generate energy-rich intermediates. In these reactions CO_2 is converted to sugar. The dark reactions involve the Calvin–Benson cycle, in which CO_2 and energy from NADPH and ATP are used to form sugars (Figure 10.4).

The components of photosynthesis are organized at the microscopic level in a unique way. In PSII chlorophyll molecules and other reactants taking part in the absorption of sunlight (antenna) and electron-transfer processes (reaction center) are organized in special assemblies. To use the entire visible-light part of the solar radiation (350–700 nm), green plants use chlorophyll a as the main light absorber along with a number of accessory pigments and a modified form of chlorophyll, called chlorophyll-a. Chlorophyll-a absorbs in the blue-violet, orange-red spectral regions while the accessory pigments cover the intermediate yellow-green-orange part. In addition, the special arrangement (antenna array) of chlorophyll molecules is made for efficient light capturing and relay to the reaction center, even when the light flux varies significantly. Photosynthesis occurs at comparable efficiency under bright and diffuse light conditions.

10.2.5   Biomimetic Design Principles for Efficient Catalytic Materials

As discussed above, metal–protein clusters play a key role in catalyzing the ORR, OER, HER and other reactions, for example the metal–perpherin clusters (metal=Fe, Mg, Co, Zn, etc.) in chlorophyll, blood cells, vitamin 12 (V_{B12}), etc. (Figure 10.5a). These metal–perpherin clusters have diverse biological functions, including the transportation of diatomic gases, chemical catalysis, diatomic gas detection, and electron transfer. Incorporating these key genetic units into the proper media could create new multifunctional materials with the desired properties for energy conversion applications.

Most biological catalytic structures do not contain only one metal in a reaction center; usually there are multiple metal centers to improve the catalytic activity, as shown in Figure 10.5b.[14] Metals play multiple roles, including binding substrate, increasing the reactivity of the substrate, preventing side reactions, and providing electrons quickly. Nature finds a clever way to avoid side reactions by using multiple metal centers. Artificial systems need to avoid passage through one-electron oxidation or reduction intermediates because such species tend to be very reactive and undergo side reactions that can destroy the light-absorbing dye molecules.

The biological key units for catalysis, for example the $CaMn_4O_5(H_2O)_4$ cluster in PSII, could be considered to be a nano-sized cluster of metal oxides (e.g., manganese oxide) with a dimension of about ~0.5 nm in a protein environment. Nano-sized metal oxides have more surface area and could be more active than the bulk materials. Two types of size effects may be distinguished at the nanoscale when compared with bulk compounds: effects that rely on increased surface-to-volume ratio and true-size effects, which also involve changes in local materials properties.

Protein in the catalytic center also plays a key role in absorbing sunlight, regulating the reaction, and transferring electrons, for example the photosynthetic reaction center is a complex of several proteins, pigments, and other co-factors assembled together to execute primary energy conversion reactions. Many amino acids in PSII are involved in proton, water, or oxygen transfer. Roles for the

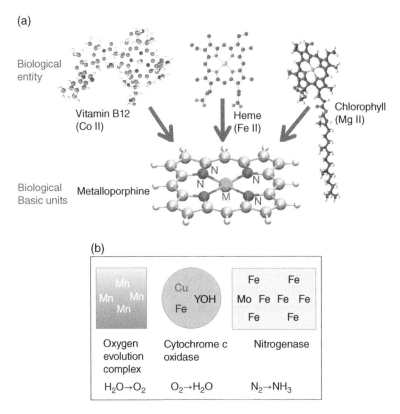

Figure 10.5 (a) Biological structures and their basic units to be considered for biomimetic basic units. (b) Multielectron redox enzymes for oxygen evolution, oxygen oxidation, and nitrogen reduction. Source: Collman & Decreau (2008).[14] Reproduced with permission of the Royal Society of Chemistry.

residues that come in direct contact with the manganese–calcium cluster could include regulation of charges and electrochemistry of the manganese cluster, and help in coordinating water molecules at appropriate metal sites and/or maintaining the stability of this cluster. It is therefore necessary to stabilize the biomimetic catalyst with additional functional groups.

Natural photosynthesis uses a series of electron-transport chains and appropriate enzymes to achieve these processes at moderate efficiencies. Artificial photosynthetic systems can stop at several intermediate steps, for example, with formation of H_2 or reduction of CO_2 to carbon-based fuels (methanol, methane, formate). Hydrogen is a key solar fuel because it permits energy storage and can be used directly in combustion engines or fuel cells.

10.3 Bioinspired Catalytic Materials and Processes

Inspired by nature, various catalysts have been developed for energy conversion and storage, including water splitting, hydrogen production, oxygen reduction, and CO_2 reduction. The following biomimetic approaches to solar energy conversion and storage have been addressed: (1) finding suitable template/substrate architectures that can mimic antennal chlorophyll function, (2) finding suitable molecular redox catalysts that permit formation of molecular/diatomic forms of hydrogen (H_2) and oxygen (O_2) in the presence of suitable oxidants and reductants, (3) total decomposition of water to H_2 and O_2 using sunlight, (4) direct conversion of sunlight to electricity, which then can be used for the production of various chemicals and other needs, and (5) reduction of CO_2 to various carbon compounds that can be used as fuels or raw materials for industry.

10.3.1 Bioinspired Catalyst for Hydrogen Fuel Cells

In a polymer electrolyte fuel cell (PEFC) the rate-limiting reaction is the ORR at the cathode.[29] The ORR pathway is mainly composed of the following reactions.

$$O_2 + 4H^+ + 4e^- \rightarrow 2H_2O \quad E_0 = 1.229\,V \tag{10.5}$$

$$O_2 + 2H^+ + 2e^- \rightarrow H_2O_2 \quad E_0 = 0.695\,V \tag{10.6}$$

$$H_2O_2 + 2H^+ + 2e^- \rightarrow 2H_2O \quad E_0 = 1.763\,V \tag{10.7}$$

Reaction (10.5) is a direct ORR pathway that involves transfer of four electrons, while reaction (10.6) follows an H_2O_2 pathway with two-electron transfer. Since a direct ORR pathway yields a higher thermodynamically reversible potential than an H_2O_2 pathway, reaction (10.5) is preferable over reaction (10.6) for the ORR in PEFC. However, all the reactions are sluggish and a large amount of platinum nanoparticles have been utilized to overcome the slow reaction of the ORR, adding to the cost and making the agglomeration problem of the catalyst during operation worse.[6,30] Searching for efficient, durable, and, most importantly, non-precious metal catalysts is of critical importance in fuel cell applications.

Since the biological catalysts are non-precious metals that efficiently catalyze the ORR with four-electron transfer, mimicking the structures of biological catalysts may dramatically enhance the performance of fuel cells. As early as in 1960s, Jasinski demonstrated that cobalt phthalocyanine could act as an electrocatalyst for oxygen reduction. Since then, transition metal macrocyclic compounds, such as porphyrin, phthalocyanine, and tetraazannulene, have been widely studied. Among them, iron-, nickel-, and cobalt-based macrocyclic compounds have been demonstrated to exhibit the highest catalytic activity for the ORR. These metal-based macrocyclic compounds contain a basic structure similar to iron porphyrins in cytochrome c oxidase. As many biomolecules contain transition-metal macrocyclic units in their skeleton for certain biological functions, the use of biomolecules with innate reactive sites for the ORR has recently been proven to be an effective approach to high-performance non-precious-metal ORR catalysts. This provides a new strategy to rationally design inexpensive and durable electrochemical oxygen reduction catalysts for metal–air batteries and fuel cells.

Transition-metal macrocyclic units have been used as ORR electrocatalysts in the following three ways: (1) as catalysts on their own, (2) as precursors for pyrolyzed $M-N_x/C$ catalysts, and (3) as a matrix for entrapping non-precious metals. Using the first approach, Collman *et al.* found the rate-determination step of the catalytic ORR in cytochrome c oxidase, and synthesized the best mimic structure of the cytochrome c oxidase active site.[14] A functional analog of the active site in the respiratory enzyme reproduces every feature in active site. When covalently attached to a liquid-crystalline self-assembled monolayer film on an Au electrode, this functional model continuously catalyzes the selective four-electron reduction of dioxygen at physiological potential and pH, under rate-limiting electron flux.

The transition-metal macrocyclic compounds were also attached to other substrates to make ORR catalysts. In this type of biomimetic catalyst, the naturally active Co–N$_x$–C center of V$_{B12}$ was directly used to catalyze the ORR. For example, a graphite electrode was prepared by absorbing non-pyrolysed V$_{B12}$ on graphite.[31,32] The electrochemical test results show that V$_{B12}$ could catalyze the ORR through a two-electron path at high pH but through a four-electron path at low pH. To improve the reactivity of the catalysts and avoid the two-electron path, carbon black was used to immobilize V$_{B12}$ with a naturally active Co–N$_x$–C center, followed by pyrolysis under nitrogen atmosphere at 700 °C. The obtained catalyst exhibits favorable ORR performance with a high durability in acid media.[33,34] In this case, the pyrolysis atmosphere can significantly change the textural structure, pore structure, and nitrogen configuration of the resultant material to improve ORR activity.[35] To further improve the activity, the catalyst with V$_{B12}$ was pyrolyzed under an ammonia atmosphere instead of a nitrogen one. After treatment, the catalyst can work as an ORR catalyst for direct methanol alkaline fuel cells in alkaline media.[34] The V$_{B12}$-based catalyst presented better ORR activity than Pt in alkaline electrolyte. X-ray photoelectron spectroscopy analyses revealed that the nitrogen configuration was changed after pyrolysis, and nitrogen species played a key role in catalyzing the ORR. The catalyst had an electron transfer number of 3.9, which was very close to the ideal theoretical value of 4. Moreover, the catalyst displayed superior methanol tolerance to Pt/C in alkaline medium, demonstrating its potential application as a cost-effective catalyst for direct methanol alkaline fuel cells.[34]

Iron (II) phthalocyanine (FePc) has a similar structure to hemeprotein and is therefore a promising catalyst for the ORR. Much effort has been put into developing FePc as ORR catalysts, but its activity and stability are not comparable to Pt-based catalysts so far.[36] The poor performance of FePc is mainly attributed to aggregation and the poor electron conductivity of FePc, which greatly decreases the active sites for the ORR and electron transfer in the ORR process. To solve these problems, FePc has been fixed on a variety of carbon materials. Chen and coworkers used a high surface area carbon substrate (Vulcan XC-72) to support FePc and found that the ORR activity of FePc was significantly increased, but its stability needed a further improvement.[37] Yuan loaded FePc on amino-functionalized multi-walled carbon nanotubes (a–MWCNTs)[38] and found that the ORR performance of FePc supported on a–MWCNTs in a microbial fuel cell was better than for FePc supported on carbon black. To further improve the activity and investigate its anticrossover effect, single-walled carbon nanotubes were used as a support for FePc.[39,40] Iron phthalocyanine with an axial ligand anchored on single-walled carbon nanotubes demonstrates higher electrocatalytic activity and better anticrossover effect for the ORR than the state-of-the-art Pt/C catalyst, as well as exceptional durability during cycling in alkaline media[40] (Figure 10.6). Theoretical calculations suggest that the rehybridization of Fe 3d orbitals with the ligand orbitals coordinated from the axial direction results in a significant change in electronic and geometric structure, which greatly increases the rate of the ORR.

Graphene, a monolayer of carbon atoms arranged in a two-dimensional honeycomb lattice, has attracted enormous attention because of its outstanding mechanical, thermal, and electrical properties. In addition, heteroatom (N, B, S, P, etc.)-doped graphene has been demonstrated to

(a)

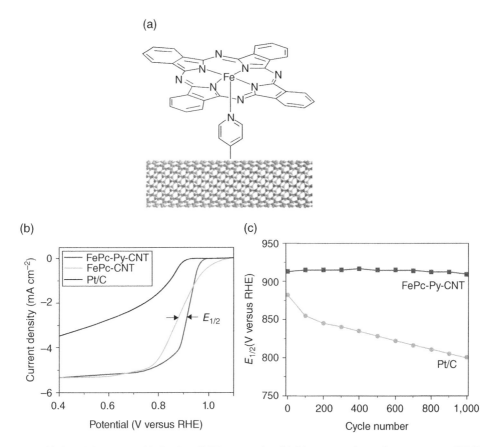

Figure 10.6 (a) Structure of FePc–Py–CNTs composite. (b) Linear scanning voltammograms of FePc–CNTs, FePc–Py–CNTs, and commercial Pt/C catalyst. (c) Half-wave potential as a function of cycle number of FePc–Py–CNTs and commercial Pt/C catalyst during durability test. Electrolyte 0.1 M KOH, scan rate 10 mV s⁻¹. Source: Cao *et al.* (2013).[40] Reproduced with permission of Nature Publishing Group.

be an excellent ORR catalyst that is free from anode crossover and shows a comparable catalytic activity and much better stability than commercial Pt/C. Graphene is also an excellent catalyst support in facilitating the ORR. Compared to CNTs or other carbon substrates, the unique two-dimensional structure and extraordinary electronic properties of graphene materials could lead to much closer contact between FePc and the graphene substrate, and thus benefit ORR performance.[41] Using graphene derivatives to enhance the ORR activity and stability of a catalyst could lead to the development of various non-precious catalysts and the bioinspired design of metal–N₄/graphene composites that hold great promise for fuel cell applications. Zhang *et al.* synthesized MPc/GX composites (M = Fe, Co; GX = nitrogen-doped graphene (NG), graphene oxide (GO), reduced graphene oxide (rGO)) as efficient non-precious catalysts for the ORR.41 Among them, the FePc/NG composite exhibits the highest onset potential for the ORR, the largest current density, and excellent tolerance to the crossover effect of methanol, for which the properties are superior to that of commercial Pt/C. In addition, although FePc alone shows very poor stability in the ORR, this can be greatly improved by attachment with NG, and the stability of the FePc/NG composite is comparable to that of the Pt/C catalyst.

Transition-metal macrocyclic compounds were also used to dope the materials to form high-performance ORR catalysts. Inspired by the hemeportein with a naturally active Fe–N_x–C center, Dai and co-workers synthesized hemin-doped poly (3,4-ethylenedioxythiophene) (PEDOT) with controllable three-dimensional nanostructures via a one-step, triphase, self-assembled polymerization routine.[42] These hierarchical structures served as a conductive medium to support the non-precious-metal macrocyclic ORR reactive centers (i.e., Fe–N_4–C) in hemin. The use of hemin is inspired by its ability to be the actual oxygen hunter and combiner in the oxygen-transport process of all vertebrate bodies, with oxygen molecules being combined directly with the iron center of Fe–N_4–C active sites for the ORR. In the hemin-doped PEDOT, the carboxyl groups within the hemin molecules are employed as doping ions to link with the PEDOT main chains by electrostatic interaction, rather than being weakly absorbed through physical sorption. PEDOT shows good electronic conductivity after doping, and thus can serve as an excellent charge collector and electron shuttle.[43] In addition, PEDOT, with its micro/nanostructures over a large surface area, can itself be an outstanding intrinsic ORR catalyst over a wide pH range, including pH 7.[42,44] These results demonstrate that the hemin-induced synergistic effect results in a very high four-electron oxygen reduction activity, better stability, and is free from methanol crossover effects even in a neutral phosphate buffer solution.

As discussed in section 10.2.1 for enzymatic ORR catalysts, nature uses bulky protein chains to separate the ORR inorganic active sites, preventing site overlap and catalytic deactivation.[45] The active sites are mostly composed of non-precious transition-metal atoms and clusters, which play a prominent role in heterogeneous catalytic transformations.[46,47] Thus, new catalytic materials can be designed that are inspired by natural ORR catalysts by using organic molecules with specific molecular backbones (adequately separating the active sites) and selected functional groups (electron withdrawing or donating) to coordinate transition-metal atoms or clusters directly on surfaces.[13] Thus, two-dimensional metal–organic coordination networks (2D-MOCNs), that is, organic molecules and metal centers self-assembled on surfaces under well-controlled conditions, constitute a promising route to fabricate functional low-dimensional bioinspired architectures.[13,48] Grumelli et al. showed that bioinspired catalytic metal centers can be effectively mimicked in 2D-MOCNs self-assembled on electrode surfaces (Figure 10.7).[13] Networks consisting of organic molecules coordinated to single iron and manganese atoms on Au(111) effectively catalyze oxygen reduction and reveal distinct catalytic activity in alkaline media. These results demonstrate the high potential of surface-engineered metal–organic networks for electrocatalytic conversions.

The above results demonstrate that the chemical activity of metal centers is determined by both the nature of the metal ion and its coordination shell, as shown in the natural design of catalysts in biological systems. The ligation separates the unsaturated metal atoms, preventing catalytic deactivation. Thus, the specific design of ligands allows the catalytic activity of metal adatoms to be tuned for a desired chemical conversion. Such work proves that surface-engineered metal–organic complexes and networks that display structural resemblance with enzyme active sites have a high potential for heterogeneous catalytic chemical conversions. The possibility to create novel and highly stable functional two-dimensional coordination complexes at surfaces using specifically designed organic molecules and transition metal centers taking inspiration from nature opens up a route for the design of a new class of nanocatalyst materials with promising applications in electrocatalysis.[13]

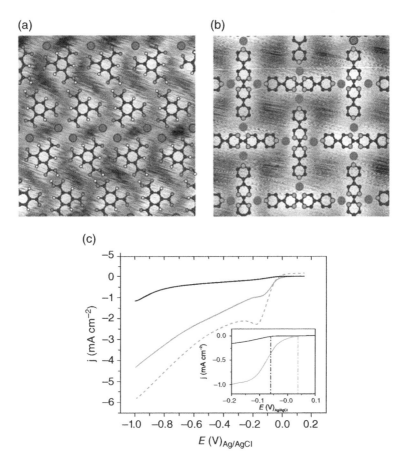

Figure 10.7 (a) STM image (8×8 nm²) of PBP–Fe network on Au(111). (b) High-resolution (3.5×3.5 nm²) STM image of PBP–Fe with the model superposed. (c) Bare Au(111) (dark grey line), TMA_Mn (light grey line), and TMA_Mn after the addition of 10 mM H_2O_2 (dashed line). The inset shows the different onset potential for the shoulder at 0.2 V. Source: Grumelli *et al.* (2013).[13] Reproduced with permission of Nature Publishing Group.

Apart from 2D-MOCNs, covalent organic frameworks (COFs) are another type of two-dimensional materials that are appealing as cost-effective catalysts due to their versatility, large surface area, excellent mechanical and electrical properties, and high stability. Using precursors that contain metal–porphyrin complexes similar to a base unit in natural photosynthesis, Xiang and co-workers synthesized a class of two-dimensional COFs with metal (e.g., Fe, Co, Mn)-incorporated macrocycles of precisely controlled nitrogen locations in the pores (Figure 10.8) by a nickel-catalyzed Yamamoto reaction.[49] Subsequent carbonization of the metal-incorporated COFs led to the formation of COF-derived graphene analogues, which acted as efficient electrocatalysts for oxygen reduction in both alkaline and acid media, with good stability and free from any methanol-crossover/CO-poisoning effects. This work highlights that COFs can be tailored by biological genetic group to create a new class of highly efficient catalysts for the ORR, thus pointing out a direction for the design and development of catalysts for energy conversion and storage devices.

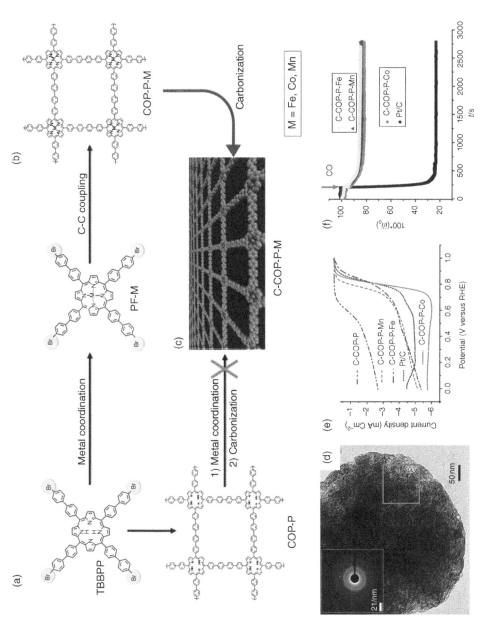

Figure 10.8 (a–c) The incorporation of non-precious metals (Fe, Co, or Mn) into C–COF, followed by carbonation. (d) TEM image of C–COF–P–Fe. Inset: the selected-area electron diffraction (SAED) pattern. (e) Electrocatalytic activity toward ORR. (f) Tolerance to crossover of fuel molecules and CO poisoning. Source: Xiang et al. (2014).[49] Reproduced with permission of John Wiley and Sons.

In cytochrome c oxidase and laccase, the active sites contain similar metal clusters of assembled Cu^{2+} complexes.[50,51] *In vitro* experiments show that laccase is more desirable and the Cu^{2+} ion is an ideal catalytic site when the energy level of the d electrons is tuned to a reasonable state.[52] However, directly mimicking the Cu^{2+} complexes in enzyme remains a challenge, and the activity of the resultant catalysts is lower than that of the natural state.[53,54] The main reasons for this inefficient catalysis are the absence of appropriate mediators for sequential electron transfer by specific amino-acid chains in apo-enzyme components and the steric variation of the coordination structures after the complexes are assembled on the electrode.[55] Wang *et al.* proposed an effective approach to tune the d electron density of the Cu^{2+} active site and simultaneously increase the electron transfer rate via the synergistic effect of electronic connection between Cu^{2+} and N, and between Cu^{2+} and metallic Cu (Cu0) in graphene formed by the pyrolysis of the mixture of graphene oxide and $Cu(phen)_2$ (CPG-900) as inspired by the catalytic sites in natural enzyme.[55] The electron density of Cu^{2+} active sites in CPG-900 is tuned by the electron donation effect from both the neighboring Cu0 of the Cu nanoparticle and the N ligand incorporated in the rigid graphene. This imperfect coordination configuration provides an optimal environment for electronic bonding of O_2 to Cu^{2+} ions. Good ORR performance on the resultant catalyst is achieved in both acidic and alkaline media. In alkaline media, superior activity to commercial Pt/C catalyst was observed with an onset potential of 0.978 V versus a reversible hydrogen electrode. This catalyst also shows excellent activity towards the OER, with the onset potential closely approaching the standard potential.

10.3.2 WOC-biomimetic Catalysts for Oxygen Evalution Reactions in Water Splitting

The electrochemical or photoelectrochemical (PEC) path to water splitting involves separating the oxidation and reduction processes into half-cell reactions, that is, the OER (Equation (10.8)) and HERs (Equation (10.9)):

$$2H_2O_{(l)} \rightarrow O_{2(g)} + 4H^+_{(aq)} + 4e^- \quad E°=1.23\,V\ vs.\ SHE \tag{10.8}$$

$$2H^+_{(aq)} + 2e^- \rightarrow H_{2(g)} \quad E°=0\,V\ vs.\ SHE \tag{10.9}$$

The free energy change for the conversion of one molecule of water into hydrogen and oxygen under standard temperature and pressure conditions is 1.23 V. In the electrolysis of water, an external voltage of approximately 2.0 V is required, which includes 1.23 V thermodynamic energy requirement and polarization losses, but in photo-assisted electrolysis of water this requirement is reduced, but the external voltage required is still called the bias voltage.[56] The semiconductor absorbs the light and produces electrons and holes. Electrons are conducted to cathode and holes oxidize water to produce oxygen and hydrogen ions. For water splitting using renewable energy to be successful, catalysts are needed to lower the overpotential and therefore also the overall voltage required to achieve water oxidation. Precious metal oxides such as IrO_2 and RuO_2 have been developed and are the most effective water-oxidation (WO) catalysts.[3,57] Considering the scarcity and high cost of precious metal oxides, the development of efficient artificial PSIIs based on low-cost Earth-abundant elements is more appealing for applications.

Since all species capable of O_2 evolution possess qualitatively identical reaction centers, and no metal element other than Mn has been identified in the catalytic cluster of PSII, many manganese compounds have been synthesized with the aim of simulating the WO catalysts of PSII not only because the WO catalyst consists of four manganese ions but also because manganese is of low cost and is environmentally friendly. Extensive research efforts have therefore been aimed at developing WO catalysts composed of the abundant element Mn. As discussed previously, the key biological unit for water oxidation can be considered as a nanoparticle. Thus, nano-sized manganese oxides are promising compounds for water oxidation because the compounds are stable, and it is easy to use, synthesize, and manufacture the catalyst. Nanometer-sized manganese oxide clusters supported on a mesoporous silica scaffold have been shown to be efficient WO catalysts in an aqueous solution at room temperature and at a pH of 5.8.[58] The high surface area silica support may be critical for the integrity of the catalytic system because it offers a perfect, stable dispersion of the nanostructured manganese oxide clusters. Other types of nanostructured manganese oxide such as α-MnO_2 nanotubes, α-MnO_2 nanowires, and β-MnO_2 nanowires have been reported as catalysts for water oxidation driven by visible light.[59] These as-prepared α-MnO_2 nanotubes are highly uniform, with a tube outer diameter of approximately 100 nm and an inner diameter of approximately 40 nm. Although the morphology and crystal structure have negligible effects on the water oxidation activity of the MnO_2 catalysts, the surface area is an important factor.

Although manganese oxides have been used as a heterogeneous catalyst for water oxidation, unlike PSII, these synthetic structural and functional models contain no amino acid groups to stabilize manganese oxide, promote proton transfer and/or decrease the activation energy for water oxidation. In PSII, there are specific intrinsic proteins that appear to be required for water oxidation. To mimic the protein environment in PSII, bovine serum albumin (BSA) was used to stabilize a nano-sized manganese oxide.[60] BSA is one of the most studied proteins and has a strong affinity for a variety of inorganic molecules binding to different sites. BSA is a soluble protein found in the plasma of the circulatory system that acts in the transport and deposition of endogenous and exogenous substances. Albumin not only induces the nucleation, but also inhibits the further growth of manganese oxide as larger particles are formed without BSA. Moreover, dispersion of manganese oxide in water is unique and may also increase the rate of reaction of particles with reactants, and could result in a high rate for many reactions.

Although manganese oxides are effective electrocatalysts under alkaline conditions, the activity of most manganese oxides is significantly reduced at neutral pH, resulting in a large electrochemical overpotential (Z) ranging from 500 to 700 mV.[61,62] This high Z is in contrast to that of the PSII tetrameric manganese cluster, which catalyzes water oxidation with Z of only 160 mV.[9,63] In PSII, charged manganese ions gradually give up electrons as they tear protons away from water molecules. This causes manganese in the 2+ and 3+ valence states to become oxidized, resulting in Mn^{4+} ions. Although the less-oxidized Mn^{3+} ions are quite stable in PSII, they are unstable in synthetic manganese oxide catalysts at neutral pH.[61,64] A strategy to overcome this instability is to speed up the regeneration of Mn^{3+} ions, which usually occurs when a water–Mn^{2+} complex loses a proton and an electron in two separate steps.[65] Yamaguchi et al. found that ring-shaped organic molecules called pyridines could help these steps to happen at the same time, a process likely promoted by amino acids in PSII. They found that the manganese oxide catalyst produced 15 times more oxygen at neutral pH when used in conjunction with a pyridine called 2,4,6-trimethylpyridine. They also tested the reaction in deuterated water, which contains a heavier isotope of hydrogen than normal water. The catalyst generated oxygen much more slowly in the presence of 2,4,6-trimethylpyridine, suggesting that removal of a proton from the water–Mn^{2+} complex is the key step that determines the overall rate of the water-splitting reaction. However, pyridines would not be suitable for large-scale water splitting because they are potential environmental pollutants. It is necessary to identify safer alternative proton-removing molecules that could be immobilized onto the surface of the manganese oxide catalyst to enhance its activity.

An effective biomimetic strategy toward ligand design was proposed for preparing highly efficient WOCs using copper-based WOCs such as the Cu-bipy, Cu-carbonate, and Cu-peptide systems[66-68] due to the high abundance and low cost of Cu and the simplicity of these systems. Although these copper-based WOCs show high current densities and good stabilities under basic conditions, difficulties in accessing high-oxidation-state copper species (Cu^{III} and/or Cu^{IV}) lead to high overpotentials for water oxidation and limit their practical utility.[69] To overcome this challenge for more active WOCs, Zhang et al. used a ligand containing suitable pendant groups to mimic the functions of tyrosine Z in facilitating the oxidation of the Cu center (Figure 10.9).[69] In PSII, a redox-active tyrosine residue, usually referred to as tyrosine Z, serves as a mediator in the electron transfer process between the catalytic center, the $CaMn_4$ cluster, and the oxidant, the photochemically generated $P680^+$. Meanwhile, tyrosine Z, as well as the adjacent histidine residue His-190, also participates in the "proton rocking" process, enabling the proton-coupled electron-transfer (PCET) mechanism for water oxidation reaction.[69] A copper-based WOC with 6,6′ -dihydroxy-2,2′ -bipyridine (H2L) as the ligand was used to mimic the role of tyrosine Z in PSII by not only providing a redox-accessible ligand but also having the hydroxyl groups participating in the PCET processes, to lower the overpotential and enhance the WOC activity. The introduction of 6,6′ -dihydroxyl groups on the bipy moiety allows the L ligand to be intimately involved in the water-oxidation catalytic cycle via ligand oxidation and significantly lowers the water-oxidation overpotential.

Cobalt-based catalysts are also promising candidates for water oxidation. Inspired by the active sites of PSII, compounds possessing Co_4O_4 cubanes were selected, such as organic cobalt complexes, inorganic polyoxometalate (POM) cobalt complexes, etc. The cubic Co_4O_4 core is considered to be the crucial structural feature for efficient catalysis of water oxidation. In addition, some cobalt solid catalysts without Co_4O_4 topology have also been reported for catalyzing water oxidation. For instance, Co-based perovskite catalyst displays electrocatalytic

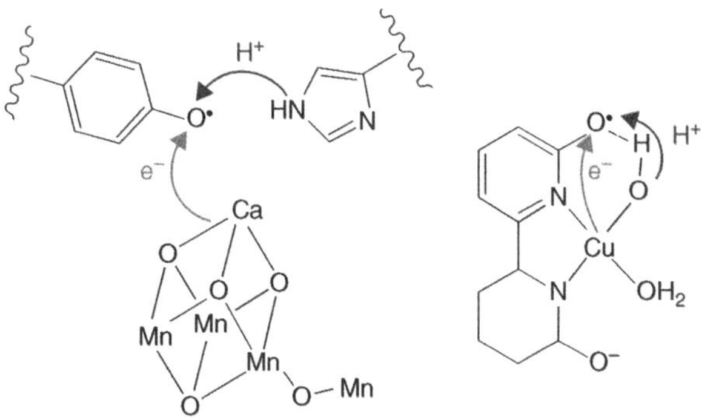

Figure 10.9 Typical PCET process in PSII (left) and the proposed ligand-assisted PCET in the Cu−L system (right). Source: Zhang et al. (2014).[69] Reproduced with permission of the American Chemical Society.

activity in a water-oxidation reaction in alkaline solution,[70,71] while a Co–Fe Prussian blue-type coordination polymer works as a bimetallic electrocatalyst for water oxidation in neutral media.[72] Cobalt(II) oxide (CoO) nanoparticles can carry out overall water splitting with a solar-to-hydrogen efficiency of around 5%.[73] This photocatalyst was synthesized from non-active CoO micropowders using two distinct methods (femtosecond laser ablation and mechanical ball milling), and the CoO nanoparticles that resulted decomposed pure water under visible-light irradiation without any co-catalysts or sacrificial reagents. However, this nanoparticle is not stable and can only work for several hours under normal conditions. In addition, various amorphous cobalt phosphate materials have also been investigated as WOCs, including Co–Pi film (CoO_x/PO_4), which can act as an efficient electrocatalyst for water oxidation with a low overpotential of 410 mV,[74] a mixture of cobalt and methylenediphosphonate (Co/M2P) that can catalyze water oxidation in an aqueous solution at pH 7.0 under visible light irradiation,[75] and cobalt metaphosphate nanoparticles $Co(PO_3)_2$, which exhibited a much lower onset overpotential than those of the Co–Pi film and Co_3O_4.[76]

In PSII, the Mn_4CaO_5 cluster adopts a distorted coordination geometry and every two octahedra are linked by di-μ-oxo (edge-shared) or mono-μ-oxo (corner-shared) bridges, which is recognized as a critical structure motif for catalytic water oxidation.[9,77] These structural features provide guidance on the biomimetic design and synthesis of new OER catalysts. Recently, an organic cobalt phosphonate crystal was demonstrated as a promising heterogeneous catalyst for photocatalytic water oxidation using $[Ru(bpy)_3]Cl_2$ as the photosensitizer and $Na_2S_2O_8$ as the sacrificial oxidant in an aqueous borate buffer solution under visible light ($\lambda > 420$ nm).[78] These cobalt phosphonates are composed of μ-oxo bridging edge-sharing CoO_6 octahedra cobalt clusters without Co_4O_4 topology (Figure 10.10), closely imitating the structure and function of the natural PSII complex, and identified as efficient molecular heterogeneous

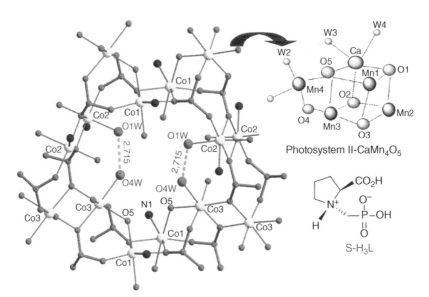

Figure 10.10 Ball-and-stick view highlighting the connectivity of the pairs of cobalt octahedra. Co, O, P, and N are shown as big grey, big dark grey, small dark, and black balls, respectively (where S-H_3L = $H_2O_3PCH_2$–NC_4H_7–CO_2H). Source: Zhou *et al.* (2015).[78] Reproduced with permission of the Royal Society of Chemistry.

catalysts for photocatalytic water oxidation under visible light irradiation. Their catalytic activity was compared to that of cobalt phosphonates with different structures in terms of O_2 evolution rate and O_2 yield under the same reaction conditions. These results provide important insight into the design of biomimetic water oxidation catalysts.

10.3.3   Hydrogenase-biomimetic Catalysts for Hydrogen Generation

Reduction of acids to molecular hydrogen as a means of storing energy is catalyzed by platinum, but its low abundance and high cost are problematic. Soon after the publication of the crystal structures of the hydrogenases, researchers started to synthesize compounds mimicking the unusual active sites of these enzymes. However, exactly mimicking the structures of hydrogenase enzymes turned out to be challenging due to the complex chemistry required to correctly arrange the different ligands around the metal core, and to finally generate functional models with catalytic activity. These considerations have led to efforts to design molecular catalysts that employ relative simple complexes. Synthetic complexes containing nickel, cobalt iron, or molybdenum have been developed recently as electrocatalysts for the production of hydrogen.

In [FeFe] hydrogenase enzymes, the amine base positioned near the iron center has been proposed to function as a proton relay that facilitates the formation or cleavage of the H–H bond.[25] To mimic the structure of the active site of hydrogenase, mononuclear complexes of Fe, Co, and Ni have been synthesized, which contain an amine base in the second coordination sphere, adjacent to a vacant coordination site or a hydride ligand on the metal center.[79,80] Some of these complexes are very effective electrocatalysts for H_2 formation or H_2 oxidation, for example a synthetic nickel complex, $[Ni(P^{Ph}_2NC_6H_4Br_2)_2]^{2+}$, with bromophenyl substituents on the amine ligand, catalyzed the formation of H_2 with turnover frequencies as high as $1040\,s^{-1}$ and an overpotential of $\sim290\,mV$.[81] Another synthetic nickel complex, $[Ni(P^{Ph}_2N^{Ph})_2](BF_4)_2$, where $P^{Ph}_2N^{Ph} = 1,3,6$-triphenyl-1-aza-3,6-diphosphacycloheptane, catalyzes the production of H_2 using protonated dimethylformamide as the proton source, with turnover frequencies of 33,000 per second (s^{-1}) in dry acetonitrile and $106,000\,s^{-1}$ in the presence of 1.2 M of water, at a potential of $-1.13\,V$ (vs. the ferrocenium/ferrocene couple).[82] The mechanistic implications of these remarkably fast catalysts point to a key role of pendant amines that function as proton relays. These results highlight the substantial promise that designed molecular catalysts hold for the electrocatalytic production of hydrogen.

One of the most effective biomimetic approaches for mimicking hydrogenase enzymes involves a family of bisdiphosphine nickel complexes.[83] These complexes combine a nickel center in an electron-rich environment with proton relays provided by a pendant base mimicking the putative azapropanedithiolato cofactor of FeFe hydrogenases.[84] The presence of basic residues at the vicinity of the catalytic metal center facilitates the formation of H_2 from a coordinated hydride ion and a proton from the ammonium group. Goff et al. used bisdiphosphine nickel complexes to assemble an electrode by binding them to carbon nanotubes (CNTs) (Figure 10.11).[85] CNTs have high surface areas (facilitating high catalyst loading), high stability and electrical conductivity, and versatile and straightforward methods for grafting molecular complexes onto their surfaces are available. The covalent attachment of a nickel bisdiphosphine-based mimic of the active site of hydrogenase enzymes onto CNTs results in a high-surface area cathode material with high catalytic activity under the strongly acidic conditions required in proton exchange

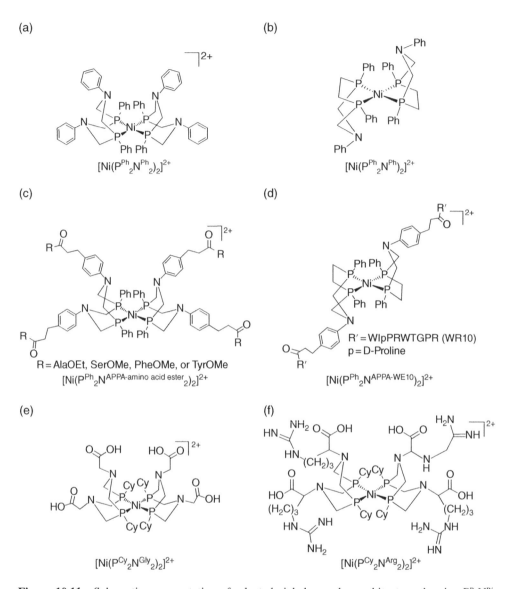

Figure 10.11   Schematic representation of selected nickel complex architectures bearing $P^R_2N^{R'}_n$ ($n = 1$ or 2) ligands. Source: Le Goff *et al.* (2009) and Caserta *et al.* (2015).[85,86] Reproduced with permission of Elsevier.

membrane technology.[85] Hydrogen evolves from aqueous sulfuric acid solution with very low overvoltages (20 mV), and the catalyst exhibits exceptional stability (more than 100,000 turn-overs). The same catalyst is also very efficient for hydrogen oxidation in this environment, exhibiting current densities similar to those observed for hydrogenase-based materials.

Mononuclear $[Ni(P^R_2N^{R'}_2)_2]^{2+}$ complexes bearing diphosphine $P^R_2N^{R'}_2$ ligands are among the most efficient bioinspired catalysts for hydrogen production, as well as H_2 oxidation.

Figure 10.11 shows some of these catalysts. Mimicking the N atom of the bridging dithiolate ligand at the active site of [FeFe] hydrogenases, the pendant amine groups were incorporated into the diphosphine ligand of the Ni ion, which acts as an essential proton relay allowing high catalytic rates for H^+/H_2 interconversion at moderate overpotentials. This demonstrates the importance of the presence of bioinspired functional groups in the second coordination sphere of the metal center in mimics of hydrogenases.[86] Moreover, complexes with acidic or basic side chains ($[Ni(P^{Ph}_2N^{APPA-Lys}_2)_2]^{2+}$, $3350 \, s^{-1}$) displayed an increase of up to five times the rate of electrocatalytic H_2 evolution in water/acetonitrile mixtures compared to the unmodified parent complex ($[Ni(P^{Ph}_2N^{Ph}_2)_2]^{2+}$, $720 \, s^{-1}$).[81] Similarly, terminal propionic acid functions were used to introduce a highly structured, sterically constrained, β-hairpin 10-mer peptide (Figure 10.11d) in the structurally related $[Ni(P^{Ph}_2N^{Ar})_2]^{2+}$ complex (the diphosphine ligands used here contain only one amine group).[86]

A structured outer coordination sphere in an Ni electrocatalyst has been demonstrated to significantly enhance H_2 oxidation activity: $[Ni^{II}(P^{Cy}_2N^{Arg}_2)_2]^{8+}$ (Arg = arginine) has a turnover frequency (TOF) of $210 \, s^{-1}$ in water with high energy efficiency (180 mV overpotential) under 1 atm H_2, and $144000 \, s^{-1}$ (460 mV overpotential) under 133 atm H_2.[87] The complex is active from pH 0–14 and is faster at low pH, the most relevant condition for fuel cells. The arginine substituents increase TOF and may engage in an intramolecular guanidinium interaction that assists in H_2 activation, while the COOH groups facilitate rapid proton movement. These results demonstrate the critical role of features beyond the active site in achieving fast, efficient catalysis.[87]

10.3.4   Artificial Photosynthesis

"Artificial leaves" that collect energy in the same way as natural ones is one of the great dreams for the use of renewable energy and sustainable development. A natural leaf integrates complex architectures and functional components to carry out photosynthesis, an amazing system that efficiently harvests solar energy and converts water and carbon dioxide into carbohydrates and oxygen.[88] The idea of the artificial leaf sounds simple enough: take a small, cheap, light-collecting device the size of a typical leaf, and use sunlight, water, and carbon dioxide to directly generate fuels by non-biological, molecular-level energy conversion.[89] This process of energy production is clean and sustainable; if realized on an industrial scale it will revolutionize our energy system. Artificial photosynthesis (APS) would provide an energy storage solution that would put solar power on the same consistent, reliable footing as any fossil fuel.

Using a biomimetic approach for natural photosynthesis, one design of an idealized molecular artificial photosynthetic system requires five essential components: a light-harvesting antenna, a photosensitized charge-separating site, an O_2-evolving catalyst, a CO_2 reduction catalyst, and a membrane isolating the catalysts, as outlined in Figure 10.12. Ideally, a triad assembly could oxidize water with one catalyst, reduce protons with another, and have a photosensitizer molecule to power the whole system. The simplest biomimetic design is that the photosensitizer is linked in tandem between a WOC and a hydrogen-evolving catalyst. The light-harvesting antenna absorbs solar radiation, which is the first step of artificial photosynthesis. To maximize absorption from the UV to the IR region, the antenna should be constructed with a variety of chromophores. This will then efficiently funnel the excitation

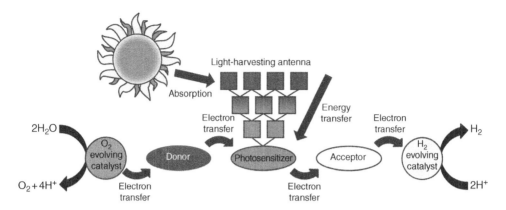

Figure 10.12 General concept for artificial photosynthesis. The light-harvesting antenna collects solar energy then funnels that energy to the photosensitizer, where initial charge separation takes place. A donor and an acceptor drive the electron and hole further apart to prevent back-electron transfer and shuttle the respective charges to the catalysts. A membrane may also be placed somewhere between the O_2 and H_2 evolving catalysts to prevent a "short circuit". When studying half-reactions either catalyst may be replaced by a sacrificial reagent. Source: Frischmann *et al.* (2013).[89] Reproduced with permission of the Royal Society of Chemistry.

energy to a photosensitizer. At the photosensitizer charge is separated by generating an electron–hole pair. Upon generating the electrons and holes, they are further separated by additional donor and acceptor molecules to increase their distance, preventing unwanted recombination and shuttling the charges to the respective catalysts.[89]

The light-harvesting antenna is the first important component for building an efficient artificial photosynthetic system. In principle photosynthesis can still occur without the participation of light-harvesting antennae; sunlight absorption and charge separation can be fulfilled by a special pair of chlorophylls alone in the reaction center. However, due to the multi-electron nature of catalytic reactions in the reaction center, excitation energy must be delivered to drive charge separation and catalysis at a rate appreciably faster than charge recombination. In natural photosynthesis, this requirement for delivering the rate of excitation energy is met by light-harvesting antennae by enlarging their absorption cross-section for the capture of sunlight and efficiently funneling the excitation energy to the reaction center. Artificial photosynthesis may therefore need to follow the rule by building a light-harvesting antenna and integrating it with other components. To integrate the system and make excitation-energy transfer and charge separation efficient, suitable chromophores are needed in artificial light-harvesting antennae and photosensitizers. There is a diversity of chromophores available to mix and match to achieve a functional system in molecular photosynthesis. A collection of selected chromophores is displayed in Figure 10.13.[89]

The second important component is a photosensitizer, in which charge separation occurs. After solar energy is collected by light-harvesting antennae, it must be delivered to photosensitizers to drive charge separation. In the process of photosensitization, back-electron transfer needs to be minimized, and it is desirable that charge-separated states are kept long enough to ensure that each photon delivered by the antenna to the photosensitizer is utilized for conversion into chemical energy. By extending the lifetime of the charge-separated state the probability

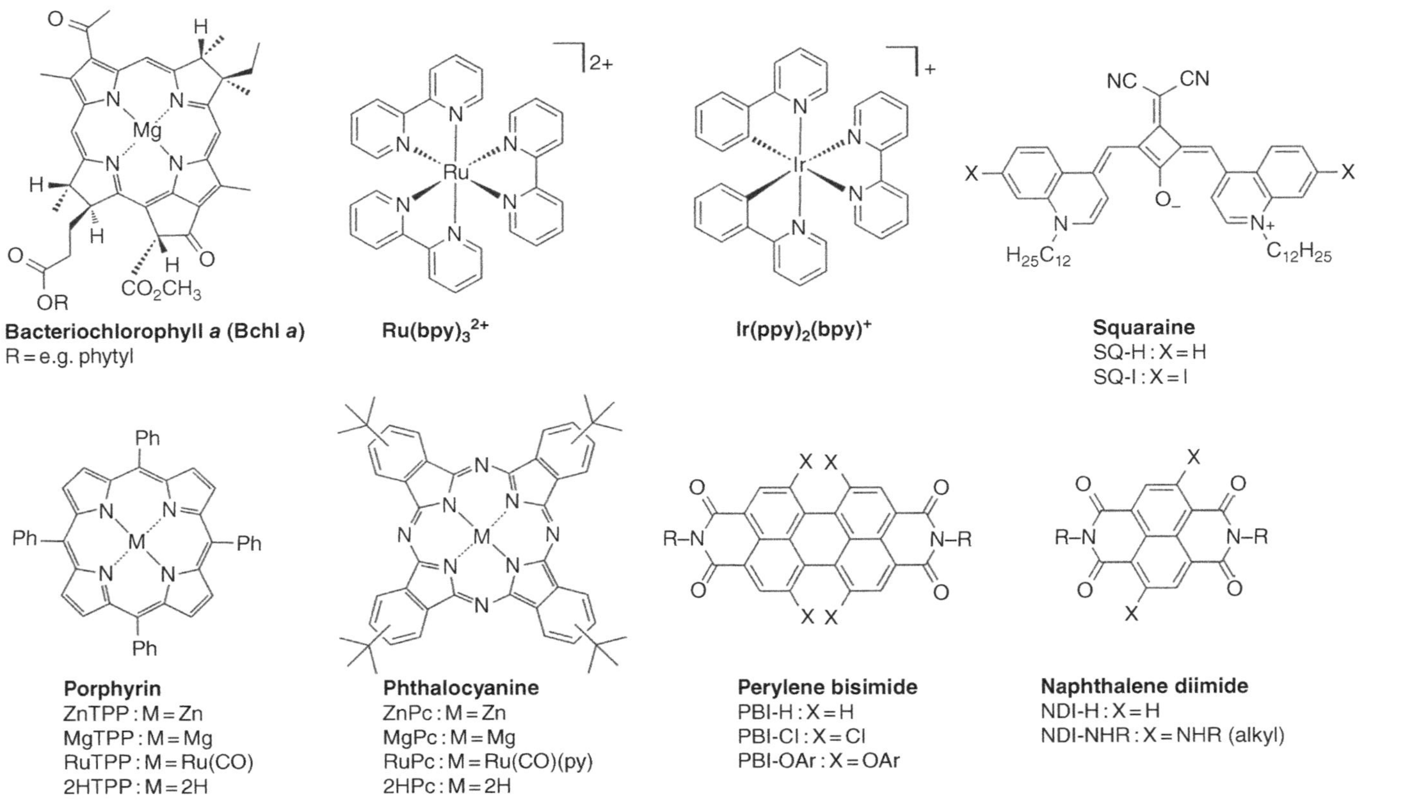

Figure 10.13 A selection of chromophores that have been utilized in energy transfer systems. Source: Frischmann *et al.* (2013).[89] Reproduced with permission of the Royal Society of Chemistry.

for productive catalytic turnover improves, as most relevant solar fuel processes require multiple redox equivalents. In natural photosynthesis, the electron and hole are rapidly separated over a large distance to keep long-lived charge-separated states through multi-step electron-transfer cascades. In this way, the electronic coupling and thus the rate of recombination are reduced by the large distance. When the free energy of the reaction is large, the rate of back-electron transfer may also decrease, resulting in long-lived charge-separated states. Inspired by the natural charge separation system, both covalent and non-covalent photosensitizing dyads, triads, and larger assemblies have been developed to generate sufficiently long-lived charge-separated states. These materials include Ru(II)(CO) porphyrins or phthalocyanines that provide kinetically inert assemblies with enhanced thermodynamic stability compared to Zn(II) analogues, metallosupramolecular self-assembly that shows a non-covalent strategy to control charge- transfer dynamics, a central Al(III) porphyrin that processes rapid electron transfer from the singlet excited state to the naphthalene diimide acceptor unit, etc.[89,90] These photo-sensitizers exhibit enhanced function upon metallosupramolecular assembly.

The third important components are catalysts for water oxidization and hydrogen reduction. These components are located at the heart of both natural and artificial photosynthesis. After charges are separated by the photosensitizers, the electrons and holes must be delivered to the relevant catalysts where solar energy is ultimately converted into chemical fuels. The basic reaction is the energetically demanding water-splitting half-reaction that describes the conversion of two H_2O into O_2, four electrons, and four protons (Equation (10.2)). Since the free energy for $2H_2O \rightarrow O_2 + 2H_2$ is $E^\circ = 1.23\,V$, the energy equivalent of photons with $\lambda = 1009\,nm$ (in the IR range) must be provided to make it happen. There are energy barriers to the reaction, the value of which depends on the type of WO catalysts, and so additional energy must be applied to drive the reaction. In PSII an overpotential of at least $0.3\,V$ is observed, limiting photo-driven catalysis to photons with a threshold wavelength $\lambda \leq 800\,nm$ (close to red). In addition, in order to drive catalytic turnover, four oxidizing equivalents must be delivered at a rate faster than non-productive back-electron transfer. Effective WO catalysts therefore play a key role in maximizing the efficiency of molecular artificial photosynthesis. Although various synthetic WO catalysts have been developed to increase the efficiency of the process, they generally do not compete with the rate, efficiency, and low overpotential exhibited by the protein-supported Mn_4Ca WO catalyst found in PSII.[89]

In PSII the protons and electrons generated during water splitting are used to produce ATP and NADPH, respectively, which provide the chemical driving force for storing solar energy in the form of carbohydrates. This natural process is too complex to mimic in synthetic systems. Instead, in artificial photosynthetic systems protons are reduced to hydrogen by catalysts, thus storing solar energy in the form of H_2 (Equation (10.3)), because hydrogen is the simplest solar fuel to synthesize and involves only the transference of two electrons to two protons. As described in section 10.3.3, many hydrogenase biomimetic catalysts have been developed for hydrogen generation with turnover frequencies of 33,000 per second.[82] In addition, a large number of non-biomimetic catalysts have also been reported based on mononuclear Co, Ni, Rh, Pd, Pt, and dinuclear Fe active sites. The non-assembled multi-component catalytic systems have been extensively optimized to improve their efficiency, resulting in a number of catalysts capable of exhibiting >1000 photocatalytic turnovers. Currently, these non-assembled multicomponent systems are the most effective systems for photocatalytic hydrogen evolution.

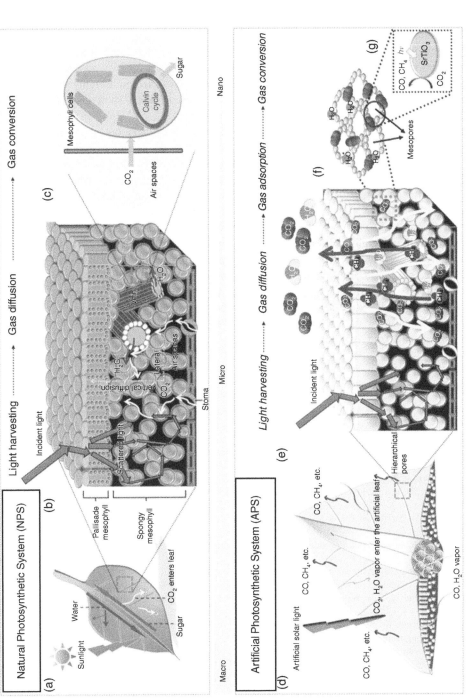

Figure 10.14 Comparison of the key processes in natural photosynthetic system (NPS) and artificial photosynthetic system (APS). (a) Basic process of photosynthesis in NPS at macroscale. (b) Light harvesting and gas diffusion processes in NPS at nanoscale. (c) Gas conversion process in mesophyll cells at microscale. (d) Basic process of artificial photosynthesis in APS at macroscale. (e) Light harvesting and gas diffusion processes in APS at microscale. (f) Gas adsorption process in APS at nanoscale. (g) Gas conversion process in APS at nanoscale. Source: Zhou *et al.* (2013).[93] Reproduced with permission of Nature Publishing Group.

In addition to hydrogen reduction, reduction of CO_2 is also a feasible process for solar energy capture. However, bifunctional catalysts are needed to catalyze reduction of CO_2 as well as H_2 evolution in artificial photosynthesis. Various bifunctional catalysts have been developed, including metal-based materials such as gold–copper bimetallic nanoparticles, MOF-253 supported active Ru carbonyl complex (MOF-253-Ru(CO)(2)Cl-2), photocatalytic CO_2 reduction under visible light, iridium coordination polymer based on a highly efficient light-harvesting Ir unit, and N-doped porous carbons.[91,92] These catalysts show greater mass activities than conventional carbon dioxide reduction catalysts.

As mentioned earlier, leaves utilize sunlight, CO_2, and water to carry out two important reactions required for the survival and growth of the global ecology: decomposition of water to molecular oxygen and reduction of CO_2 to carbohydrates and other carbon-rich products (Figure 10.14a).[88] The whole process of natural photosynthesis has been mimicked to generate fuels. In such artificial leaves, the complex morphological details of the architecture must be preserved to photoreduce CO_2 into hydrocarbon fuels in the APS of perovskite titanates ($ATiO_3$, A = Sr, Ca, and Pb) (Figure 10.14). This process imitates the method utilized by real leaves by using artificial sunlight as the energy source, water as an electron donor, and CO_2 as the carbon source to create hydrocarbon fuels (mainly carbon monoxide and methane). Various co-catalysts (Au, Ag, Cu, Pt, RuO_2, and NiO_x) are inserted for enhanced activity; of these, Au has been shown to be the most effective. Thus, a close reproduction of a leaf's structure will result in a promising APS with efficient mass flow and light absorbing ability. Eventually, these structures can also pave the way for new methods to improve the mass flow network for gas diffusion and enhanced light harvesting.

References

1. Zhou, X., Qiao, J., Yang, L. & Zhang, J. A review of graphene-based nanostructural materials for both catalyst supports and metal-free catalysts in PEM fuel cell oxygen reduction reactions. *Advanced Energy Materials* **4**, 1301523 (2014).
2. Wang, Z.L., Xu, D., Xu, J.J. & Zhang, X.B. Oxygen electrocatalysts in metal–air batteries: from aqueous to nonaqueous electrolytes. *Chemical Society Reviews* **43**, 7746–7786 (2014).
3. Lee, Y., Suntivich, J., May, K.J., Perry, E.E. & Shao-Horn, Y. Synthesis and activities of rutile IrO_2 and RuO_2 nanoparticles for oxygen evolution in acid and alkaline solutions. *Journal of Physical Chemistry Letters* **3**, 399–404 (2012).
4. Morales-Guio, C.G., Stern, L.A. & Hu, X. Nanostructured hydrotreating catalysts for electrochemical hydrogen evolution. *Chemical Society Reviews* **43**, 6555–6569 (2014).
5. Su, D.S. *et al.* Metal-free heterogeneous catalysis for sustainable chemistry. *ChemSusChem* **3**, 169–180 (2010).
6. Sealy, C. The problem with platinum. *Materials Today* **11**, 65–68 (2008).
7. Olson, J.M. Photosynthesis in the Archean era. *Photosynthesis Research* **88**, 109–117 (2006).
8. McEvoy, J.P. & Brudvig, G.W. Water-splitting chemistry of photosystem II. *Chemical Reviews* **106**, 4455–4483 (2006).
9. Umena, Y., Kawakami, K., Shen, J.R. & Kamiya, N. Crystal structure of oxygen-evolving photosystem II at a resolution of 1.9 angstrom. *Nature* **473**, 55–U65 (2011).
10. Heinekey, D.M. Hydrogenase enzymes: Recent structural studies and active site models. *Journal of Organometallic Chemistry* **694**, 2671–2680 (2009).
11. Mulder, D.W. *et al.* Insights into [FeFe]-hydrogenase structure, mechanism, and maturation. *Structure* **19**, 1038–1052 (2011).
12. Silverstein, T.P. Photosynthetic water oxidation vs. mitochondrial oxygen reduction: distinct mechanistic parallels. *Journal of Bioenergetics and Biomembranes* **43**, 437–446 (2011).

13. Grumelli, D., Wurster, B., Stepanow, S. & Kern, K. Bio-inspired nanocatalysts for the oxygen reduction reaction. *Nature Communications* **4**, 2904 (2013).

14. Collman, J.P. & Decreau, R.A. Functional biomimetic models for the active site in the respiratory enzyme cytochrome c oxidase. *Chemical Communications*, 5065–5076 (2008).

15. Berg, J.M., Tymoczko, J.L. & Stryer, L. *Biochemistry*, 5th edn (W.H. Freeman, New York; 2002).

16. Wang, C.H., Wang, C.T., Huang, H.C., Chang, S.T. & Liao, F.Y. High stability pyrolyzed vitamin B12 as a non-precious metal catalyst of oxygen reduction reaction in microbial fuel cells. *RSC Advances* **3**, 15375–15381 (2013).

17. Choi, C.H., Park, S.H. & Woo, S.I. Phosphorus–nitrogen dual doped carbon as an effective catalyst for oxygen reduction reaction in acidic media: effects of the amount of P-doping on the physical and electrochemical properties of carbon. *Journal of Materials Chemistry* **22**, 12107 (2012).

18. Yu, D., Xue, Y. & Dai, L. Vertically aligned carbon nanotube arrays co-doped with phosphorus and nitrogen as efficient metal-free electrocatalysts for oxygen reduction. *Journal of Physical Chemistry Letters* **3**, 2863–2870 (2012).

19. Zhang, J.Z.Z., Xia, Z.H. & Dai, L.M. Nitrogen–phosphorus co-doped 3D mesoporous nanocarbon foams as efficeint metal-free bifunctional electrocatalysts for high-performance rechargeable Zn–air batteries. *Nature Nanotechnology* **10**, 444–452 (2015).

20. Masa, J. *et al.* Trace metal residues promote the activity of supposedly metal-free nitrogen-modified carbon catalysts for the oxygen reduction reaction. *Electrochemistry Communications* **34**, 113–116 (2013).

21. Najafpour, M.M., Moghaddam, A.N., Allakhverdiev, S.I. & Govindjee, S.I. Biological water oxidation: lessons from nature. *Biochimica et Biophysica Acta* **1817**, 1110–1121 (2012).

22. Najafpour, M.M. A soluble form of nano-sized colloidal manganese(IV) oxide as an efficient catalyst for water oxidation. *Dalton Transactions* **40**, 3805–3807 (2011).

23. Lubitz, W., Ogata, H., Rudiger, O. & Reijerse, E. Hydrogenases. *Chemical Reviews* **114**, 4081–4148 (2014).

24. Chenevier, P. *et al.* Hydrogenase enzymes: Application in biofuel cells and inspiration for the design of noble-metal free catalysts for H₂ oxidation. *Comptes Rendus Chimie* **16**, 491–505 (2013).

25. Fontecilla-Camps, J.C., Volbeda, A., Cavazza, C. & Nicolet, Y. Structure/function relationships of [NiFe]- and [FeFe]-hydrogenases. *Chemical Reviews* **107**, 4273–4303 (2007).

26. Florin, L., Tsokoglou, A. & Happe, T. A novel type of iron hydrogenase in the green alga *Scenedesmus obliquus* is linked to the photosynthetic electron transport chain. *Journal of Biological Chemistry* **276**, 6125–6132 (2001).

27. Cammack, R. Bioinorganic chemistry – Hydrogenase sophistication. *Nature* **397**, 214–215 (1999).

28. Allakhverdiev, S.I. *et al.* Photosynthetic hydrogen production. *Journal of Photochemistry and Photobiology C – Photochemistry Reviews* **11**, 101–113 (2010).

29. Appleby, A.J. Electrocatalysis of aqueous dioxygen reduction. *Journal of Electroanalytical Chemistry* **357**, 117–179 (1993).

30. Li, L., Hu, L.P., Li, J. & Wei, Z.D. Enhanced stability of Pt nanoparticle electrocatalysts for fuel cells. *Nano Research* **8**, 418–440 (2015).

31. Zagal, J.H. & Paez, M.A. Electro-oxidation of hydrazine on electrodes modified with vitamin B₁₂. *Electrochimica Acta* **42**, 3477–3481 (1997).

32. Zagal, J.H., Aguirre, M.J. & Paez, M.A. O₂ reduction kinetics on a graphite electrode modified with adsorbed vitamin B₁₂. *Journal of Electroanalytical Chemistry* **437**, 45–52 (1997).

33. Chang, S.-T. *et al.* Vitalizing fuel cells with vitamins: pyrolyzed vitamin B12 as a non-precious catalyst for enhanced oxygen reduction reaction of polymer electrolyte fuel cells. *Energy and Environmental Science* **5**, 5305–5314 (2012).

34. Liu, S., Deng, C., Yao, L., Zhong, H. & Zhang, H. Bio-inspired highly active catalysts for oxygen reduction reaction in alkaline electrolyte. *International Journal of Hydrogen Energy* **39**, 12613–12619 (2014).

35. Zhong, H.X., Zhang, H.M., Liu, S.S., Deng, C.W. & Wang, M.R. Nitrogen-enriched carbon from melamine resins with superior oxygen reduction reaction activity. *ChemSusChem* **6**, 807–812 (2013).

36. Morozan, A., Jousselme, B. & Palacin, S. Low-platinum and platinum-free catalysts for the oxygen reduction reaction at fuel cell cathodes. *Energy & Environmental Science* **4**, 1238–1254 (2011).

37. Chen, R.R., Li, H.X., Chu, D. & Wang, G.F. Unraveling oxygen reduction reaction mechanisms on carbon-supported Fe-phthalocyanine and Co-phthalocyanine catalysts in alkaline solutions. *Journal of Physical Chemistry C* **113**, 20689–20697 (2009).

38. Yuan, Y. *et al.* Iron phthalocyanine supported on amino-functionalized multi-walled carbon nanotube as an alternative cathodic oxygen catalyst in microbial fuel cells. *Bioresource Technology* **102**, 5849–5854 (2011).

39. Dong, G., Huang, M. & Guan, L. Iron phthalocyanine coated on single-walled carbon nanotubes composite for the oxygen reduction reaction in alkaline media. *Physical Chemistry Chemical Physics* **14**, 2557–2559 (2012).

40. Cao, R. *et al.* Promotion of oxygen reduction by a bio-inspired tethered iron phthalocyanine carbon nanotube-based catalyst. *Nature Communications* **4** (2013).

41. Zhang, C.Z., Hao, R., Yin, H., Liu, F. & Hou, Y.L. Iron phthalocyanine and nitrogen-doped graphene composite as a novel non-precious catalyst for the oxygen reduction reaction. *Nanoscale* **4**, 7326–7329 (2012).

42. Guo, Z. *et al.* Biomolecule-doped PEDOT with three-dimensional nanostructures as efficient catalyst for oxygen reduction reaction. *Small* **10**, 2087–2095 (2014).

43. Ahmad, S. *et al.* Towards flexibility: metal free plastic cathodes for dye sensitized solar cells. *Chemical Communications* **48**, 9714–9716 (2012).

44. Guo, Z.Y. *et al.* Self-assembled hierarchical micro/nano-structured PEDOT as an efficient oxygen reduction catalyst over a wide pH range. *Journal of Materials Chemistry* **22**, 17153–17158 (2012).

45. Li, W.M., Yu, A.P., Higgins, D.C., Llanos, B.G. & Chen, Z.W. Biologically inspired highly durable iron phthalocyanine catalysts for oxygen reduction reaction in polymer electrolyte membrane fuel cells. *Journal of the American Chemical Society* **132**, 17056–17058 (2010).

46. Barth, J.V. Fresh perspectives for surface coordination chemistry. *Surface Science* **603**, 1533–1541 (2009).

47. Norskov, J.K., Bligaard, T., Rossmeisl, J. & Christensen, C.H. Towards the computational design of solid catalysts. *Nature Chemistry* **1**, 37–46 (2009).

48. Barth, J.V., Costantini, G. & Kern, K. Engineering atomic and molecular nanostructures at surfaces. *Nature* **437**, 671–679 (2005).

49. Xiang, Z. *et al.* Highly efficient electrocatalysts for oxygen reduction based on 2D covalent organic polymers complexed with non-precious metals. *Angewandte Chemie International Edition* **53**, 2433–2437 (2014).

50. Kim, E. *et al.* Superoxo, mu-peroxo, and mu-oxo complexes from heme/O_2 and heme-Cu/O_2 reactivity: Copper ligand influences in cytochrome c oxidase models. *Proceedings of the National Academy of Sciences of the United States of America* **100**, 3623–3628 (2003).

51. Muramoto, K. *et al.* Bovine cytochrome c oxidase structures enable O_2 reduction with minimization of reactive oxygens and provide a proton-pumping gate. *Proceedings of the National Academy of Sciences of the United States of America* **107**, 7740–7745 (2010).

52. Blanford, C.F., Heath, R.S. & Armstrong, F.A. A stable electrode for high-potential, electrocatalytic O_2 reduction based on rational attachment of a blue copper oxidase to a graphite surface. Chemical Communications, 1710–1712 (2007).

53. McCrory, C.C.L., Ottenwaelder, X., Stack, T.D.P. & Chidsey, C.E.D. Kinetic and mechanistic studies of the electrocatalytic reduction of O_2 to H_2O with mononuclear Cu complexes of substituted 1,10-phenanthrolines. *Journal of Physical Chemistry A* **111**, 12641–12650 (2007).

54. Thorum, M.S., Yadav, J. & Gewirth, A.A. Oxygen reduction activity of a copper complex of 3,5-diamino-1,2, 4-triazole supported on carbon black. *Angewandte Chemie International Edition* **48**, 165–167 (2009).

55. Wang, J., Wang, K., Wang, F.B. & Xia, X.H. Bioinspired copper catalyst effective for both reduction and evolution of oxygen. *Nature Communications* **5**, 5285 (2014).

56. Jones, R.H. *Materials for the hydrogen economy* (Taylor & Francis Ltd, Hoboken; 2007).

57. Stoerzinger, K.A., Qiao, L., Biegalski, M.D. & Shao-Horn, Y. orientation-dependent oxygen evolution activities of rutile IrO_2 and RuO_2. *Journal of Physical Chemistry Letters* **5**, 1636–1641 (2014).

58. Jiao, F. & Frei, H. Nanostructured manganese oxide clusters supported on mesoporous silica as efficient oxygen-evolving catalysts. *Chemical Communications* **46**, 2920–2922 (2010).

59. Boppana, V.B. & Jiao, F. Nanostructured MnO_2: an efficient and robust water oxidation catalyst. *Chemical Communications* **47**, 8973–8975 (2011).

60. Najafpour, M.M., Sedigh, D.J., King'ondu, C.K. & Suib, S.L. Nano-sized manganese oxide-bovine serum albumin was synthesized and characterized. It is promising and biomimetic catalyst for water oxidation. *RSC Advances* **2**, 11253–11257 (2012).

61. Takashima, T., Hashimoto, K. & Nakamura, R. Mechanisms of pH-dependent activity for water oxidation to molecular oxygen by MnO_2 electrocatalyst. *Journal of the American Chemical Society* **134**, 1519–1527 (2012).

62. Mohammad, A.M., Awad, M.I., El-Deab, M.S., Okajima, T. & Ohsaka, T. Electrocatalysis by nanoparticles: Optimization of the loading level and operating pH for the oxygen evolution at crystallographic ally oriented manganese oxide nanorods modified electrodes. *Electrochimica Acta* **53**, 4351–4358 (2008).

63. Yano, J. *et al.* Where water is oxidized to dioxygen: Structure of the photosynthetic Mn_4Ca cluster. *Science* **314**, 821–825 (2006).

64. Takashima, T., Hashimoto, K. & Nakamura, R. Inhibition of charge disproportionation of MnO_2 electrocatalysts for efficient water oxidation under neutral conditions. *Journal of the American Chemical Society* **134**, 18153–18156 (2012).

65. Yamaguchi, A. *et al.* Regulating proton-coupled electron transfer for efficient water splitting by manganese oxides at neutral pH. *Nature Communications* **5**, 4256 (2014).

66. Chen, Z & Meyer, T.J. Copper(II) catalysis of water oxidation. *Angewandte Chemie International Edition* **52**, 700–703 (2013).

67. Barnett, S.M., Goldberg, K.I. & Mayer, J.M. A soluble copper–bipyridine water-oxidation electrocatalyst. *Nature Chemistry* **4**, 498–502 (2012).

68. Zhang, M.T., Chen, Z.F., Kang, P. & Meyer, T.J. Electrocatalytic water oxidation with a copper(II) polypeptide complex. *Journal of the American Chemical Society* **135**, 2048–2051 (2013).

69. Zhang, T., Wang, C., Liu, S., Wang, J.L. & Lin, W. A biomimetic copper water oxidation catalyst with low overpotential. *Journal of the American Chemical Society* **136**, 273–281 (2014).

70. Suntivich, J., May, K.J., Gasteiger, H.A., Goodenough, J.B. & Shao-Horn, Y. A perovskite oxide optimized for oxygen evolution catalysis from molecular orbital principles. *Science* **334**, 1383–1385 (2011).

71. Yamada, Y., Yano, K., Hong, D.C. & Fukuzumi, S. $LaCoO_3$ acting as an efficient and robust catalyst for photocatalytic water oxidation with persulfate. *Physical Chemistry Chemical Physics* **14**, 5753–5760 (2012).

72. Pintado, S., Goberna-Ferron, S., Escudero-Adan, E.C. & Galan-Mascaros, J.R. Fast and persistent electrocatalytic water oxidation by Co–Fe Prussian blue coordination polymers. *Journal of the American Chemical Society* **135**, 13270–13273 (2013).

73. Liao, L.B. *et al.* Efficient solar water-splitting using a nanocrystalline CoO photocatalyst. *Nature Nanotechnology* **9**, 69–73 (2014).

74. Kanan, M.W. & Nocera, D.G. In situ formation of an oxygen-evolving catalyst in neutral water containing phosphate and Co^{2+}. *Science* **321**, 1072–1075 (2008).

75. Shevchenko, D., Anderlund, M.F., Thapper, A. & Styring, S. Photochemical water oxidation with visible light using a cobalt containing catalyst. *Energy & Environmental Science* **4**, 1284–1287 (2011).

76. Ahn, H.S. & Tilley, T.D. Electrocatalytic water oxidation at neutral pH by a nanostructured $Co(PO_3)_2$ anode. *Advanced Functional Materials* **23**, 227–233 (2013).

77. Tachibana, Y., Vayssieres, L. & Durrant, J.R. Artificial photosynthesis for solar water-splitting. *Nature Photonics* **6**, 511–518 (2012).

78. Zhou, T. *et al.* Bio-inspired organic cobalt(ii) phosphonates toward water oxidation. *Energy and Environmental Science* **8**, 526–534 (2015).

79. DuBois, D.L. & Bullock, R.M. Molecular electrocatalysts for the oxidation of hydrogen and the production of hydrogen – The role of pendant amines as proton relays. *European Journal of Inorganic Chemistry* 1017–1027 (2011).

80. Dubois, M.R. & Dubois, D.L. Development of molecular electrocatalysts for CO_2 reduction and H_2 production/oxidation. *Accounts of Chemical Research* **42**, 1974–1982 (2009).

81. Kilgore, U.J. *et al.* $[Ni((P_2N_2C_6H_4X)-N-Ph)_2]^{2+}$ complexes as electrocatalysts for H_2 production: effect of substituents, acids, and water on catalytic rates. *Journal of the American Chemical Society* **133**, 5861–5872 (2011).

82. Helm, M.L., Stewart, M.P., Bullock, R.M., DuBois, M.R. & DuBois, D.L. A synthetic nickel electrocatalyst with a turnover frequency above $100,000\,s^{-1}$ for H_2 production. *Science* **333**, 863–866 (2011).

83. Wilson, A.D. *et al.* Nature of hydrogen interactions with Ni(II) complexes containing cyclic phosphine ligands with pendant nitrogen bases. *Proceedings of the National Academy of Sciences of the United States of America* **104**, 6951–6956 (2007).

84. Silakov, A., Wenk, B., Reijerse, E. & Lubitz, W. N^{14} HYSCORE investigation of the H-cluster of [FeFe] hydrogenase: evidence for a nitrogen in the dithiol bridge. *Physical Chemistry Chemical Physics* **11**, 6592–6599 (2009).

85. Le Goff, A. *et al.* From hydrogenases to noble metal-free catalytic nanomaterials for H_2 production and uptake. *Science* **326**, 1384–1387 (2009).

86. Caserta, G., Roy, S., Atta, M., Artero, V. & Fontecave, M. Artificial hydrogenases: biohybrid and supramolecular systems for catalytic hydrogen production or uptake. *Current Opinion in Chemical Biology* **25**, 36–47 (2015).

87. Dutta, A., Roberts, J.A.S. & Shaw, W.J. Arginine-containing ligands enhance H_2 oxidation catalyst performance. *Angewandte Chemie International Edition* **53**, 6487–6491 (2014).

88. Barber, J. Biological solar energy. *Philosophical Transactions of the Royal* Society A **365**, 1007–1023 (2007).

89. Frischmann, P.D., Mahata, K. & Wurthner, F. Powering the future of molecular artificial photosynthesis with light-harvesting metallosupramolecular dye assemblies. *Chemical Society reviews* **42**, 1847–1870 (2013).

90. Rodriguez-Morgade, M.S., Torres, T., Atienza-Castellanos, C. & Guldi, D.M. Supramolecular bis(rutheniumphthalocyanine)-perylenediimide ensembles: Simple complexation as a powerful tool toward long-lived radical ion pair states. *Journal of the American Chemical Society* **128**, 15145–15154 (2006).

91. Artero, V., Chavarot-Kerlidou, M. & Fontecave, M. Splitting water with cobalt. *Angewandte Chemie International Edition* **50**, 7238–7266 (2011).

92. Costentin, C., Drouet, S., Robert, M. & Saveant, J.-M. A local proton source enhances CO_2 electroreduction to CO by a molecular Fe catalyst. *Science* **338**, 90–94 (2012).

93. Zhou, H. *et al.* Leaf-architectured 3D hierarchical artificial photosynthetic system of perovskite titanates towards CO_2 photoreduction into hydrocarbon fuels. *Scientific Reports* **3**, 1667 (2013).

Part III

Biomimetic Processing

11

Biomineralization and Biomimetic Materials Processing

11.1 Introduction

Modern material manufacturing technology has evolved to be able to fabricate various materials, including ceramics, metals, polymers, and their composites from nano to macro scale. However, most traditional approaches to synthesis of materials are energy inefficient, require stringent conditions (e.g., elevated temperature, pressure, or pH), and often produce toxic byproducts. For example, in Czochralski crystal growth, high-purity, semiconductor-grade silicon is melted in crucible at high temperature, and a seed crystal is dipped into the molten silicon and then pulled upwards to form large single crystals that are needed in semiconductor industry. In particular for nanoscale materials, the quantities produced are small, and the resultant material is usually irreproducible because of the difficulties in controlling agglomeration. Similarly, the vast bulk of minerals found on Earth are formed under extreme temperatures and pressures, and required long times for their various transformations.

Living organisms make materials in a different way from traditional materials processing. Nature creates materials based on biogenic origin at moderate ambient temperatures, pressures, and neutral pH. These biological materials are assembled in aqueous environments under the directing genetic codes; most of them are formed using elements that are abundant in the Earth's crust. In the biosynthesis, macromolecules play a key role in forming various materials, including polymers, ceramics, metals, and their composites. The organic macromolecules, generated by cells, collect and transport raw materials and assemble them into short- and long-range ordered materials. In addition to biopolymers, minerals, metals, and their composites are formed with control over the crystalline and hierarchical structure from the nano to the macro scale. The materials produced by living organisms have properties that usually surpass those of analogous synthetically manufactured materials with similar phase composition.[1]

Biomimetic Principles and Design of Advanced Engineering Materials, First Edition. Zhenhai Xia.
© 2016 John Wiley & Sons, Ltd. Published 2016 by John Wiley & Sons, Ltd.

Minerals are almost ubiquitous components of living organisms, in which they serve a multitude of crucial structural and biochemical functions. Through millions of years of evolution, bacteria, plants, and animals have created diversified inorganic material of, perhaps, 60 different kinds, including hydroxyapatite (HA), calcium carbonate, and silica. In biologically controlled mineralization, or biomineralization, living organisms deposit inorganic minerals in a highly regulated manner to produce mineralized structures such as bones, shells, and teeth. These biominerals often exhibit impressive morphologies, structures, and compositions, such as uniform grain and/or particle sizes, oriented crystallographic structures, and hierarchical structures, resulting in complex architectures that have multifunctional properties. For example, HA in the bones and teeth of mammals and calcium carbonate in molluscan shells have elaborate hierarchical structures, giving them conspicuously high mechanical hardness and flexibility, which are not provided by conventional synthetic materials.[2,3] A number of single-celled organisms (bacteria and algae) also produce inorganic materials (e.g., magnetite and silica) either intracellularly or extracellularly. These hard tissues are normally mechanical devices (e.g., skeletal, cutting, grinding) or serve a physical function (e.g., magnetic, optical, piezoelectric).[4] The impressive mechanical, physical, and chemical properties of biominerals stem from their complex structures consisting of organic and inorganic constituents. These interesting features and unique biomineralization processes have inspired considerable efforts in biomimetic material synthesis and assembly.[5]

The bioinspired synthesis approach has been used for generating a variety of technologically relevant materials. Although this approach is still in its initial stage of research, some of the work has shown potential for synthesizing high-quality ceramic and metal nanoparticles with low cost and efficient and environmentally friendly approaches. For example, metals, metal oxides, chalcogenides, and hydroxides have been grown with controlled structures, composition, and morphologies through bioinspired synthesis approaches under mild conditions.[6,7] Various specialized proteins extracted from organisms or their biomimics are applied to the biomimetic synthetic routes, through which the crystal size, morphology, surface structures, composition, crystallinity, and even hierarchical organization of materials can be controlled. The applications of these principles will create environmentally benign routes for the fabrication of structural and functional materials with multiple combinations of outstanding properties, including light weight, high flexibility, mechanical strength, dynamic function, and structural hierarchy.

In this chapter, the mechanisms of typical biomineralization in different living systems will be discussed and the biomimetic synthesis of inorganic materials will be examined. Emphasis will be placed on material formation mechanisms and bioinspired process principles with the use of protein/peptide molecules that are extracted from the organisms or identified through the molecular evolution process, showing specific affinity to an arbitrary inorganic material.

11.2   Materials Processing in Biological Systems

11.2.1   Biomineralization

Biomineralization is the process by which mineral crystals are deposited in an organized fashion in the matrix (either cellular or extracellular) of living organisms. Many organisms construct structural ceramic (biomineral) composites from seemingly mundane materials. In biomineralization, cell-mediated processes control both the nucleation and growth of the mineral and the development of composite microarchitecture.[8] A general approach of natural biomineralization systems for the formation of minerals is the use of an organized protein matrix as an overall framework. A general biomineralization process involves:

1. *nucleation*: specific protein nucleators provide a template or sequester the ions and direct them in an ordered orientation to form the initial crystal structure
2. *crystal expansion*: ions add into specific sites on the initial crystals, and crystals aggregate
3. *organic molecules regulate crystal expansion*: the molecules (e.g., matrix proteins) can bind to the crystal to inhabit or shape crystal formation, and can transport or sequester ions.

Biomineralization is a complex process in which proteins with special binding ability and functions play a key role in determining the microstructure and properties. Although the biomachinery facilitating biomineralization is complex – involving signaling transmitters, inhibitors, and transcription factors – many elements of this "toolkit" are shared between phyla as diverse as corals, molluscs, and vertebrates.[9] These shared components tend to perform quite fundamental tasks, such as nucleation and initial growth of minerals, whereas genes control more finely tuned aspects that occur later in the biomineralization process. During the growth of minerals, the genes direct the precise alignment and structure of the crystals in different lineages.[10] So far, a number of proteins that control biomineralization processes have been identified, although the detailed functions of proteins at the molecular level have remained largely unknown. According their functions in the mineralization processes, these proteins can be classified into several groups: promotion of crystal formation,[11,12] matrix-assisted orientation of crystals,[13,14] growth inhibition by face-selective surface adsorption,[15] and control of the crystal phase[16] (Table 11.1).

The schematics of protein-controlled processes of biomineralization are illustrated in Figure 11.1. Silicatein, for example, which exists in material synthesis in sponges, plays a key role in the formation of the skeletal structures of marine sponges during their

Table 11.1 Important proteins directly involved in biomineralization.[5,20]

Mineral	Protein	Bioprototypes	Function
Calcium carbonate	Perlucin	The nacreous layer of the shell	Calcite precipitation
	MSI31,60	The nacreous layer of the shell	Framework of the prismatic layer
	Pif	The nacreous layer of the shell	Aragonite crystal formation
	Ansocalcin	Goose egg shell matrix	Template for calcite nucleation
	CAP-1	The exoskeleton of crayfish	Crystal growth regulation
Silica	Silicatein	Sponge spicules	Silica polymerization
	Silaffin	Diatom shells	Silica precipitation
Iron oxide	Mms6	Bacterial magnetites	Crystal size and shape control of magnetites
Metal	Ferritin	All living organisms	Iron homeostasis and storage
	Bovine serum albumin	Cows	Formation of nanoparticles

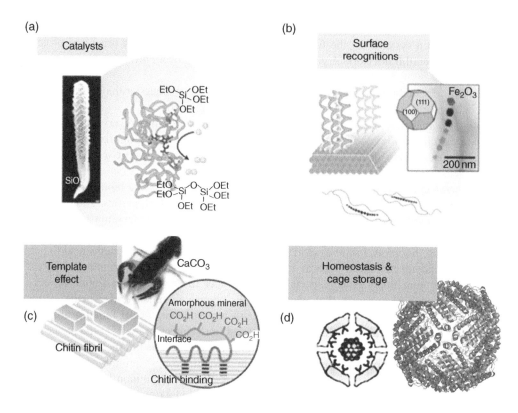

Figure 11.1 Molecularly controlled processes of biomineralization: (a) biosilicification catalyzed by silicatein, (b) magnetosomes by surface-directed biomineralization, (c) organic matrix-templated biomineralization, and (d) iron homeostasis and storage by ferritin. Source: (a–c) Arakaki *et al.* (2015)[5] and (d) Suzuki *et al.* (2009).[21] Reproduced with permission of the Royal Society of Chemistry.

biomineralization.[17] The nucleation and growth of silica are directly catalyzed by the protein. This provides us with a powerful tool to synthesize new materials with controllable composition, microstructure, shape, and properties under ambient conditions.[18] Compared with traditional process of silica, this approach is more effective in controling microstructure and compositions, and thus the properties of the minerals. Another protein, Mms6, identified in magnetotactic bacteria, regulates the surface structure of magnetite nanoparticles.[16] The protein regulation leads to the formation of nano-sized single-domain magnetite crystals with precise shape and microstructure. In bone growth, proteins form the preform to regulate the growth direction and size of the minerals between the proteins.

In addition to biomineralization, specialized proteins can also reduce and/or store metals in the form of nanoparticles. Ferritin, for example, is a ubiquitous intracellular protein that stores iron and releases it in a controlled fashion. This protein is produced by almost all living organisms, including algae, bacteria, higher plants, and animals.[19] Ferritin is found in most tissues as a cytosolic protein, but small amounts are secreted into the serum where it functions as an iron carrier. This protein can be used as a scaffold for producing metal nanoparticles such as Co, Ni, Cr, Au, Ag, Pt, nanoparticles, etc.

11.2.2 Surface-directed Biomineralization

Surface-directed biomineralization is a process in which certain crystal surfaces directly grow from the protein surfaces. With certain proteins as growth base, a variety of nanoparticles can be synthesized in similar way to in nature, and their size, morphology, and composition can be precisely controlled. For example, magnetotactic bacteria are known to have the ability to synthesize nano-sized single-domain magnetite crystals (Fe_3O_4) that are aligned in a chain, enabling the cells to swim or migrate along magnetic field lines (Figure 11.2a–c). During formation of magnetite crystals, the cells form a specialized intracellular membranous compartment, also called magnetosome, the diameter of which is typically in the range 20–100 nm.[22] The biomineralization process in magnetosomes is strictly controlled, leading to a highly regular and specific crystal size, shape, number, and assembly for a given bacterial strain. Such exquisite control over magnetite structure provides an excellent biomimetic model for the design and synthesis of nanoparticles with highly controlled size and crystal structures, useful for many applications, including nanotechnology-based therapeutics and nanoscale devices.

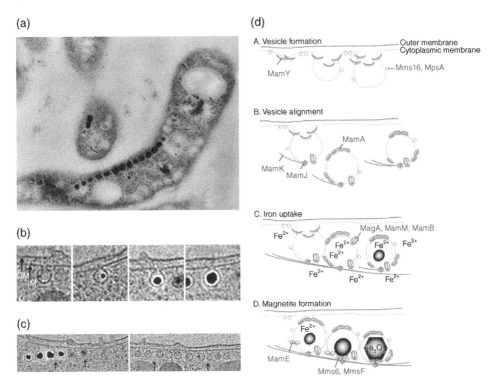

Figure 11.2 (a) Transmission electron microscopy (TEM) image of sectioned *Magnetospirillum magneticum AMB*-1 reveals the presence of a chain of electron-dense magnetite crystals. Reproduced with permission of the National Academy of Sciences. (b) Sections from an electron cryotomographic image of AMB-1 show that magnetosomes are invaginations of the inner cell membrane at various stages of biomineralization. (c) The same study shows the presence of filaments parallel to the magnetosome chain. Source: Komeili (2012).[23] Reproduced with permission of Oxford University Press. (d) Schematic for the hypothesized mechanism of magnetite biomineralization in magnetotactic bacteria. Source: Arakaki *et al.* (2015).[5] Reproduced with permission of the Royal Society of Chemistry.

The assembly of magnetosomes can be divided into several stages, including membrane biogenesis, magnetosome protein localization, and biomineralization, each of which is strictly controlled by genes.[23] Figure 11.2d illustrates the biological process of biomineralization based on magnetosome protein. In the first stage of the biomineralization process, a series of vesicles are formed in the inner cell membrane, which serves as the precursor for magnetite crystals (Figure 11.2d(A)). These vesicles are then assembled into a linear chain along with cytoskeletal filaments (Figure 11.2d(B)). After the vesicles are formed, ferrous ions are accumulated into the vesicles through the transmembrane iron transporters (Figure 11.2d(C)). Internal iron is strictly controlled by an oxidation–reduction system. In the final stage, accumulated iron ions are crystallized within the vesicle (Figure 11.2d(D)).[5]

Several proteins are involved in the formation and maintenance of membrane vesicles; each of them has its own specific functions for mineralizing. Among them, Mms6 protein is one of the important proteins for regulating the morphology of cubo-octahedral magnetite crystals. Three other proteins, designated Mms5 (MamG), Mms7 (MamD), and Mms13 (MamC), have a common amphiphilic characteristic containing hydrophobic N-terminal and hydrophilic C-terminal regions. The C-terminal region is considered to be an iron-binding site,[16] while the N-terminal region is responsible for the self-aggregation of this protein.[24] These proteins spatially distribute in specific positions and work together to guide crystal growth.

11.2.3 Enzymatic Biomineralization

Some organisms, including diatoms, radiolaria, choanoflagellates, sponges, and higher plants, use specialized proteins to build up their skeletons with silica. During the nucleation and growth of silica these proteins serve as catalysts and usually matrix as well, and the mineral is formed at neutral pH and ambient temperature. In this process, or biosilicification, inorganic silicon is incorporated into living organisms as silica, which occurs on the scale of gigatons.[25] A well-known example is siliceous sponges, where the specific proteins, silicatein, enzymatically regulate the biosilicification process and at the same time serve as the organic matrix for the silica product. The process mainly occurs in the unicellular diatoms and multicellular sponges, but silica also deposits in plants and even in higher mammals, having been reported in the electric organs of the fish *Psammobatis extenta*.[26] Several proteins work together to produce silica structures (Figure 11.3). These silicatein proteins can be classified into three different types: silicatein α, β, and γ. Among them, silicatein α is most seen, occupying about 70% of the mass of the proteinaceous filament.[27] The biosilicification process can be continued *ad infinitum*, leading to long insoluble biosilica molecules.

One of the key steps in silica biosynthesis is Si–O–Si bond formation from the condensation of two Si–OH groups in either silica-forming organisms or *in vitro* silicification.[29] In this process, orthosilicate $Si(OH)_4$ is condensed into long polymers with the elimination of water:

$$\underset{\underset{OH}{|}}{\overset{\overset{OH}{|}}{HO-Si-OH}} + \underset{\underset{OH}{|}}{\overset{\overset{OH}{|}}{HO-Si-OH}} \longrightarrow \underset{\underset{OH}{|}}{\overset{\overset{OH}{|}}{HO-Si}}-O-\underset{\underset{OH}{|}}{\overset{\overset{OH}{|}}{Si-OH}} + H_2O \qquad (11.1)$$

The rate-limiting step in this process is the hydrolysis of alkoxysilanes, which typically uses acids or bases as catalysts. In sponge spicule formation, the growth of silica is catalyzed

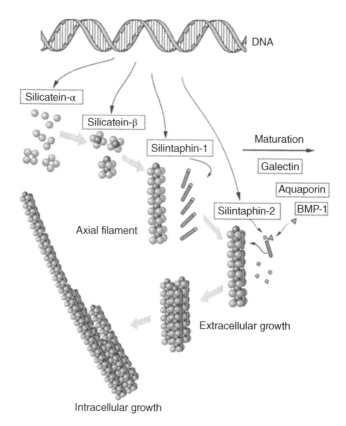

Figure 11.3 The self-assembly process of the organic layers in siliceous spicules. Source: Wang *et al.* (2012).[28] Reproduced with permission of the Royal Society of Chemistry.

by the axial filaments in the hydrolysis of various silicon alkoxides such as tetraethoxysilane (TEOS) and methyl- and phenyl-triethoxysilanes. This catalytic activity was retained in silicatein monomers dissociated from axial filaments. Silicatein not only works in the natural environment, but also catalyzes silsesquioxanes *in vitro* at ambient temperature, pressure, and neutral pH. The process was reproduced even with a recombinant form synthesized in genetically engineered *Escherichia coli*.[30] Thus, silicatein should be very useful in creating new types of silica materials for various applications.

In addition to being catalysts, silicateins serve as the template for constructing axial filaments within the structure. As shown in Figure 11.3, the silicateins, mediated by the interaction of hydrophobic patches on the surfaces of the silicatein isoforms, self-assemble into filamentous structures.[31] The polycondensed silica are then deposited on the surface of the filaments along the entire filament length.[32]

11.2.4 Organic Matrix-templated Biomineralization

In organic matrix-templated biomineralization, a macromolecule matrix is usually formed as a preform of the minerals. The mineralization occurs within the pores or gaps of the organic template. For example, in bone growth the collagen assembles itself into fibrils and fibers. These fibrils further assemble into ordered fibril bundles with the end of one collagen fibril connected to the beginning of the next, linked with special bonds, and these porous structures are mechanically stable due to crosslinking. In the mineralization process HA, a mineral phase of calcium phosphate, deposits inside the cavities in between the fibrils (interfibrillar) as well as inside the fibrils themselves (intrafibrillar), forming parallel mineralized collagen fibrils.[33] The overlapping region of the fibrils in the fiber bundles is less mineralized. Confined by collagen templating, the mineral platelets are formed with a specific size just occupying the space that corresponds to the distance between one gap and the next, across one overlapping region (Figure 11.4a).[34] Another well-known example is the nacre of the abalone shell.[35] Plate-like aggregates of submicrometer thickness are formed in nanocrystalline aragonite with uniaxial crystallographic orientation between the pre-organized macromolecular sheets (Figure 11.4c). The exoskeleton of crayfish is also an example of a $CaCO_3$-based biominerals formed in organic templating. The exoskeleton is

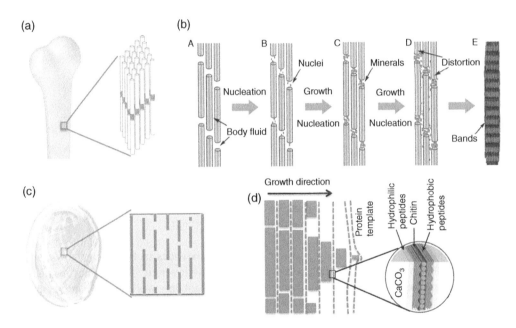

Figure 11.4 Schematic depicting a proposed mechanism of intralar mineralization of collagen in (a) bone and (b) the process of bone biomineralization, A→B, nucleation: Within the aqueous mineralizing solution containing the polymeric process-directing agent (i.e., polyaspartate), the negatively charged polymer sequesters ions and at some critical ion concentration generates liquid–liquid-phase separation within the crystallizing solution, forming nanoscopic droplets of a highly hydrated, amorphous calcium phosphate phase. B→C, growth: The nanoscopic droplets of this polymer-induced liquid precursor (PILP) phase adsorb to the collagen fibril and due to their fluidic character become pulled up and into the whole zones and interstices of the collagen fibril by capillary action. C→D, in-plane growth: The collagen fibril becomes fully imbibed with the amorphous mineral precursor, which then solidifies as the hydration waters are excluded. The amorphous precursor phase crystallizes, leaving the collagen fibril embedded with nanoscopic crystals of hydroxyapatite. D→E, thickness growth. Source: Weiner & Addadi (1997).[35] Reproduced with permission of Elsevier. (c) Nacre. (d) Process of shell biomineralization.[5,36] Source: Sarikaya (1999).[4] Reproduced with permission of the Royal Society of Chemistry.

mainly composed of α-chitin/protein microfibril frameworks, with both calcite and amorphous calcium carbonate deposited within the fibril structures.[5]

There are several stages in the biomineralization of bone, including vesicle mineralization, subsequent collagen mineralization, and secondary mineralization. The first stage is so-called matrix vesicle mineralization, in which small extracellular vesicles referred to as matrix vesicles are formed. These matrix vesicles contain many enzymes and transporters, which synthesize and incorporate Ca^{2+} and PO_4^- into the vesicles. These ions deposit inside the vesicles, and crystalline calcium phosphates are then nucleated on the matrix. Calcium phosphates, that is, HA crystals, grow and eventually break through the membrane to get out of the matrix vesicles. On being exposed, these ribbon-like calcium phosphates assemble themselves radially into a spherical mineralized structure, referred to as a mineralized nodule or calcifying globule. After this matrix vesicle mineralization process, the mineralized nodules deposit in a specific orientation within collagen fibrils, extending mineralization along with their longitudinal axis from the contact points of collagen fibrils. As the nodules form and deposit, the HA continues growing as a plate, with the elongated c axis of the crystal aligned parallel to the fiber axis. This in-plane growth is quick, followed by a much slower growth in thickness until the whole cavity is filled.[37] Matrix vesicle mineralization and subsequent collagen mineralization are grouped as primary mineralization associated with osteoblastic bone formation. After primary mineralization, secondary mineralization occurs, gradually increasing mineral density of bone matrix.[38]

Another interesting protein-templated biomineralization is the basins of mollusks, the nacreous part. The outer layer of nacre is composed of different proteins and calcite, the thermodynamically most stable calcium carbonate species. It is worth noting that the acidic proteins bound to the chitin matrix could also significantly influence the mineralization because they can bind to the side of an aragonite crystal, regulating the crystallites to orient along this face in the crystal growth. During biomineralization, amorphous calcium carbonate fills the space within the organized matrix of chitin, and then mineralizes to aragonite. In this process, several proteins work together to form the fine structure of mollusks. While the chitin matrix controls size and shape, solvated proteins change the mineral phase to the thermodynamically unstable aragonite as well as the shape of the crystal.[39] Many other proteins from the nacreous layer lead to the growth of aragonite instead of calcite, and some of these proteins display interesting and promising properties for biomimicks, for example the protein N16N, extracted from nacre, was shown to lead a sandwich-like structure of protein and mineral in the growth of aragonite. The protein itself does not orient the mineral, instead the mineral orients itself by mineral bridges through the protein layers. N16N coexists with chitin in nature and might play an important role in the formation of unique brick-and-mortar structures.[40]

The biomineralization of the lobster's armor is also protein templated, but its structure is crystalline α-chitin. In the chitin matrix, the partly crystalline chitin fibers mostly orient with the c axis (which is the fiber axis) parallel to the surface, forming a plywood-like structure, but a small fraction of the fibers align orthogonally with respect to the surface and interpenetrate the chitin layers. The cavities of the matrix are filled with calcite localized in the fibers orthogonal to the surface and preferentially in the outer layers.[41] Amorphous calcium carbonate is the dominant phase in the layers far below the surface.[42] Such structure provides hardening on the outer layer and toughening from less mineralized fibers with inner layers against mechanical attacks. Apart from chitin, many other proteins are involved in the biomineralization. Among them, an acidic protein from the crayfish is interesting.

This type of protein not only directs the growth of calcite from square particles to round particles, but also changes the structure of chitin.[38]

11.2.5   Homeostasis and Storage of Metallic Nanoparticles

Living organisms have the unique ability to control necessary but potentially toxic metal in their bodies. Among the many proteins capable of forming metal nanoparticles, ferritin is a well-studied family of proteins that plays an important role in iron homeostasis and storage. Ferritin is composed of 24 subunits of two types, heavy chain (H-ferritin) and light chain (L-chain), self-assembled into a hollow cage-like architecture with an external diameter of 12 nm and an 8 nm diameter cavity.[19] An iron oxide core comprising up to ca. 4500 atoms of iron is located in the cavity of ferritin. At the subunit junctions, some channels are formed for the transport of iron and other metal ions into and out of protein shell. The superior stability of the ferritin cage architecture over a wide range of pH (3–10) and temperatures (up to 80–100 °C) makes it an excellent nanoreactor for metal nanoparticles.[43]

Another protein that is widely used as a protein template for synthesizing metal nanoparticles is bovine serum albumin, the most abundant protein in the circulatory system. Human serum albumin consists of 585 amino acids forming a single polypeptide of known sequence. The albumin molecule has an overall ellipsoidal shape (about 140×40 Å) and is composed of domains. In its native state, the molecule presents a heart shape and consists of three homologous domains (I, II, and III). Each domain is made up of a sequence of large–small–large loops forming a triplet. The most important physiological function of serum albumin is to maintain the osmotic pressure and pH of blood, and transport a wide variety of endogenous and exogenous compounds, including fatty acids, metal, amino acids, steroids, and drugs.[44] Because of these extraordinary characteristics, albumins from various sources have gained extensive biomedical and industrial applications as well as research interest. This protein is readily available and relatively cheap for large-scale production. The molecule can simultaneously bind cationic and anionic metal complexes, and has superior stabilizing and modifying abilities for further biological/biomedical application.[45]

11.2.6   Bioinspired Strategies for Synthesizing Processes

As discussed above, many organisms are able to control mineralization with specialized proteins, generating a range of inorganic materials with desired microstructures. Such biominerals can be further assembled into exquisite hierarchical composite structures by organisms, many with superior properties to those that we are able to form synthetically. Although biominerals themselves are limited for engineering uses, their unique processes are appealing for fabricating bioinspired materials that have the potential to be used in current and future technology. To do this, several levels of biomimetic process strategies can be applied to create materials that are useful for engineering applications.

11.2.6.1   Microorganism-mediated Synthesis

Directly using the mineralizing ability of live microorganisms can generate various biominerals or metals useful for many applications. In the large family of microorganisms, bacteria and fungi represent two main groups, prokaryotic and eukaryotic microorganisms, and are considered to be the main bioinspired candidates for the synthesis of nanoparticles. These microorganisms not only exhibit complicated biological activity but also possess a complicated hierarchical structure in a controlled way. Besides a large variety of minerals, metal nanoparticles can be found in the periplasmic space, on the cell wall and outside the cells. With more insights into metal resistance in the bioreduction process mediated by microorganisms, it is believed that various metal particles can be synthesized by using different microorganisms in which various enzymes participate in the bioreduction process of transporting electrons from certain electron donors to metal electron acceptors.[20]

11.2.6.2   Biomolecule-mediated Synthesis

This approach uses the proteins identified from living organisms. It is common to use proteins for the synthesis of biominerals, which are also used by organisms for this purpose. The functions of these proteins include promotion of crystal formation, matrix-assisted orientation of crystals, growth inhibition by face-selective surface adsorption, and control of the crystal phases. Many natural molecules, such as silicatein, have been identified to function well in synthesizing minerals under non-bioenvironments.

11.2.6.3   Biopanning

There are many organic synthetic molecules that can template inorganic materials, functioning like those proteins identified from the living organisms. One technique for identifying such bioinspired sequences involves rational design. This uses existing knowledge and/or computational modeling of protein–mineral interactions to design peptide sequences that are able to template the formation of a desired material under mild reaction conditions.[46] This approach uses libraries containing between 10^7 and 10^{15} randomly generated peptide sequences for mineral synthesis.[47,48] These sequences are genetically encoded within organisms so that a particular peptide sequence is displayed on the surface of each different bacterium or virus in the library. However, not all sequences are useful for mineral synthesis. Only those sequences that bind strongly to the target are retained after washing. To synthesize a specific mineral, the selected sequences are amplified and re-exposed to the inorganic material, thus enriching the sample in the sequences that have an affinity for the desired mineral.[20,49]

The protein-assisted synthesis has several advantages for the formation of minerals compared with other biotemplates.[20] First, the biomineralization processes are eco friendly and inherently "green". Reactions occur at room temperature, in aqueous solutions with neutral pH, and use benign reducers (some reactions even occur without reducers), which are normally much milder than traditional chemical synthesis pathways. Second, the size, morphology, component, and crystal structure of minerals can be finely controlled by choosing different kinds of proteins and/or variable experimental conditions. This kind of approach through protein may be more effective to control these characteristics, and thus the properties of the minerals. Third, the surfaces of the minerals usually consist of a layer of protein molecules containing a plethora of chemical functional groups. This inherent organic coating can conveniently conjugate the targeting agents and drug molecules for various applications such as surface modification and drug delivery. Fourth, the protein-capped minerals are very stable in various chemical solutions such as salt, phosphate and MES buffer. Last, the protein template is relatively cheap and can be easily obtained (via natural extraction or protein engineering), which enables commercial production of the minerals.

11.3   Biomimetic Materials Processes

11.3.1   Synthesis of Mineralized Collagen Fibrils with Macromolecular Templates

Biomimetic synthetic systems have been developed to synthesize minerals by macromolecular templates. Bone is a typical biocomposite constituted of mineralized collagen fibrils. The mineral is formed under the regulation of non-collagenous proteins (NCPs); several NCPs work together lead to the intrafibrillar mineralization. Inspired by bone growth, a multi-functional protein, named (MBP)–BSP–HA, based on bone sialoprotein (BSP) and HA binding protein, was designed to mimic the intrafibrillar mineralization process *in vitro*.[50] The three functional domains of (MBP)–BSP–HA provide the artificial protein with multiple designated functions for intrafibrillar mineralization, including binding calcium ions, collagen, and HA. If nothing is added in the mineralization system, flake-like minerals grow along the surface of reconstituted collagen fibrils (Figure 11.5a). When both (MBP)–BSP–HA and polyacrylic acid are present in the system, although fibril surfaces remain smooth, an obscure band pattern of collagen fibrils is formed due to a high degree of intrafibrillar mineralization. As shown in Figure 11.5b, the microstructures exhibit highly organized alignment of minerals inside the fibril. The minerals are periodically arranged along the long axis of collagen fibrils. The morphology of crystals inside a fibril changes from needle to platelet. The size of crystals in the tip of a fractured fibril is 30–50 nm in length and 15–20 nm in width. The size and shape of the HA crystals are similar to those of the bone. In addition, the high-resolution transmission electron microscope image reveals the growth orientation of the apatite mineral, in which the crystal lattice, with an interplanar spacing of 0.34 nm, is parallel to the (002) plane of HA (Figure 11.5c).[50]

To understand the biomimetic synthetic process, a mechanism of the intrafibrillar mineralization process directed by (MBP)–BSP–HA is proposed in Figure 11.5d. In the growth process, (MBP)–BSP–HA first specifically binds to the hole zones of collagen fibrils through hydrophobic interaction between hydrophobic residues of BSP and collage fibrils.[51] HA grows on the (MBP)–BSP–HA template within the hole zones. During the growth, polyacrylic acid acts as a cationic absorbing base, accumulating calcium ions from the solution to form droplets of amorphous calcium phosphate (ACP) in the fluid phase. The negatively charged ACP is attracted to positively charged regions of the collagen fibrils through electrostatic interactions. Afterwards, ACP infiltrates and diffuses throughout the interior of the collagen fibrils, owing to the highly hydrated and fluid characteristics of the amorphous phase. The (MBP)–BSP–HA within the collagen fibrils is then surrounded by ACP. The size of the Ca^{2+} equilateral triangle matches the distribution of calcium atoms in the (002) plane of the HA crystal lattice. The BSP fragment can interact with the Ca^{2+} equilateral triangle. The transformation energy of the amorphous phase changing to the crystalline phase is decreased due to the size match between the Ca^{2+} equilateral triangle within ACP and the (002) plane of HA, so the crystals prefer to grow in alignment with the [002] axis, perpendicular to the (002) plane, and the space confinement of the nearby tropocollagen molecules restricts the transverse growth of crystals.

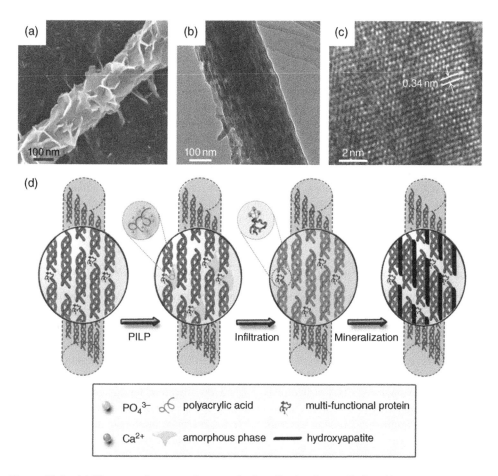

Figure 11.5 (a) Electron microscopy images of mineralized collagen fibrils without any additive for 72 h. (b) TEM image of an isolated collagen fibril, platelet-like mineral crystals with uniform size grown inside the fibril. (c) An HRTEM image of the apatite crystal inside the fibril. (d) Intrafibrillar mineralization process directed by (MBP)–BSP–HAP. Source: Ping *et al.* (2015).[50] Reproduced with permission of the Royal Society of Chemistry.

Due to the high affinity of HA fragments to HA, HA fragments can bind to mineral surfaces and restrict the growth of minerals when they reach another protein located in a neighboring hole zone. Finally, the size of platelet-like nanocrystals closely matches the spacing supplied by the collagen fibril and the protein template.[50]

In addition to the biomimetic intrafibrillar mineralization process, synthetic systems for the formation of $CaCO_3$/polymer hybrids have also been developed. In abalone shell growth, well-controlled oriented aragonite nanocrystals are formed in brick-and-mortar structures.[39] Although precise duplication of such biological crystal structures is still challenging, synthetic approaches have been explored to build similar structures. In a synthetic approach mimicking shell biomineralization, aragonite crystallization was induced by a crystalline poly(vinyl alcohol) (PVA) matrix in the presence of poly(acrylic acid) (PAA).[52] The polycrystalline PVA matrix was prepared by annealing a spin-coated PVA film. Since the lattice of the PVA (100) face closely matches the aragonite lattice, carboxylic acid groups are aligned and adsorbed on the crystalline PVA through the PAA interacting with the hydroxy groups. These aligned carboxylates locally regulate the deposition of the calcium ions, leading to the nucleation of aragonite. The presence of the PAA is important in regulating the structures since the distance between the functional groups on the PAA polymer backbone is the same as that for PVA, leading to the selective formation of aragonite. On the other hand, if poly(glutamic acid) (PGA) is used as a soluble additive, thin-film vaterite will be formed on the PVA matrix because distances between the functional groups are different from the aragonite lattice.[52] Furthermore, unidirectionally oriented polymer/$CaCO_3$ hybrids can be synthesized using macroscopically oriented polymer matrices through macromolecular templating in the presence of simple acidic polymers.[53] These biomimetic processes provide new ways of developing mineral/polymer composites with well-controlled structures.

11.3.2   Synthesis of Nanoparticles and Films Catalyzed with Silicatein

Living organisms such as diatoms and sponges can build fascinating glass nanoarchitectures, but the ways in which this process is controlled by nature is more appealing for the biomimetic synthesis of advanced engineering materials. A key protein, silicatein, has been identified as being responsible for the growth of silica. Since silicatein serves as catalyst as well as a templating matrix, it has been widely used for the synthesis of silica and other inorganic materials under mild conditions.[26,54,55] For example, in a *Monorhaphis chuni* giant spicule, a three-dimensional mesoporous silica structure is formed on silicatein templates with a perfect lattice.[56] This capability of silicateins has been used to synthesize cristobalite and remarkable flexible rods of aligned calcite nanocrystals.[57,58] In the presence of silicatein, nanocrystalline anatase TiO_2 was prepared from titanium(IV) bis(ammonium lactato)dihydroxide (Ti(BALDH)), a water-stable alkoxide-like conjugate of titanium.[59] Oriented aggregates of Ga_2O_3 nanocrystals grow from $Ga(NO_3)_3$,[60] and nanostructured $BaTiOF_4$,[61] an intermediate for production of $BaTiO_3$, was formed from $BaTiF_6$ under ambient conditions.

Silicateins can be modified by genetic engineering to manufacture inorganic materials through environmentally benign routes. The modified silicateins can catalyze various synthetic processes, depending on the sequences inserted in the molecules. The recombinant silicatein catalyzes polymerization of alkoxysilanes at neutral pH and ambient temperature to form silica as well as silicones such as straight-chained poly(dimethylsiloxane) (PDMS).[62] Immobilized silicateins regulate the growth on gold-coated surfaces, polystyrene, and silicon wafers, yielding uniform silica films with controlled thickness, roughness, and hydrophilicity.[63] The thin films are electrically insulative and have potential for device applications. Recombinant silicateins can be modified with the fusion of additional peptide sequences, including polyHis (His-tag), bearing a strong affinity to Ni^{2+} or Co^{2+} chelated with nitrilotriacetic acid (NTA), and polyglutamate (Glu-tag), with affinity to HA, making them remarkably useful tools for immobilization. After modification with His-tag bound to the gold surface with Ni-NTA, the recombinant silicatein gains biocatalytic activity, which catalyzes reactions with Ti(BALDH) and hexafluorozirconate to form layered structures of titania and zirconia, respectively.

The silicatein also promotes the formation of titania coating from Ti(BALDH) on the surface of WS_2 nanotubes. As shown in Figure 11.6, silicatein-immobilized WS_2 nanotubes

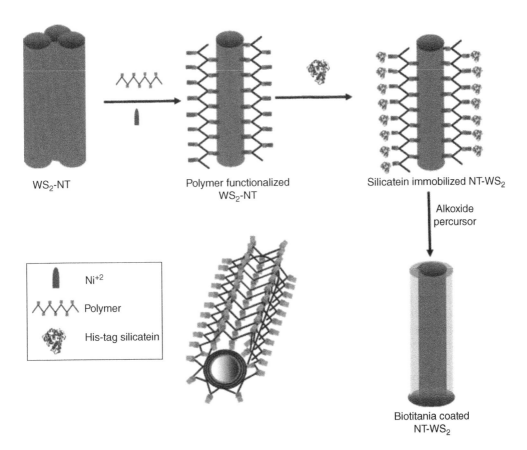

Figure 11.6 Fabrication of the biotitania/vNT-WS2 nanocomposite. In the first step the WS$_2$ nanotube is functionalized with the multifunctional polymer ligand (gray) by complexation through Ni^{2+} groups. The NTA tripod ligand is bound to the side groups of the polymer. In the next step, the silicatein-containing His-tag is attached to the NTA ligand by complexation of Ni^{+2} ions through the His-tag. Finally, the water stable precursor of titanium is hydrolyzed by the immobilized silicatein. Source: Tahir *et al.* (2009).[64] Reproduced with permission of John Wiley and Sons.

were prepared in several steps. First, the nanotubes were coated with polymers with NTA side-chains, and then a His-tagged silicatein was bound to NTA on the surface of the nanotubes.[64] His-tagged silicatein was also used to catalyze the synthesis of nanostructured cassiterite SnO$_2$ on glass surfaces from sodium hexafluorostannate. In this process, His-tagged silicatein was immobilized on glass surfaces using an NTA anchor.[65] The immobilized His-tagged silicatein was also coated on magnetite or gold nanoparticles to retain its native hydrolytic activity to catalyze the growth of silica.[66] In the case of gold nanoparticles, His-tagged silicatein acts as a nanoreactor to synthesize and immobilize gold nanoparticles obtained from auric acid onto the core-shell polymer colloids.[67]

Because of its unique functions in the synthesis of various nanomaterials at ambient conditions, recombinant silicatein opens new doors for biomaterial and medical applications, for example the activity of bone-mineralizing cells can be enhanced when the tissue culture plates are modified with silicatein-mediated silica. Glu-tag-immobilized recombinant silicateins catalyze the growth of biosilica coatings from sodium metasilicate on both synthetic HA nanofibrils and dental HA.[68] This work indicates that Glu-tagged silicateins have considerable potential in broad biomedical applications, including regenerative and prophylactic implementations.[5]

11.3.3   Synthesis of Magnetite using Natural and Synthetic Proteins

Magnetite nanoparticles (MNPs) are one of the key components in the development of many novel bio- and nanotechnological applications such as nanomotors, nanogenerators, nanopumps, and other similar nanometer-scale devices. These MNPs could be used as nano-magnetite in films, magnetic inks in the form of ferrofluids, magnetic recording media, liquid sealing, dampers in motors and shock absorbers, and for heat transfer in loudspeakers. Furthermore, in the biotechnological and biomedical fields magnetite nanoparticles are needed for magnetic separation of biomolecules, magnetic resonance imaging (MRI), tissue repair, drug delivery, hyperthermia treatment of tumor cells, and magnetofection.[69-71] In many of these applications, especially in biotechnological applications, it is necessary to tailor and functionalize the particles; mimicking the natural processes of bacterial magnetosomes provides an attractive alternative to chemically synthesized iron oxide particles.

   Biomimetic approaches have been used in the preparation of magnetite crystal nanopar-ticples. Two synthetic methods have been developed for producing regular nanoparticles; both involve a specialized protein, Mms6, that is found in natural magnetosomes.[16,72,73] One method is the co-precipitation of ferrous and ferric ions in the presence of Mms6 protein to produce uniform magnetite crystals with sizes ranging from 20 to 30 nm. In the absence of the protein morphologies and size of crystals are irregular. Other types of nanoparticles, such as cobalt ferrite ($CoFe_2O_3$), can be produced with the same synthetic approach.[74] The other approach to producing magnetic nanoparticles involves partial oxidation of ferrous hydroxide in the presence of Mms6 protein (Figure 11.7).[72] Similar to the first approach, this method produces magnetites with uniform size, narrow size distribution (approximately 20 nm in diameter), and a cubo-octahedral morphology consisting of crystal faces (Figure 11.7a,c). The synthetic crystal surfaces are similar to those of the magnetosomes observed in the *Magnetospirillum magneticum* strain AMB-1. However, in the absence of Mms6 the crystals were octahedral, consisting of crystal faces that were larger and had increased size distribution (approximately 30 nm in diameter) (Figure 11.7b,d). Thus, protein plays an important role in the synthesized crystals. During the formation of the nanoparti-cles, the Mms6 protein regulates crystal growth of the nanoparticles by binding precursors and/or magnetite crystals.[16]

   Apart from the magnetite-regulating proteins Mms6, biomimetic Mms6 was also devel-oped for the synthesis of magnetite crystals. The key functional group in Mms6 that regu-lates magnetite growth is the C-terminal acidic region. Short synthetic peptides were used to mimic the characteristic amino acid sequences of Mms6 (Figure 11.7e).[73] This M6A peptide contains the active group and leads to the formation of regular particles with a cubo-octahedral morphology similar to those of the magnetosomes and particles formed in the presence of the Mms6 protein. Cobalt-doped magnetite nanoparticles with similar morphology were also produced with this method.[75] These nanoparticles synthesized with

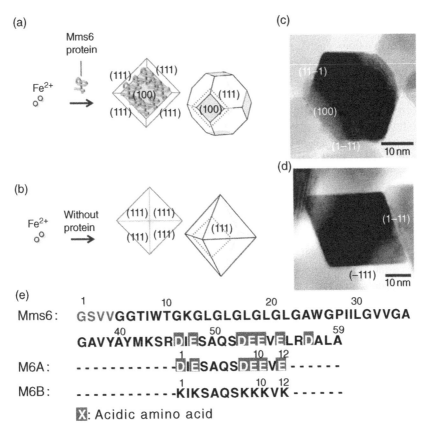

(e)

```
         1              10                20           30
Mms6: GSVVGGTIWTGKGLGLGLGLGLGLGAWGPIILGVVGA
         40             50                59
       GAVYAYMKSRDIESAQSDEEVELRDALA
                                1       10  12
M6A:   - - - - - - - - - -DIESAQSDEEVE- - - - - -
                                1       10  12
M6B:   - - - - - - - - - -KIKSAQSKKKVK- - - - - -
```

X: Acidic amino acid

Figure 11.7 Schematic illustrations of magnetic particles synthesized (a) with or (b) without Mms6. TEM images of magnetic particles synthesized (c) in the presence of Mms6 and (d) in the absence of the Mms6 protein. Source: Amemiya *et al.* (2007).[72] Reproduced with permission of Elsevier. (e) Amino acid sequences of Mms6, M6A, and M6B. D, aspartic acid; E, glutamic acid. Source: Arakaki *et al.* (2015).[5] Reproduced with permission of the Royal Society of Chemistry.

biomimetic proteins have high magnetic saturation, high coercivity, and narrow size distribution in the single-domain size.

Mediated with Mms6, magnetic particles can also form on planar substrates.[76] To this end, a silicon substrate was first modified with a self-assembled monolayer of octadecyltrimethoxysilane and then the Mms6 protein through hydrophobic interactions between the protein molecules and the monolayer, so that the active carboxylic groups of the Mms6 protein are present on the surface of the substrates. The magnetic crystals can then grow on the surface of the substrate with the Mms6 protein. The particle film can also be made into patterns by applying a soft-lithographic technique to fabricate the surface patterning of the Mms6 protein. The subsequent crystal growth mediated with the proteins forms biomimetic magnetite nanoparticle arrays.[76] Nanocrystals arrays of magnetites may have many applications, such high-density data storage. These methods offer an alternative strategy for magnetite formation under mild conditions.

11.3.4   Nanofabrication of Barium Titanate using Artificial Proteins

Complex metal oxides such as $BaTiO_3$ are widely sought materials in the ceramic and electronic industries due to their piezoelectric, ferroelectric, and dielectric properties. The synthesis of $BaTiO_3$ typically requires harsh reaction conditions, such as high temperature and pressure, that provide poor control over composition, crystal structure, and nanoscale morphology. In contrast, living organisms are able to maintain impressive regulation over the crystallization of inorganic materials during naturally occurring mineralization processes. The ability to translate the advantages of natural biomineralization processes to the reality for the synthesis of $BaTiO_3$ could have major impacts in applications such as batteries, solar cells, optoelectronics, and sensor technologies.

A bioinspired approach has been demonstrated in the fabrication of $BaTiO_3$ nanostructures.[77] Because there are no known biogenic perovskite materials in nature, ceramics such as $BaTiO_3$ are typically produced by bioinspired approaches incorporating the advantages found in the biological processes of calcification or silification, as shown in Figure 11.8a. In these processes, minerals are formed through slow growth processes by precipitating or condensing the dissolved metal ions or molecular conjugates from the organism's environment. In general, calcium-containing biomineralization (calcification) is more common in nature. About a half of the biomineralization processes studied thus far involve calcium-containing materials through the highly evolved ability of living organisms to manipulate soluble Ca^{2+}, concentrating this ion from the environment (in the case of marine organisms) or nutrient supplies (in terrestrial animal bone formation).[77] These calcium-containing biominerals take the form of crystalline phases. In contrast, biogenic silica structures are amorphous at the atomic level, typically forming through polycondensation reactions in which silicic precursors must be actively concentrated and sequestered for the silification process to occur.[1] Although the biomineralization processes for different materials vary, they all involve specialized proteins that play a key role in the formation of the materials. These biogenic systems provide inspiration for the rationally controlled synthesis of nanostructured $BaTiO_3$.

$BaTiO_3$ crystals are synthesized through biomimic approaches involving biomineralizing proteins extracted or adopted from biological systems. Specifically, as soon as metal oxides such as SiO_2, TiO_2, and Ga_2O_3 are formed through polycondensation under ambient conditions, their complex binary phases are crystallized by native silicatein filaments from the sponge *Tethya aurantia*.[55] To gain control over the size of the $BaTiO_3$ nanoparticles, biological macromolecules were used as molecular templates during the hydrolysis of a Ba/Ti double metal alkoxide precursor. Similar to microemulsion methods, these templates can serve as soft materials "reactors" that control the flux of reactants to and from the nucleation site.[77] In the presence of peptides, monodisperse $BaTiO_3$ nanoparticles are formed inside peptide templates at room temperature and ambient pressure (Figure 11.8b).[78] In this process, Bis(N-a-amido-glycylglycine)heptane-1,7-dicarboxylate and bolaamphiphile peptide were used as monomers. They self-assemble into ring-like structures, facilitating $BaTiO_3$ nucleation. On addition of the soluble metal alkoxide, Ti-peptide chelation occurs within the core of the peptide ring structures as the template for $BaTiO_3$ nucleation and growth. Interestingly, the inner cavity size (~44 nm) of the self-assembled peptide rings can be controlled by varying the pH of the solution, where an increase in pH results in a decrease in cavity size. Thus, the size of the monodisperse $BaTiO_3$ nanoparticles can be adjusted between 6 and 12 nm by varying the pH value of the solution. Although the process is slow (1–4 day synthesis), nanoparticle

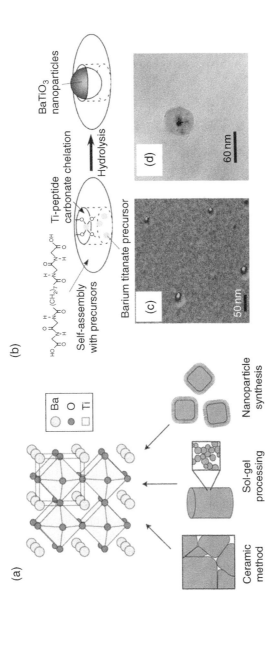

Figure 11.8 (a) BaTiO$_3$ perovskite crystal structure and conventional synthetic strategies. (b) Biomimetic process for the formation of ferroelectric BaTiO$_3$ nanoparticles inside peptide-ring templates. (c) and (d) SEM images of BaTiO$_3$ nanoparticles. Source: Tao *et al.* (2010).[77] Reproduced with permission of the Royal Society of Chemistry.

aggregation does not occur during this templating process. The as-prepared nanoparticles are tetragonal in crystal structure and show switching behavior under an applied electric field. Because of the unique protein interactions with the inorganic surface, this type of ferroelectric (i.e., non-centrosymmetric crystal phase) is different from the cubic-phase nanoparticles that most synthetic methods produce. In addition, the reaction temperature for the synthesis of $BaTiO_3$ by bioinspired synthetic approaches is far below that required for ceramic, sol-gel, and hydrothermal methods.[77] These bioinspired synthetic approaches bypass the difficulties of traditional binary metal oxide synthesis reactions and provide impressive control over $BaTiO_3$ particle size and aggregation.

11.3.5   Protein-assisted Nanofabrication of Metal Nanoparticles

Metal nanomaterials (MNMs) exhibit unique catalytic, electrical, magnetic, and thermal properties, and show potential for a variety of applications. Inspired by the biological paradigm for the formation of MNMs, biomimetic syntheses have emerged as innovative and alternatively attractive synthetic protocols for MNMs. Among many proteins capable of forming metal nanomaterials, ferritin is a well-studied family of proteins that plays an important role in iron homeostasis and storage. This protein has a hollow cage-like architecture with an external diameter of 12 nm and an 8 nm diameter cavity with an iron core located in it.[19] Before ferritin can be used for synthesizing nanoparticles, its iron oxide core must be removed from the cage. The empty ferritin shell forms a scaffold structure (referred to as apo-Ft) for the synthesis of various metal nanoparticles due to its ability to adsorb various metal ions.

For apo-Ft templated synthesis, the first step is to introduce a certain amount of metal ions into its aqueous solution, followed by the addition of reducer (sodium borohydride or ascorbic acid). A variety of metal–protein hybrids (Co, Ni, Cr, Au, Ag, Pt, nanoparticles etc.) can be synthesized by using apo-Ft as scaffolding. The paired Au cluster, for example, was synthesized within the apo-Ft nanoreactor, where ferrous ion is oxidized by the ferroxidase center of H-ferritin.[79] The ferroxidase center is composed of six amino acid residues: one histidine (His), one aspartic acid (Asp), one glutamine (Gln), and three glutamic acids (Glu). Here, Au clusters grow under "points of control" by residue. In the process, NaOH is added to lower the reaction rate in a controlled manner. Since every apo-Ft has two H-ferritin subunits, a pair of Au clusters can be assembled in each part. The final size of the resulting Au clusters is determined by the molar ratio of Au ions to apo-Ft, therefore the sizes of Au clusters can be adjusted by controling the Au(III)/apo-Ft ratio, increased. In principle, other metal clusters can also be synthesized through judiciously tuning the reaction conditions.

Apo-Ft can also be used as the template to synthesize bimetallic clusters or nanoparticles, for example bimetallic Au/Pd core-shell and alloy AuPd NPs were prepared by using two completely different strategies in apo-Ft aqueous solution, as shown in Figure 11.9.[21] Because of the different coordination properties of Pd(II) and Au(III) ions to apo-Ft, for the synthesis of Au + Pd core–shell nanoparticles, the Au core is first prepared as a monometallic nanoparticle in apo-Ft, followed by introduction and reduction of Pd ions to form the shell. The Au/Pd alloy nanoparticles were obtained by mixing aqueous solutions of $KAuCl_4$ and K_2PdCl_4, and subsequent co-reduction by $NaBH_4$. In principle, a series of bimetallic, even tri-metallic, NPs could be prepared by adopting these methods. However, the size of the particles is limited to 8 nm due to the apo-Ft size.

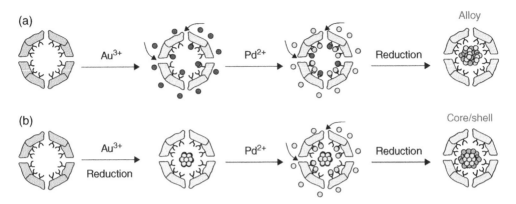

Figure 11.9 The two different synthesis methods of Au/Pd NPs in apo-Ft solution: (a) synthesis of Au/Pd alloy structure and (b) synthesis of Au/Pd core–shell structure. Au(III) and Pd(II) atoms are represented by dark grey and light grey, and Au(0), and Pd(0) are represented by light grey and dark grey, respectively. Source: Suzuki et al. (2009).[21] Reproduced with the permission of the Royal Society of Chemistry.

Bovine serum albumin (BSA) is also as an excellent protein template for the synthesis of metal nanoparticles because of its ability to simultaneously bind cationic and anionic metal complexes. Unlike ferritin, which has a cage shape, BSA has a heart shape and consists of large numbers of reactive functional groups, and thus can work as a reducer and a stabilizer for the synthesis of metal nanoparticles with different sizes and morphologies. Single-crystalline Au nanoplates (0.6–3 μm in size along their longest edge and ~19 nm in thickness) have been synthesized in aqueous solution of BSA at physiological temperature.[80] In this process BSA provides the dual functions of catalysis and templating: metal ion reduction and directing the anisotropic growth of Au nanocrystals. The nucleation and crystalline growth of Au nanoparticles largely depends on environmental factors such as temperature and pH. At physiological temperature, BSA retains its native configuration and no particles are formed. The configuration change of BSA is reversible between 40 and 50 °C, but becomes irreversible between 50 and 60 °C. When the temperature rises from 60 °C to 70 °C, unfolding of BSA with β-aggregation occurs followed by gel formation.[81] Thus, a mild temperature is preferable for the formation of high-yield Au nanoplates (80%). However, no Au nanoplates were formed at high temperature (100 °C). The sizes of Au nanoplates can be tuned, ranging from a few micrometers to tens of nanometers, by adding different amounts of Ag ions.[80]

Apart from three-dimensional noble metal microspheres, some porous structures, such as mesoflowers and macroporous film, can also be synthesized by using BSA as the template. To obtain these flower-like structures, Ag mesoflowers composed of multi-layer Ag nanosheets are first prepared using BSA. Pt and its alloy structures are synthesized via the galvanic replacement reaction using the as-prepared Ag mesoflowers as the sacrificial template. Due to their three-dimensional porous structure and small crystal sizes, these Pt mesoflowers exhibit enhanced electrocatalytic activity toward electrical oxidation of methanol.[82] The three-dimensional macroporous Au film was also synthesized using BSA based on a bottom-up biomineralization approach.[83] In preparing the Au thin films, BSA and metals ions are first mixed at room temperature to form a solution. Ascorbic acid was then rapidly added to solution to reduce metal ions, leading to the nucleation, growth, and formation of primary clusters and small nanoparticles that finally self-assemble into supraspheres. In the final step, macroporous Au films are formed by carrying out calcination at high temperature to remove proteins. Similar procedures can be used to synthesize other macroporous metals. These porous structures have promising potential applications in catalysis, sensing, biofuel cells, and tissue engineering.

References

1. Mann, S. *Biomineralization: Principles and Concepts in Bioinorganic Materials Chemistry* (Oxford University Press, Oxford; 2001).
2. Meyers, M.A., McKittrick, J. & Chen, P.Y. Structural biological materials: Critical mechanics-materials connections. *Science* **339**, 773–779 (2013).
3. Weaver, J.C. *et al.* The stomatopod dactyl club: A formidable damage-tolerant biological hammer. *Science* **336**, 1275–1280 (2012).
4. Sarikaya, M. Biomimetics: Materials fabrication through biology. *Proceedings of the National Academy of Sciences of the United States of America* **96**, 14183–14185 (1999).
5. Arakaki, A. *et al.* Biomineralization-inspired synthesis of functional organic/inorganic hybrid materials: organic molecular control of self-organization of hybrids. *Organic & Biomolecular Chemistry* **13**, 974–989 (2015).
6. Faivre, D. & Godec, T.U. From bacteria to mollusks: the principles underlying the biomineralization of iron oxide materials. *Angewandte Chemie International Edition* **54**, 4728–4747 (2015).
7. Politi, Y. & Weaver, J.C. Biomineralization. Built for tough conditions. *Science* **347**, 712–713 (2015).
8. Heuer, A.H. *et al.* Innovative materials processing strategies – a biomimetic approach. *Science* **255**, 1098–1105 (1992).
9. Westbroek, P. & Marin, F. A marriage of bone and nacre. *Nature* **392**, 861–862 (1998).
10. Jackson, D.J. *et al.* Parallel evolution of nacre building gene sets in molluscs. *Molecular Biology and Evolution* **27**, 591–608 (2010).
11. Weiss, I.M., Kaufmann, S., Mann, K. & Fritz, M. Purification and characterization of perlucin and perlustrin, two new proteins from the shell of the mollusc *Haliotis laevigata*. *Biochemical and Biophysical Research Communications* **267**, 17–21 (2000).
12. Kroger, N., Deutzmann, R. & Sumper, M. Polycationic peptides from diatom biosilica that direct silica nanosphere formation. *Science* **286**, 1129–1132 (1999).
13. Sudo, S. et al. Structures of mollusc shell framework proteins. *Nature* **387**, 563–564 (1997).
14. Suzuki, M. *et al.* An acidic matrix protein, Pif, is a key macromolecule for nacre formation. *Science* **325**, 1388–1390 (2009).
15. Lakshminarayanan, R., Kini, R.M. & Valiyaveettil, S. Investigation of the role of ansocalcin in the biomineralization in goose eggshell matrix. *Proceedings of the National Academy of Sciences of the United States of America* **99**, 5155–5159 (2002).
16. Arakaki, A., Webb, J. & Matsunaga, T. A novel protein tightly bound to bacterial magnetic particles in *Magnetospirillum magneticum* strain AMB-1. *Journal of Biological Chemistry* **278**, 8745–8750 (2003).
17. Shimizu, K., Cha, J., Stucky, G.D. & Morse, D.E. Silicatein alpha: Cathepsin L-like protein in sponge biosilica. *Proceedings of the National Academy of Sciences of the United States of America* **95**, 6234–6238 (1998).
18. Kisailus, D., Truong, Q., Amemiya, Y., Weaver, J.C. & Morse, D.E. Self-assembled bifunctional surface mimics an enzymatic and templating protein for the synthesis of a metal oxide semiconductor. *Proceedings of the National Academy of Sciences of the United States of America* **103**, 5652–5657 (2006).
19. Watt, R.K., Petrucci, O.D. & Smith, T. Ferritin as a model for developing 3rd generation nano architecture organic/inorganic hybrid photo catalysts for energy conversion. *Catalysis Science & Technology* **3**, 3103–3110 (2013).
20. Huang, J. *et al.* Bio-inspired synthesis of metal nanomaterials and applications. *Chemical Society Reviews* **44**, 6330–6374 (2015).
21. Suzuki, M. *et al.* Preparation and catalytic reaction of Au/Pd bimetallic nanoparticles in Apo-ferritin. *Chemical Communications*, 4871–4873 (2009).
22. Blakemore, R. Magnetotactic bacteria. *Science* **190**, 377–379 (1975).

23. Komeili, A. Molecular mechanisms of compartmentalization and biomineralization in magnetotactic bacteria. *Fems Microbiology Reviews* **36**, 232–255 (2012).

24. Wang, L. *et al.* Self-assembly and biphasic iron-binding characteristics of Mms6, a bacterial protein that promotes the formation of superparamagnetic magnetite nanoparticles of uniform size and shape. *Biomacromolecules* **13**, 98–105 (2012).

25. Treguer, P. *et al.* The silica balance in the world ocean – A reestimate. *Science* **268**, 375–379 (1995).

26. Otzen, D. The role of proteins in biosilicification. *Scientifica* **2012**, 867562–867562 (2012).

27. Scheffel, A. *et al.* An acidic protein aligns magnetosomes along a filamentous structure in magnetotactic bacteria. *Nature* **440**, 110–114 (2006).

28. Wang, X., Schroeder, H.C., Wang, K., Kaandorp, J.A. & Mueller, W.E.G. Genetic, biological and structural hierarchies during sponge spicule formation: from soft sol-gels to solid 3D silica composite structures. *Soft Matter* **8**, 9501–9518 (2012).

29. Arakaki, A., Yamagishi, A., Fukuyo, A., Tanaka, M. & Matsunaga, T. Co-ordinated functions of Mms proteins define the surface structure of cubo-octahedral magnetite crystals in magnetotactic bacteria. *Molecular Microbiology* **93**, 554–567 (2014).

30. Zhou, Y., Shimizu, K., Cha, J.N., Stucky, G.D. & Morse, D.E. Efficient catalysis of polysiloxane synthesis by silicatein alpha requires specific hydroxy and imidazole functionalities. *Angewandte Chemie International Edition* **38**, 780–782 (1999).

31. Murr, M.M. & Morse, D.E. Fractal intermediates in the self-assembly of silicatein filaments. *Proceedings of the National Academy of Sciences of the United States of America* **102**, 11657–11662 (2005).

32. Cha, J.N. *et al.* Silicatein filaments and subunits from a marine sponge direct the polymerization of silica and silicones in vitro. *Proceedings of the National Academy of Sciences of the United States of America* **96**, 361–365 (1999).

33. Olszta, M.J. *et al.* Bone structure and formation: A new perspective. *Materials Science & Engineering R – Reports* **58**, 77–116 (2007).

34. Jager, I. & Fratzl, P. Mineralized collagen fibrils: A mechanical model with a staggered arrangement of mineral particles. *Biophysical Journal* **79**, 1737–1746 (2000).

35. Weiner, S. & Addadi, L. Design strategies in mineralized biological materials. *Journal of Materials Chemistry* **7**, 689–702 (1997).

36. Kakisawa, H. & Sumitomo, T. The toughening mechanism of nacre and structural materials inspired by nacre. *Science and Technology of Advanced Materials* **12**, 064710 (2011).

37. Fratzl, P., Gupta, H.S., Paschalis, E.P. & Roschger, P. Structure and mechanical quality of the collagen-mineral nano-composite in bone. *Journal of Materials Chemistry* **14**, 2115–2123 (2004).

38. Schulz, A., Wang, H., van Rijn, P. & Böker, A. Synthetic inorganic materials by mimicking biomineralization processes using native and non-native protein functions. *Journal of Materials Chemistry* **21**, 18903 (2011).

39. Cartwright, J.H.E. & Checa, A.G. The dynamics of nacre self-assembly. *Journal of the Royal Society Interface* **4**, 491–504 (2007).

40. Killian, C.E. *et al.* Mechanism of calcite co-orientation in the sea urchin tooth. *Journal of the American Chemical Society* **131**, 18404–18409 (2009).

41. Raabe, D. *et al.* Microstructure and crystallographic texture of the chitin-protein network in the biological composite material of the exoskeleton of the lobster *Homarus americanus. Materials Science & Engineering A: Structural Materials* **421**, 143–153 (2006).

42. Al-Sawalmih, A. *et al.* Microtexture and chitin/calcite orientation relationship in the mineralized exoskeleton of the American lobster. *Advanced Functional Materials* **18**, 3307–3314 (2008).

43. Uchida, M. *et al.* Biological containers: Protein cages as multifunctional nanoplatforms. *Advanced Materials* **19**, 1025–1042 (2007).

44. Sekula, B., Zielinski, K. & Bujacz, A. Crystallographic studies of the complexes of bovine and equine serum albumin with 3,5-diiodosalicylic acid. *International Journal of Biological Macromolecules* **60**, 316–324 (2013).

45. Xavier, P.L., Chaudhari, K., Baksi, A. & Pradeep, T. Protein-protected luminescent noble metal quantum clusters: an emerging trend in atomic cluster nanoscience. *Nano Reviews* **3** (2012).

46. Wang, F., Li, D. & Mao, C. genetically modifiable flagella as templates for silica fibers: From hybrid nanotubes to 1D periodic nanohole arrays. *Advanced Functional Materials* **18**, 4007–4013 (2008).

47. Sarikaya, M., Tamerler, C., Jen, A.K.Y., Schulten, K. & Baneyx, F. Molecular biomimetics: nanotechnology through biology. *Nature Materials* **2**, 577–585 (2003).

48. Reiss, B.D. *et al.* Biological routes to metal alloy ferromagnetic nanostructures. *Nano Letters* **4**, 1127–1132 (2004).

49. Tamerler, C. & Sarikaya, M. Molecular biomimetics: nanotechnology and bionanotechnology using genetically engineered peptides. *Philosophical Transactions of the Royal Society A* **367**, 1705–1726 (2009).

50. Ping, H. *et al.* Organized intrafibrillar mineralization, directed by a rationally designed multi-functional protein. *Journal of Materials Chemistry B* **3**, 4496–4502 (2015).

51. Baht, G.S., Hunter, G.K. & Goldberg, H.A. Bone sialoprotein-collagen interaction promotes hydroxyapatite nucleation. *Matrix Biology* **27**, 600–608 (2008).

52. Hosoda, N., Sugawara, A. & Kato, T. Template effect of crystalline poly(vinyl alcohol) for selective formation of aragonite and vaterite $CaCO_3$ thin films. *Macromolecules* **36**, 6449–6452 (2003).

53. Nishimura, T., Ito, T., Yamamoto, Y., Yoshio, M. & Kato, T. Macroscopically ordered polymer/$CaCO_3$ hybrids prepared by using a liquid-crystalline template. *Angewandte Chemie International Edition* **47**, 2800–2803 (2008).

54. Morse, D.E. Silicon biotechnology: harnessing biological silica production to construct new materials. *Trends in Biotechnology* **17**, 230–232 (1999).

55. Brutchey, R.L. & Morse, D.E. Silicatein and the translation of its molecular mechanism of biosilicification into low temperature nanomaterial synthesis. *Chemical Reviews* **108**, 4915–4934 (2008).

56. Zlotnikov, I. *et al.* A perfectly periodic three-dimensional protein/silica mesoporous structure produced by an organism. *Advanced Materials* **26**, 1682–1687 (2014).

57. Fuchs, I., Aluma, Y., Ilan, M. & Mastai, Y. induced crystallization of amorphous biosilica to cristobalite by silicatein. *Journal of Physical Chemistry B* **118**, 2104–2111 (2014).

58. Natalio, F. *et al.* Flexible minerals: self-assembled calcite spicules with extreme bending strength. *Science* **339**, 1298–1302 (2013).

59. Sumerel, J.L. *et al.* Biocatalytically templated synthesis of titanium dioxide. *Chemistry of Materials* **15**, 4804–4809 (2003).

60. Kisailus, D., Choi, J.H., Weaver, J.C., Yang, W.J. & Morse, D.E. Enzymatic synthesis and nanostructural control of gallium oxide at low temperature. *Advanced Materials* **17**, 314–318 (2005).

61. Brutchey, R.L., Yoo, E.S. & Morse, D.E. Biocatalytic synthesis of a nanostructured and crystalline bimetallic perovskite-like barium oxofluorotitanate at low temperature. *Journal of the American Chemical Society* **128**, 10288–10294 (2006).

62. Wolf, S.E. *et al.* Formation of silicones mediated by the sponge enzyme silicatein-alpha. *Dalton Transactions* **39**, 9245–9249 (2010).

63. Rai, A. & Perry, C.C. Mussel adhesive protein inspired coatings: a versatile method to fabricate silica films on various surfaces. *Journal of Materials Chemistry* **22**, 4790–4796 (2012).

64. Tahir, M.N. *et al.* Enzyme-mediated deposition of a TiO_2 coating onto biofunctionalized WS2 chalcogenide nanotubes. *Advanced Functional Materials* **19**, 285–291 (2009).

65. Andre, R., Tahir, M.N., Schroeder, H.C.C., Mueller, W.E.G. & Tremel, W. Enzymatic synthesis and surface deposition of tin dioxide using silicatein-alpha. *Chemistry of Materials* **23**, 5358–5365 (2011).

66. Shukoor, M.I. *et al.* Fabrication of a silica coating on magnetic gamma-Fe_2O_3 nanoparticles by an immobilized enzyme. *Chemistry of Materials* **20**, 3567–3573 (2008).

67. Lange, B. *et al.* Functional polymer-opals from core-shell colloids. *Macromolecular Rapid Communications* **28**, 1987–1994 (2007).

68. Natalio, F. *et al.* Bioengineering of the silica-polymerizing enzyme silicatein-alpha for a targeted application to hydroxyapatite. *Acta Biomaterialia* **6**, 3720–3728 (2010).

69. Gupta, A.K. & Gupta, M. Synthesis and surface engineering of iron oxide nanoparticles for biomedical applications. *Biomaterials* **26**, 3995–4021 (2005).

70. Bain, J. & Staniland, S.S. Bioinspired nanoreactors for the biomineralisation of metallic-based nanoparticles for nanomedicine. *Physical Chemistry Chemical Physics* **17**, 15508–15521 (2015).

71. Stanicki, D., Elst, L.V., Muller, R.N. & Laurent, S. Synthesis and processing of magnetic nanoparticles. *Current Opinion in Chemical Engineering* **8**, 7–14 (2015).

72. Amemiya, Y., Arakaki, A., Staniland, S.S., Tanaka, T. & Matsunaga, T. Controlled formation of magnetite crystal by partial oxidation of ferrous hydroxide in the presence of recombinant magnetotactic bacterial protein Mms6. *Biomaterials* **28**, 5381–5389 (2007).

73. Arakaki, A., Masuda, F., Amemiya, Y., Tanaka, T. & Matsunaga, T. Control of the morphology and size of magnetite particles with peptides mimicking the Mms6 protein from magnetotactic bacteria. *Journal of Colloid and Interface Science* **343**, 65–70 (2010).

74. Prozorov, T. *et al.* Cobalt ferrite nanocrystals: Out-performing magnetotactic bacteria. *ACS Nano* **1**, 228–233 (2007).

75. Galloway, J.M. *et al.* Magnetic bacterial protein Mms6 controls morphology, crystallinity and magnetism of cobalt-doped magnetite nanoparticles in vitro. *Journal of Materials Chemistry* **21**, 15244–15254 (2011).

76. Galloway, J.M. *et al.* Biotemplated magnetic nanoparticle arrays. *Small* **8**, 204–208 (2012).

77. Tao, A.R., Niesz, K. & Morse, D.E. Bio-inspired nanofabrication of barium titanate. *Journal of Materials Chemistry* **20**, 7916 (2010).

78. Nuraje, N. *et al.* Room temperature synthesis of ferroelectric barium titanate nanoparticles using peptide nanorings as templates. *Advanced Materials* **18**, 807–811 (2006).

79. Sun, C. *et al.* controlling assembly of paired gold clusters within apoferritin nanoreactor for in vivo kidney targeting and biomedical imaging. *Journal of the American Chemical Society* **133**, 8617–8624 (2011).

80. Xie, J., Lee, J.Y. & Wang, D.I.C. Synthesis of single-crystalline gold nanoplates in aqueous solutions through biomineralization by serum albumin protein. *Journal of Physical Chemistry C* **111**, 10226–10232 (2007).

81. Bakshi, M.S. *et al.* Protein films of bovine serum albumen conjugated gold nanoparticles: A synthetic route from bioconjugated nanoparticles to biodegradable protein films. *Journal of Physical Chemistry C* **115**, 2982–2992 (2011).

82. Zhuang, L. *et al.* Porous platinum mesoflowers with enhanced activity for methanol oxidation reaction. *Journal of Solid State Chemistry* **191**, 239–245 (2012).

83. Hou, C., Yang, D., Liang, B. & Liu, A. Enhanced performance of a glucose/O_2 biofuel cell assembled with laccase-covalently immobilized three-dimensional macroporous gold film-based biocathode and bacterial surface displayed glucose dehydrogenase-based bioanode. *Analytical Chemistry* **86**, 6057–6063 (2014).

Index

Printed and bound by CPI Group (UK) Ltd, Croydon, CR0 4YY

16/04/2025

14658472-0002